GIANT PANDAS

GIANT PANDAS

Biology and Conservation

Edited by Donald Lindburg
and Karen Baragona

With a Foreword by George B. Schaller

UNIVERSITY OF CALIFORNIA PRESS
Berkeley   Los Angeles   London

University of California Press
Berkeley and Los Angeles, California

University of California Press, Ltd.
London, England

© 2004 by the Regents of the University of California

Library of Congress Cataloging-in-Publication Data

Giant pandas : biology and conservation / Edited by Donald Lindburg and Karen Baragona ; with a foreword by George B. Schaller.
    p.   cm.
  Includes bibliographical references (p.).
  ISBN 0-520-23867-2 (cloth: alk. paper)
    1. Giant panda.   2. Giant panda—Conservation.   I. Lindburg, Donald, 1932–   II. Baragona, Karen, 1967–
QL737.C214P365   2004
599.789—dc22   2003064565

Manufactured in the United States of America
10   09   08   07   06   05   04
10   9   8   7   6   5   4   3   2   1

The paper used in this publication meets the minimum requirements of ANSI/NISO Z39.48-1992(R 1997) ∞

TO LINDA
DEVOTED FRIEND OF CHINA,
ITS PEOPLE, AND ITS WILDLIFE

When westerners first visited the home of the panda we were at the primitive stage of shooting or grabbing them. Understandably, we were banned from the locality, and gradually we learned to value and appreciate pandas. Since then they have received adoration a-plenty, but their natural history was almost completely unstudied and there seemed a considerable danger that they might be lost forever. Now, at last, they are starting to get the serious attention they deserve. Chinese and western scientists are working together to investigate the ecology of the giant panda, its life in the wild and its natural behaviour. These are vital steps in devising a programme for its conservation, a programme in which captive pandas may play an important part as a sort of backstop against the extinction of this rare and extraordinary animal. For the pressure is still on the wild panda: predators, the disappearance of bamboo from large areas, and the burgeoning human population of the Far East all work against them. The Chinese authorities, however, have made an outstanding commitment to the preservation of the pandas, have already protected vast areas, and are fighting the threat of panda food shortages. The future now begins to look hopeful and there is a good chance that the world will salvage one of its most exotic and enigmatic natural treasures: *Ailuropoda melanoleuca,* the fabulous giant panda.

MORRIS AND MORRIS, *THE GIANT PANDA* (1981)

CONTENTS

Foreword *George B. Schaller* / xi

Preface / xiii

Acknowledgments / xv

INTRODUCTION: A NEW DAWNING FOR GIANT PANDAS? / 1
Donald Lindburg and Karen Baragona

PART ONE · *Evolutionary History of Giant Pandas* / 7

1 · **PHYLOGENETIC POSITION OF THE GIANT PANDA: HISTORICAL CONSENSUS THROUGH SUPERTREE ANALYSIS** / 11
Olaf R. P. Bininda-Emonds

BRIEF REPORT 1.1
Phylogenetic Placement of the Giant Panda Based on Molecular Data / 36
Lisette P. Waits

2 · **WHAT IS A GIANT PANDA? A QUERY ABOUT ITS PLACE AMONG THE URSIDAE** / 38
Lee R. Hagey and Edith A. MacDonald

3 · **A PALEONTOLOGIST'S PERSPECTIVE ON THE ORIGIN AND RELATIONSHIPS OF THE GIANT PANDA** / 45
Robert M. Hunt, Jr.

4 · **VARIATION IN URSID LIFE HISTORIES: IS THERE AN OUTLIER?** / 53
David L. Garshelis

BRIEF REPORT 4.1
Life History Traits and Reproduction of Giant Pandas in the Qinling Mountains of China / 74
Dajun Wang, Xiaojian Zhu, and Wenshi Pan

PART TWO · *Studies of Giant Panda Biology* / 77

5 · **FUTURE SURVIVAL OF GIANT PANDAS IN THE QINLING MOUNTAINS OF CHINA** / 81
Wenshi Pan, Yu Long, Dajun Wang, Hao Wang, Zhi Lü, and Xiaojian Zhu

---

To ensure proper referencing, this volume follows the Western convention of listing given names ahead of family names for all Chinese authors on chapter opening pages, in the table of contents, and in reference lists. The Eastern convention of family names ahead of given names is followed by some authors in the acknowledgments to their chapters and in the contributors list.

WORKSHOP REPORT 5.1
Genetic Studies of Giant Pandas in Captivity and in the Wild / 88
Yaping Zhang and Oliver A. Ryder

6 · **NUTRITIONAL STRATEGY OF GIANT PANDAS IN THE QINLING MOUNTAINS OF CHINA** / 90
Yu Long, Zhi Lü, Dajun Wang, Xiaojian Zhu, Hao Wang, Yingyi Zhang, and Wenshi Pan

BRIEF REPORT 6.1
Spatial Memory in the Giant Panda / 101
Loraine R. Tarou, Rebecca J. Snyder, and Terry L. Maple

7 · **CHEMICAL COMMUNICATION IN GIANT PANDAS: EXPERIMENTATION AND APPLICATION** / 106
Ronald R. Swaisgood, Donald Lindburg, Angela M. White, Hemin Zhang, and Xiaoping Zhou

BRIEF REPORT 7.1
Chemical Composition of Giant Panda Scent and Its Use in Communication / 121
Lee R. Hagey and Edith A. MacDonald

8 · **REPRODUCTION IN GIANT PANDAS: HORMONES AND BEHAVIOR** / 125
Rebecca J. Snyder, Dwight P. Lawson, Anju Zhang, Zhihe Zhang, Lan Luo, Wei Zhong, Xianming Huang, Nancy M. Czekala, Mollie A. Bloomsmith, Debra L. Forthman, and Terry L. Maple

PART THREE · *Pandas and Their Habitat* / 133

9 · **COMPARATIVE ECOLOGY OF GIANT PANDAS IN THE FIVE MOUNTAIN RANGES OF THEIR DISTRIBUTION IN CHINA** / 137
Jinchu Hu and Fuwen Wei

PANEL REPORT 9.1
Assessing the Habitat and Distribution of the Giant Panda: Methods and Issues / 149
Colby Loucks and Hao Wang

BRIEF REPORT 9.1
Using DNA from Panda Fecal Matter to Study Wild-Living Populations / 155
Yunwu Zhang, Hemin Zhang, Guiquan Zhang, Oliver A. Ryder, and Yaping Zhang

10 · **GIANT PANDA MIGRATION AND HABITAT UTILIZATION IN FOPING NATURE RESERVE, CHINA** / 159
Yange Yong, Xuehua Liu, Tiejun Wang, Andrew K. Skidmore, and Herbert H. Prins

BRIEF REPORT 10.1
Noninvasive Techniques for Monitoring Giant Panda Behavior, Habitat Use, and Demographics / 170
Matthew E. Durnin, Jin Yan Huang, and Hemin Zhang

11 · **MAPPING HABITAT SUITABILITY FOR GIANT PANDAS IN FOPING NATURE RESERVE, CHINA** / 176
Xuehua Liu, M. C. Bronsveld, Andrew K. Skidmore, Tiejun Wang, Gaodi Dang, and Yange Yong

PANEL REPORT 11.1
Restoring Giant Panda Habitat / 187
Chunquan Zhu and Zhiyun Ouyang

12 · **SYMPATRY OF GIANT AND RED PANDAS ON YELE NATURAL RESERVE, CHINA** / 189
Fuwen Wei, Ming Li, Zuojian Feng, Zuwang Wang, and Jinchu Hu

13 · **BALANCING PANDA AND HUMAN NEEDS FOR BAMBOO SHOOTS IN MABIAN NATURE RESERVE, CHINA: PREDICTIONS FROM A LOGISTIC-LIKE MODEL** / 201
Fuwen Wei, Guang Yang, Jinchu Hu, and Stephen Stringham

PANEL REPORT 13.1
Management of Giant Panda Reserves
in China / 210
Changqing Yu and Xiangsui Deng

PART FOUR · *Giant Panda Conservation* / 213

14 • A NEW PARADIGM FOR PANDA RESEARCH AND CONSERVATION: INTEGRATING ECOLOGY WITH HUMAN DEMOGRAPHICS, BEHAVIOR, AND SOCIOECONOMICS / 217
Jianguo Liu, Zhiyun Ouyang, Hemin Zhang, Marc Linderman, Li An, Scott Bearer, and Guangming He

PANEL REPORT 14.1
China's National Plan for Conservation of the Giant Panda / 226
Zhi Lü and Yongfan Liu

15 • BIOLOGICAL FRAMEWORK FOR EVALUATING FUTURE EFFORTS IN GIANT PANDA CONSERVATION / 228
Eric Dinerstein, Colby Loucks, and Zhi Lü

PANEL REPORT 15.1
National Survey of the Giant Panda / 234
Changqing Yu and Shaoying Liu

16 • THE LEGACY OF EXTINCTION RISK: LESSONS FROM GIANT PANDAS AND OTHER THREATENED CARNIVORES / 236
John L. Gittleman and Andrea J. Webster

PANEL REPORT 16.1
Reintroduction of Giant Pandas: An Update / 246
Sue Mainka, Wenshi Pan, Devra Kleiman, and Zhi Lü

17 • BIOMEDICAL SURVEY OF CAPTIVE GIANT PANDAS: A CATALYST FOR CONSERVATION PARTNERSHIPS IN CHINA / 250
Susie Ellis, Anju Zhang, Hemin Zhang, Jinguo Zhang, Zhihe Zhang, Mabel Lam, Mark Edwards, JoGayle Howard, Donald Janssen, Eric Miller, and David Wildt

BRIEF REPORT 17.1
Conservation Education Initiatives in China: A Collaborative Project among Zoo Atlanta, Chengdu Zoo, and Chengdu Research Base of Giant Panda Breeding / 264
Sarah M. Bexell, Lan Luo, Yan Hu, Terry L. Maple, Rita McManamon, Anju Zhang, Zhihe Zhang, Li Song Fei, and Yuzhong Tian

WORKSHOP REPORT 17.1
International Coordination and Cooperation in the Conservation of Giant Pandas / 268
J. Craig Potter and Kenneth Stansell

CONCLUSION: CONSENSUS AND CHALLENGE: THE GIANT PANDA'S DAY IS NOW / 271
Donald Lindburg and Karen Baragona

APPENDIX A. KEYNOTE ADDRESS BY FU MA / 279

APPENDIX B. KEYNOTE ADDRESS BY MARSHALL JONES / 282

APPENDIX C. MEMORANDUM OF CONSENSUS / 286

Contributors / 289

Index / 297

FOREWORD

When, in 1978, Hu Jinchu and his colleagues erected a hut and several tents in the forests of the Qionglai Mountains in Wolong Natural Reserve to study giant pandas, it was the beginning of an intensive effort to save a species that was much adored but little known. I was privileged to become a member of the Chinese research team on behalf of World Wildlife Fund. On May 15, 1980, I first examined and measured panda droppings and feeding sites. Four and one-half years later, in January 1985, I left the project. We had by then collected a baseline of information about the life of pandas. To this, Pan Wenshi and his team added superbly during a thirteen-year study at another research site located in the Qinling Mountains. Natural history remains the cornerstone of knowledge about species and their habitat, providing information, defining problems, and suggesting solutions on which realistic conservation plans depend. However, given the economic, social, and political pressures in today's world, knowledge alone will not lead to conservation. Between 1975 and 1989, while these studies were in progress, the panda lost half of its habitat in Sichuan Province to logging and agriculture. The survivors remain in small, fragmented populations, isolated in about twenty-four forest patches—a blueprint for extinction.

Conservation requires a vision beyond pandas to include the whole forest ecosystem, human land use, and community development, all within a cultural context. The 1990s have seen a consistent change in that direction. A national conservation plan began to be implemented in 1993. An up-to-date census of pandas was completed. New reserves were established, and they now number forty, encompassing half the panda's habitat. Reserve staff were trained to patrol and monitor wildlife. The number and survival of captive-born young increased greatly. And in a dramatic policy shift, the government banned logging of old-growth forest and initiated a vast reforestation program to turn steep hillsides from "grain to green."

Giant Pandas: Biology and Conservation, based on a conference held in 2000, presents a valuable collection of reports that place the species into its biological, ecological, and political context. For the first time in a quarter century, panda research and thought have been summarized, including material until now available only in Chinese. This volume also shows the value of international cooperation—of sharing ideas and techniques—to produce new data and insights leading to better management in both the wild and captivity. I am impressed by the large number of Chinese authors—many of them former students of Hu Jinchu and Pan Wenshi—who contributed their knowledge.

However, the volume is more than a progress report. It also emphasizes what must now be done to assure the panda a future as a national treasure of China and an international icon of our natural world. Conservation is a dynamic process. Plans must constantly be revised and adapted to changing conditions and new ideas. During the 1980s, the government of China was reluctant to permit us to collect socioeconomic information, thereby preventing the integration of panda ecology and human demography. This attitude has changed. There is now appreciation for the idea that panda conservation means more than protecting reserves, that it means dealing with the whole landscape. New concepts, such as buffer zones, linkages between forest fragments, multiple-use areas, and cooperation with the local people who have a stake in the resources have been officially accepted. To implement a conservation program requires scientific understanding, local commitment, and official involvement—a complex, never-ending task of conflict resolution. Conservation is a mission, not a job, requiring sustained emotional investment in those who strive to achieve it. This volume offers essential guidelines based on an ideal combination of fact and eloquent advocacy. But it still is important to heed Charles Darwin's admonition, expressed in *The Origin of Species:* "We need not marvel at extinction; if we must marvel, let it be at our own presumption in imagining for a moment that we understand the many complex contingencies on which the existence of each species depends."

The first step in solving a problem is to define it with clarity and wisdom. This has largely been done, and panda conservation is now high on China's agenda. Yet much still remains to be implemented in the field and laboratory and on the policy level. The panda's habits differ somewhat from area to area, but only in the Qinling Mountains has the species been studied sufficiently to offer details on community structure, birth and death rates, and trends in population size. Pandas and their habitat must be monitored with remote sensing, radiotracking, and ground surveys. Captive reproduction has improved so markedly that a surplus of animals for reintroduction to the wild will soon be available. A comprehensive search for a safe and suitable habitat for such reintroductions has yet to be made, and focused research on the best method of giving captives their full freedom is needed. Management and survival of the fragmented populations is a long-term challenge, one that must be measured not in terms of a few years, but in centuries. For obscure reasons, the government of China has at present banned radiotracking, with the result that basic research, monitoring, and the potential of reintroductions are severely hampered. Even more troublesome is the shortage of permanent government funds to support and strengthen reserves. International funds, mostly derived from panda loans to zoos, are a main reason why some critical reserves are not languishing.

In the 1980s, I was filled with creeping despair, as the panda seemed increasingly shadowed by fear of extinction. But now, in this new millennium, *Giant Pandas: Biology and Conservation* rightly projects hope, optimism, and opportunity. As the volume editors note, the prospects for saving the giant panda are today unequaled. The species impinges on the mind of all who behold its wondrous image. National pride, ethics, aesthetics, and resource economics combine to place international value on the panda. Recent years have illuminated the animal's life as never before, but so much more needs to be done on its behalf. The panda cannot compromise its needs, whereas humanity can use its knowledge, self-restraint, and compassion to offer the species a secure wilderness home. If the measures proposed in this book are implemented, if the correct choices are soon made, the panda will surely endure as a living symbol of conservation and luminous wonder of evolution.

GEORGE B. SCHALLER
Wildlife Conservation Society

PREFACE

The giant panda first came under scientific study during the 1970s, following gifts of juvenile pairs from the Chinese government to selected countries, including the United States. Without exception, however, in these exposures of giant pandas to the outside world, propagation and exhibitry took precedence over in-depth biological inquiry. In the late 1970s, a massive die-off of bamboo in many parts of the panda's range prompted authorities in China to send teams into the field, including Professor Hu Jinchu, who, in 1978, set up the first of several bases for monitoring the wild population at Wolong Nature Reserve. When China and the World Wildlife Fund initiated a collaborative study of pandas at Wolong in 1980, one of North America's foremost field biologists, Dr. George Schaller, was selected to lead the team. Among his collaborators were Hu Jinchu, Pan Wenshi, and Zhu Jing. Professors Hu and Pan have continued their studies on wild pandas until quite recently, and several of their students remain active in panda research today. Their efforts have also sparked studies by colleagues in other parts of the giant panda's range. Both professors have great knowledge of giant pandas and are internationally respected as China's most distinguished panda specialists. The Wolong project also led to the establishment of a breeding center at the lower end of the reserve that has provided an important venue for captive research in recent years.

China has taken significant steps in the past few years to advance conservation and study of the giant panda. These include the establishment of new panda reserves, the restriction of logging across much of the panda's range, sponsorship of a comprehensive survey of the wild population and its habitat, the initiation of international loans to raise revenues for conservation, and the approval of several collaborative research endeavors between Chinese investigators and researchers from abroad. The World Wildlife Fund has been supporting capacity-building activities and providing technical assistance on panda reserve management and community-based conservation in China since the early 1980s. Under the aegis of the Conservation Breeding Specialist Group of the International Union for the Conservation of Nature, and in cooperation with local experts, an evaluation of the reproductive health and potential of China's captive population has been realized. Several U.S. and Chinese zoos are today involved in collaborative studies of captive giant pandas in China.

As a result of this increased collaboration between East and West, indications of heightened societal awareness of the giant panda's rareness and plight became evident. As forces were joined, a new spirit of cooperation and strong friendships developed. Sensing the building momentum, we organized an international conference, held in San Diego, California, in October 2000. Entitled *Panda 2000: Conservation Priorities for the New Millennium,* this event provided a forum for eighty-five presentations on various aspects of giant panda conservation, captive breeding, and scientific study. Of the 230 registrants, more than fifty were from China. Delegates from England, Germany, Japan, Mexico, The Netherlands, Spain, and Switzerland also attended. The China Wildlife Conservation Association and the Chinese Association of Zoological Gardens acted as cosponsors of the conference. The success of *Panda 2000* rested on the judicious selection of panda specialists from around the world as participants and a meeting format of workshops, panel discussions, and paper and poster sessions that led to lively debate and consensus building.

The main objectives of *Panda 2000* were to utilize newly available information in achieving agreement on the most critical goals for panda conservation and to fortify existing collaborations and develop new ones for accomplishing those goals. Another objective was to enhance international understanding in areas of policy. Achievement in these areas entailed discussions of priorities in the conservation of the wild population of pandas and of captive adjuncts to such efforts. An opportunity for sharing recent scientific findings was seen as essential to the conservation process.

This volume aims, therefore, to bring together evidence of the recent growth in knowledge of giant panda biology and of expanded conservation initiatives. Through its publication, a trove of information heretofore available primarily in the Chinese language becomes available to the non-Chinese community, as do the viewpoints of Chinese colleagues who are leading the conservation endeavors of the present. The thirty-two contributions from eighty-three authors and coauthors represent only a portion of the topics covered at *Panda 2000*. We have opted to include primarily those that deal directly with giant panda biology and conservation, although we realize that many of the reports on captive breeding also contribute to these domains. Some of the contributions shed new light on longstanding issues, whereas the majority inform of more recent developments. We have attempted to present an integrated message by supplying cross-references and section introductions that allow readers to see the connecting threads of the four parts of the volume. The result is to bring to the fore a more complete picture of one of Nature's most compelling species.

DONALD LINDBURG
San Diego, California
KAREN BARAGONA
Washington, D.C.

ACKNOWLEDGMENTS

We thank each of the authors for sharing the most recent products of their endeavors, first through participation in *Panda 2000*, and now in printed form. We are especially grateful for their patience in seeing their work go through review and revision, and in many cases, rather heavy editing to achieve uniformity in format and style. *Panda 2000* could not have taken place without the generous financial support of the Zoological Society of San Diego and the World Wildlife Fund. Mabel Lam is gratefully acknowledged for her assistance in organizing the conference and in recruiting outstanding translation services.

Pamela Case, administrative assistant to the first editor, invested many hours and considerable expertise in bringing about this work. Before her, Kristin Abbott helped to organize the conference and shepherd the early stages of book preparation. Dr. Yu Changqing and Eugene Lee, colleagues of the second editor, provided valuable assistance in organizing the conference and contributions to the third section of the book, respectively. From the beginning, Doris Kretschmer at the University of California Press expressed a keen interest in publishing a volume on giant pandas, and we appreciate the assistance, always cheerfully given, of press staff Nicole Stephenson, Jenny Wapner, and Marilyn Schwartz in bringing this work to fruition. Peter Strupp and Cyd Westmoreland of Princeton Editorial Associates brought outstanding professionalism and efficiency to the final editing process.

Our strategy has been to include paper presentations as chapters and follow these, where relevant, with the reports of panel and workshop leaders; that is, brief summaries of the hard work that went on in their sessions. A smaller number of papers have been included as brief reports because of their preliminary, yet promising, findings.

Our goal was to seek peer review of each scientific contribution, and we are grateful to those who have cooperated with us in this activity, specifically Randy Champeau, Eric Dinerstein, Mark Edwards, Devra Kleiman, Colby Loucks, William McShea, Eric Miller, Michael Mooring, Michael Pelton, John (Andy) Phillips, Richard Reading, Donald Reid, Bruce Schulte, Steven Stringham, Ronald Swaisgood, David Wildt, and Kevin Willis.

Finally, we acknowledge the assistance of Chia Tan, postdoctoral fellow at the Zoological Society of San Diego, in the translation of chapters 5 and 9, originally submitted in Chinese.

Introduction

A NEW DAWNING FOR GIANT PANDAS?

Donald Lindburg and Karen Baragona

NEW INITIATIVES devoted to the conservation and study of giant pandas *(Ailuropoda melanoleuca)* during the past decade have resulted from a convergence of activity in several quarters. The history of this secretive denizen of the dense bamboo forests of western China—first as a trophy for the hunter's gun, then as a spectacle for the entertainment of Westerners—has endowed the panda with a mystique that will not go away. As a candidate for conservation, the giant panda elicits a level of public concern that is rarely equaled by other wild forms (Lü et al. 2000).

Biological inquiry only adds to this mystique. Consider, for example, two of the panda's most widely recognized traits—a carnivore that is nearly entirely herbivorous in its diet (Sheldon 1937) and has evolved a pseudothumb to aid in feeding (Wood-Jones 1939). More recent work on panda biology has thrown into sharp relief other traits that evoke curiosity and wonder. An example is the discovery that females will have no more than a single, short period of heat during an annual mating season (Xu et al. 1981; Kleiman 1983), seemingly enhancing greatly the cost of reproductive failure in a given year. Another is the finding that the term fetus is smallest in size relative to dam size among all placental mammals (Gittleman 1994), and although twins are produced in nearly half of all litters, dams routinely abandon one twin at parturition (Schaller et al. 1985).

It is within this milieu of wonderment that eager championing of pandas on the one hand and the desire for exclusivity in their promotion on the other have the potential for exacerbating differences among those involved with the panda. *Panda 2000: Conservation Priorities for the New Millennium,* an international conference on which this volume is based, was remarkable for its coalescence around a widely shared view that the giant panda may yet have its day. Participants from several academic disciplines and from various levels of involvement engaged in spirited debate and evolved agreement on action plans and priorities. Perhaps the clearest consensus drawn from this event was that the panda's day is now.

WHY ANOTHER BOOK ON GIANT PANDAS?

This volume is, in one sense, the proceedings of the *Panda 2000* conference, an event that brought together scientists and conservationists selected as participants on the basis of the skill and cogency of their efforts. It is much more than a "proceedings," however, given this selection process and that all scientific contributions have been subject to peer review. As a scholarly book on a little-studied mammal, it represents but a beginning in an important field of endeavor, embracing as it does a renewed analysis of a taxon that is symbolic of progress in conservation and research. A brief review of its history reveals how the panda attained this symbolic position.

Following in the footsteps of their adventurous father and former U.S. president, Kermit and Theodore Roosevelt Jr., were apparently the first hunters from the West to shoot a wild panda (Roosevelt and Roosevelt 1929). In so doing, they "set off an avalanche of Western hunting expeditions, often sponsored by museums eager for specimens to grace their exhibition halls" (Catton 1990: 12). Shortly thereafter, however, attention shifted to bringing the first live panda to the West, an objective first achieved in 1936 by Ruth Harkness, the socialite widow of an adventurer and animal collector (Harkness 1938). Over the next four years, another twelve pandas were captured and exported to zoos in London and the United States (Zhao et al. 1991). Trafficking ended during World War II, although not until the panda had become a bartering chip in the political arena. Demonstrating once again the powerful emotions stirred by the panda, two were presented to the Bronx Zoo in 1941 as an inducement to the U.S. public to support the Nationalist Party government in China during the early stages of World War II (Morris and Morris 1966; Perry 1969). Such high-stakes dealing were possible only because of the frenzied response to pandas by the U.S. public in the prewar era.

As China experienced political upheaval after the war years, we know only in retrospect that pandas began to show up in captivity again in the 1950s, mainly in Chinese zoos, but also in a limited trickle to zoos in London and Moscow (Morris and Morris 1966). Remarkably, within a few years of their reappearance in captivity, the Beijing Zoo announced (in 1963) the first birth of a giant panda (Ouyang and Tung 1964). Additional births were to follow, in very low numbers, over the next three decades or so, including a small number produced by artificial means.

The Western world had its next opportunity to see live pandas up close when, in 1972, China's government presented then U.S. President Richard Nixon with a gift of two juveniles, placed in the National Zoo in Washington, D.C. (Reed 1972). Gifts to England, France, Germany, Japan, Mexico, and Spain were soon to follow. Although reports from this period indicated great hopes for captive breeding, as of this writing, only a single elderly male and four aging offspring survive from these seven pairs. Pandas thus acquired a reputation as being difficult to breed in captivity and to keep alive, but this conclusion has been shown by recent events to be more a matter of human failure and bad luck than any problem with the species itself (Zhu et al. 2001).

Field studies began with the massive flowering and die-off of bamboo, the panda's primary food source, in various mountain ranges in China in the 1970s, soon to be followed by the first in-depth study of a wild panda population (Schaller et al. 1985). A relatively small number of ecological studies by scientists from the West, often with Chinese collaborators, were subsequently carried out as extensions of the Schaller team's effort. Starting from its joint sponsorship of Schaller's project, World Wildlife Fund (WWF) spearheaded a range of giant panda conservation programs that continue to the present time. These initiatives include collaborating with China's State Forestry Administration to lead the "Third National Survey of the Giant Panda and Its Habitat"; providing training and technical assistance in wildlife monitoring and law enforcement patrolling on panda reserves; designing and implementing innovative community-based conservation programs, including ecotourism development and cooperation with the Shaanxi Province Forestry Department in delineating five dispersal corridors for reconnection of fragmented panda habitat.

Yet another phase in the history of the charismatic giant panda began in the 1980s, when short-term loans to North American zoos proliferated. This capitalization on a popular animal for commercial gain violated international law and was eventually banned by the American Zoo and Aquarium Association (AZA) and the U.S. Fish and Wildlife Service (see summary in Schaller [1993]). However, revived interest in an appropriately structured captive program soon led to the development by several North American zoos of an AZA Conservation Action Plan and Species Survival Plan (SSP). Dr. Devra Kleiman of the Smithsonian's National Zoological Park became the first species coordinator for North America and a keeper of the newly established international studbook.

Against this important background, a long-term United States–China international loan program, designed to generate financial resources for use in implementing China's conservation plans for the giant panda, was launched in 1996. These funds are being applied to improving habitat protection primarily through an upgrading of infrastructure. On the U.S. side, loan policy promulgated by its Fish and Wildlife Service (1998) mandates research that will enhance the conservation of wild-living pandas as the sole justification for an importation. Significantly, commercial benefit from the loans is prohibited, and captive breeding is, at best, a secondary activity. In general, this new era of international loans has sparked a resurgence in international collaborations and in resolve to ensure the survival of giant pandas.

With few notable exceptions, the themes developed in this volume have a short, recent history. Beginning in the 1930s, when the first live pandas came to the United States, one finds a fairly voluminous popular and semipopular literature that offers, at best, anecdotal information about the nature of the species. From these early arrivals and from pairs gifted to the outside world in the 1970s, descriptive accounts of anatomical features, diseases, pathology, behavior, and genetics were published. Many of these were opportunistic (e.g., whenever a panda died) and usually appeared as notes in a range of semipopular and clinical publications. Our search of the literature indicates that more substantive articles appearing in scientific journals and books averaged about 2.2 per year from 1980 through 1999. Taken together, the information garnered from this period provided an important baseline for work with larger samples and refined methodologies that have appeared in the more recent scientific literature.

Since 1990, a number of popular books on giant pandas have appeared (Catton 1990; Priess and Gao 1990; Maple 2000; Lü 2001; Kiefer 2002; Lumpkin and Seidensticker 2002). In addition, reports on captive pandas appeared in the proceedings of two international symposia (Asakura and Nakagawa 1990; Klos and Frädrich 1990), but for readers of English, only one truly scientific book (Schaller et al. 1985) has ever been published. A volume that describes over 10 years of study of radiocollared pandas in the Qinling Mountains is at this writing available only in the Chinese language (Pan et al. 2001).

Against this background of adulation but too little in the way of concrete advancements, we come to the modern era of work with giant pandas. China's increasingly resolute embrace of responsibility for saving its wildlife from extinction and the change in international loan policy by the U.S. government have launched an era in which biological study receives its primary impetus from concern for the giant panda's survival. We now have long-term research loans, and much of the growing body of knowledge of giant panda biology arises from studies of captive-held individuals. Fortunately, in the tradition of Schaller, Hu, and Pan, other Chinese workers have maintained a presence in the field, and this volume offers a first exposure to their work.

Although the panda's future is still far from certain, sweeping changes in China's environmental policies let us dare to hope that we may be witnessing a sea change in the way panda conservation can be accomplished. Two policies that grew out of a need to stem soil erosion and control flooding are having enormous favorable ramifications for giant pandas and their habitat. These include a ten-year ban on commercial logging throughout the panda's entire range, beginning in 1998, and a massive decade-long reforestation program—each with a $10 billion bill to be footed largely by the Chinese central government. Another watershed policy initiative is a thirty-year, $30 billion plan to dramatically expand China's "protected areas" system. These three breakthroughs, made possible by China's economic boom and motivated by its concern for the economic and human consequences of pillaging the environment, present an unprecedented opportunity to protect and restore panda habitat on a scale that would have been pure fantasy until now. Lest we get carried away with optimism, though, it is sobering to consider the countervailing force exerted by the Western China Development Program, the goal of which is to close the yawning gap between the rapidly modernizing and prosperous eastern half of the country and the still relatively backward, impoverished western half. This is to be accomplished in large part through major infrastructure development (e.g., highways, hydropower projects); expanding such extractive industries as mining; and promoting large-scale tourism, which, if not managed prudently, could visit havoc upon pandas and their forests.

We see three intertwined strands emerging in this renewed interest in giant pandas: (1) the surmounting of cultural and language differences in communication about the species, (2) the wedding of governmental authority and resources to programmatic proposals for setting goals, and (3) the significance of organismic biology in shaping those proposals. Given that local communities, governments, nongovernmental organizations (NGO), philanthropists, and many more have a stake in future developments affecting pandas and their habitats, respect for the national sovereignty of the host country and for different ways of doing things is, in this case, profoundly important in communicating about objectives. Very little conservation can occur without the concurrence and support of range country governments, and we applaud China for new policies recently promulgated at the highest levels of government for conserving its unique biotic heritage. The third strand, scientific inquiry into the biology of the species, raises questions of a particulate nature, but also places the panda in an ecosystem that must be appreciated by governments and conservers alike. Although the desire to protect the giant panda and its habitat is a driving force in policy development, the animal's flagship role is undoubted (Leader-Williams and Dublin 2000).

The editors come to this volume as Westerners involved with NGOs committed to conservation and its linkage to scientific inquiry. We have spent a considerable amount of time in China and in the habitats of the giant panda, gaining firsthand knowledge of the cultural, logistic, and financial issues. We recognize a discipline of endeavor that is known today as "conservation biology," but here we use these terms separately to give voice to those whose primary activity is conservationist as opposed to scientific. Irrespective of these two domains, the work presented in this volume derives from scientists with a strong conservation ethic and conservationists with a keen appreciation of the importance of science to their work. Our background and experience with animals, partially in captivity, but especially in their natural ranges, enable us to appreciate the convergence of these emphases into a common agenda aimed at ensuring a future for giant pandas.

REFERENCES

Asakura, S., and S. Nakagawa, eds. 1990. *Giant panda: Proceedings of the second international symposium on giant pandas, Tokyo, November 10–13, 1987*. Tokyo: Tokyo Zoological Park Society.

Catton, C. 1990. *Pandas*. Bromley, Kent: Christopher Helm.

Gittleman, J. L. 1994. Are the pandas successful specialists or evolutionary failures? *BioScience* 44: 456–64.

Harkness, R. 1938. *The lady and the panda*. London: Nicholson and Watson.

Kiefer, M. 2002. *Chasing the panda: How an unlikely pair of adventurers won the race to capture the myth-*

ical *"white bear."* New York: Four Walls Eight Windows.

Kleiman, D. 1983. Ethology and reproduction of captive giant pandas *(Ailuropoda melanoleuca)*. *Z Tierpsychol* 62:1–46.

Klos, H.-G., and H. Frädrich, eds. 1990. *Giant Panda: Proceedings of the international symposium on giant panda, September 28–October 1, 1984, Berlin.* Bongo 10.

Leader-Williams, N., and H. Dublin. 2000. Charismatic megafauna as "flagship species." In *Priorities for the conservation of mammalian diversity*, edited by A. Entwistle and N. Dunstone, pp. 53–81. Cambridge: Cambridge University Press.

Lü, Z. 2001. *Giant pandas in the wild*. Gland, Switzerland: World Wildlife Fund International.

Lü, Z., W. Pan, X. Zhu, D. Wang, and H. Wang. 2000. What has the giant panda taught us? In *Priorities for the conservation of mammalian diversity*, edited by A. Entwistle and N. Dunstone, pp. 325–34. Cambridge: Cambridge University Press.

Lumpkin, S., and J. Seidensticker. 2002. *Smithsonian book of giant pandas*. Washington, D.C.: Smithsonian Institution Press.

Maple, T. L. 2000. *Saving the giant panda*. Atlanta: Longstreet.

Morris, R., and D. Morris. 1966. *Men and pandas*. New York: McGraw-Hill.

Ouyang, K., and S.-H. Tung. 1964. In the Peking Zoo—The first baby giant panda. *Anim Kingdom* 67:45–46.

Pan, W., Z. Lü, X. Zhu, D. Wang, H. Wang, Y. Long, D. Fu, and X. Zhou. 2001. *A chance for lasting survival*. (In Chinese.) Beijing: Peking University Press.

Perry, R. 1969. *The world of the giant panda*. New York: Taplinger (Bantam).

Priess, G., and X. Gao, eds. 1990. *The secret world of pandas*. New York: Abrams.

Reed, T. H. 1972. What's black and white and loved all over? *Nat Geogr* 142:803–15.

Roosevelt, T., and K. Roosevelt. 1929. *Trailing the giant panda*. New York: Charles Scribner's Sons.

Schaller, G. B. 1993. Rent-a-panda. Chapter 14 in *The last panda*, pp. 235–49. Chicago: University of Chicago Press.

Schaller, G. B., J. Hu, W. Pan, and J. Zhu. 1985. *The giant pandas of Wolong*. Chicago: University of Chicago Press.

Sheldon, W. 1937. Notes on the giant panda. *J Mammal* 18:13–19.

U.S. Fish and Wildlife Service. 1998. Policy on giant panda permits. *Fed Reg* 63(166):45839–54.

Wood-Jones, F. 1939. The "thumb" of the giant panda. *Nature* 3613:157.

Xu, G., G. He, and Z. Ye. 1981. Reproduction and hand-rearing of the giant panda. (In Chinese.) *China Zoo Yearbook* 4:10–16.

Zhao, Q., Z. Fan, D. Kleiman, and J. Gipps. 1991. *The giant panda studbook*. Beijing: Chinese Association of Zoological Gardens.

Zhu, X., D. Lindburg, W. Pan, K. Forney, and D. Wang. 2001. The reproductive strategy of giant pandas *(Ailuropoda melanoleuca)*: Infant growth and development and mother-infant relationships. *J Zool Lond* 253:141–55.

PART ONE

Evolutionary History of Giant Pandas

In a spirited essay on the zoological placement of the giant panda, George Schaller (1993: 267) concluded his review of the evidence by saying "[t]he giant panda still pseudothumbs his nose at us." Can it be that opinions on this topic remain divided even today? And does it matter? We believe both questions must be answered in the affirmative, and for this reason, in part one we present results from more recent work on the phylogeny of the giant panda and its relatives. Our objective is not, as Schaller would say, "to prod heretics into becoming true believers" in any one school of thought (1993: 262), but to let new facts have whatever impact they may on those who read them.

Obviously, where the pandas belong in phylogenetic terms is important, not just from the intellectual pleasures derived from settling a debate or satisfying our curiosity, but in understanding its unusual constellation of traits and how these may affect its prospects for survival (Gittleman 1994). Were there a consensus that the panda's affinities lie with the ursids, for example, then conservation programs modeled after the Ursidae are more likely to be successful. Pandas, furthermore, live in sympatry with members of two of the putative branches of their family tree, the Asiatic black bear *(Ursus thibetanus)* and the red panda *(Ailurus fulgens)*, and this reality has potential significance in the competition for common resources, particularly as living space continues to decrease.

To satisfy these ends, and many more, we note that the phylogenetic affinities of the giant panda were first addressed by its missionary/naturalist discoverer Père Armand David in 1869, when he assigned it to the bear genus *Ursus*. Within a year, however, it was reclassified by Milne-Edwards as belonging with the raccoons, based on the very same anatomical materials. In this initial divergence of opinion, a long-lived controversy was generated. With the advent of molecular techniques as identifiers in the 1970s, it may be thought that questions of phylogeny have been irrefutably resolved (O'Brien et al. 1985). Yet, as Schaller notes (1993), even among molecular specialists there are differing conclusions. Whether the giant panda will ever be categorized to the satisfaction of all is not known, but we suggest that a broadened perspective that includes genetics, anatomy, life history, and behavior data may bring us closer to that goal.

In chapter 1, Bininda-Emonds reviews comprehensively a vast amount of historical data (105 studies) on the giant panda and on *Ailurus*, the red or lesser panda, with a view to producing a consensus on their phylogenetic affinities. It is his contention that supertree analysis overcomes limitations arising from disparate data sources that are fundamentally incompatible and cannot be analyzed simultaneously. The giant panda is conclusively identified as the sister species of the extant ursids, a placement that has been increasingly accepted and only occasionally questioned over the past twenty years. Separation of the panda lineage from other ursids appears to have taken place nearly 22 million years ago. As for the red panda, Bininda-Emonds concludes that supertree analysis suggests affinities with the Procyonidae, although this result is not strongly supported. Red pandas may, in his opinion, represent the last survivor of an ancient arctoid lineage that separated from the Procyonidae about 29 million years ago.

Brief report 1.1 complements Bininda-Emonds's analysis by focusing solely on molecular phylogenetics in addressing the giant panda's placement. Waits reports that ten of twelve studies published since the original work of Sarich (1973) place the giant panda with the Ursidae. Waits also summarizes unpublished results from two mitochondrial DNA analyses of bears and other carnivores that provide strong molecular support for including giant pandas with bears.

Yet another line of evidence bearing on phylogeny is presented by Hagey and MacDonald in chapter 2. In a novel approach to the problem, these authors examined the profile of bile salts (the breakdown products of cholesterol metabolism) of the giant panda and related carnivores. The bile salts of carnivores are characterized by the presence of two taurine conjugated bile acids, taurochenodeoxycholic acid and taurocholic acid. Among the Ursidae, an additional taurine conjugated bile salt, tauroursodeoxycholic acid, is also present. When compared with published phylogenetic relationships among ursids, the amount of tauroursodeoxycholic acid in bile was found to be directly proportional to species' evolutionary age. An examination of the bile of the giant panda showed that it has the lowest proportion of tauroursodeoxycholic acid of all species examined. Hagey and MacDonald conclude that the near absence of tauroursodeoxycholic acid in giant panda's biliary bile acids suggests that, in terms of evolutionary age, the giant panda is older than any extant bear.

Any discussion of giant panda phylogeny would be incomplete without a summary of the fossil record. In chapter 3, Hunt notes that we are the beneficiaries of a much improved record for Cenozoic carnivorous mammals, and this record places the giant panda in the arctoid family Ursidae. Recent discoveries near Lufeng, Yunnan Province, extend the fossil record of the giant panda into the late Miocene (~7 million years ago). These fossils—isolated teeth—are among the oldest that can be confidently attributed to the ancestry of *Ailuropoda,* and represent an early stage in the development of its unique dentition. The Lufeng fossils suggest that the lineage leading to the living giant panda separated from those of other ursids during the Miocene. Living ursine bears can be traced back to the early Miocene *Ursavus;* however, the ursine radiation that led to the diverse species of living bears did not occur until the Plio-Pleistocene and, hence, is a relatively recent event. Thus the giant panda is believed to be closely related to the ancestral ursine bear *Ursavus,* and may have diverged from that genus as recently as the late Miocene.

In chapter 4, Garshelis reviews the all-important topic of ursid life histories and how these may clarify the interrelatedness of extant species. The eight species of bears are a small family, but collectively and individually, they span a wide geographical and elevational range. As such, the bears must be adaptable to highly diverse conditions. Bears show considerable variation in diet, habitat use, home range size, arboreality, den selection, hibernation, age of sexual maturity, litter interval, litter size, and cub survival. Variation within species often exceeds variation between species. Garshelis concludes that the giant panda is most divergent taxonomically and has some distinctive life history traits, such as a specialized but nonnutritive diet, exceedingly small cubs at birth, and, despite frequently producing twins (like other bears), generally raising only one cub. However, each of the other bears also has distinctive biological and behavioral peculiarities that set them apart, and in that context, no one species can be regarded as a life history outlier.

Because so much of what we know about giant pandas has derived from observing them in captivity, the information from the wild that is provided by Wang, Zhu, and Pan in brief report 4.1 provides a long-awaited glimpse into the nature of giant panda reproduction. Adding to our knowledge of their life history, fifteen giant pandas (twelve with radiocollars) in the Qinling Mountains of Shaanxi Province were observed for nearly a decade, starting in 1992. Mating events, the length of gestation, interbirth intervals, the timing of births, and even the sex ratio of newborn cubs were documented. These data enable us to make the comparisons that further elucidate questions of phylogeny and to begin the process of estimating viability of wild populations during a time of severe habitat contraction.

REFERENCES

Gittleman, J. 1994. Are the pandas successful specialists or evolutionary failures? *BioScience* 44:456–64.

O'Brien, S. J., W. G. Nash, D. E. Wildt, M. E. Bush, and R. E. Benveniste. 1985. A molecular solution to the riddle of the giant panda's phylogeny. *Nature* 317:140–44.

Sarich, V. M. 1973. The giant panda is a bear. *Nature* 245:218–20.

Schaller, G. B. 1993. The panda is a panda. Appendix B in *The last panda,* pp. 261–267. Chicago: University of Chicago Press.

1

Phylogenetic Position of the Giant Panda

HISTORICAL CONSENSUS THROUGH SUPERTREE ANALYSIS

Olaf R. P. Bininda-Emonds

PERHAPS NO QUESTION in mammalian systematics has engendered such long-term controversy and uncertainty as the phylogenetic placement of the giant panda *(Ailuropoda melanoleuca)*. Although its formal introduction to Western science placed it as a member of the bear family (Ursidae) (David 1869), similarities to the lesser or red panda *(Ailurus fulgens)* and, by extension, raccoons and allies (Procyonidae) were quickly noted (Milne-Edwards 1870). Since that time, a variety of evidence has been used to ally *Ailuropoda* with ursids, procyonids, or *Ailurus* (either within the previous two families or as the separate family Ailuridae), or to place it in a family by itself (Ailuropodidae). Fueled largely by molecular evidence, there is perhaps finally a growing consensus that *Ailuropoda* represents the sister group to the remaining ursids.

In this chapter, I approach the question of panda phylogeny from a historical perspective to examine trends in the placement of *Ailuropoda* through time. My work follows on from that of O'Brien et al. (1991), although it differs in two key respects. First, I employ a phylogenetic rather than a taxonomic perspective. The taxonomic status of any species is highly subjective (e.g., is *Ailuropoda* "sufficiently distinct" to justify being placed in its own family?), whereas a study of its phylogenetic or sister-group relationships is much more objective and concrete. Second, through the use of supertree construction (sensu Sanderson et al. 1998), I am able to infer the consensus estimate of the affinity of *Ailuropoda* for any given time period. Thus, I can demonstrate how consensus opinion has shifted over time and by how much. This resembles work done previously with *Ailurus* and the pinnipeds (Bininda-Emonds 2000a). Because *Ailurus* has played a critical historical role in the controversy surrounding the origin of *Ailuropoda,* I also perform similar analyses for it here.

ISSUES OF EVIDENCE AND CONVERGENCE

The uncertainty surrounding the placement of *Ailuropoda* derives from the numerous similarities that it shows to each of ursids, procyonids, and especially to *Ailurus*. In fact, were it not for the existence of the procyonid-like *Ailurus,* the acceptance of *Ailuropoda* as an ursid would likely be considerably less—if at all—in dispute. The

TABLE 1.1
Shared Features between Ailuropoda *and Ursids, Procyonids, and* Ailurus

FEATURE OF *AILUROPODA*	URSIDAE	PROCYONIDAE	*AILURUS*
Morphology			
Size and gross morphology	+		
Skull robustness	−		
Brain morphology	+		
Auditory region and ossicles	+		
Epipharyngeal bursa		+	
Ridges on hard palate	+		
Dentition (especially massive size)	−		+
Skeletal robustness	−	+	
Specialized sesamoid on forepaw			? (unique)
Respiratory tract	+		
Intestines (shorter and less complex)	−		
External (soft) morphology	? (unique)		
Genitalia			+
Coloration		+	+
Hair structure	+		
Molecular			
Karyotype	+	+	+
Serology/immunology	+		
Sequence data	+		
Behavioral			
Vocalizations	−	+	+
Life history traits	+		?
Scent marking (and glands)	−		+
Feeding behavior			+
Mating behavior			+
Other			
Fossil affinities	+		

NOTES: A plus sign indicates that the feature is similar between the two taxa. A minus sign indicates that the feature has been used to argue against a relationship between the two taxa, even if *Ailuropoda* does not share the feature with another taxon.

key lines of evidence used to infer the ancestry of *Ailuropoda* historically are given in table 1.1. The majority of evidence, both morphological and molecular, allies *Ailuropoda* with ursids. Fossil evidence, particularly the inferred close relationship with the Pliocene fossil ursid *Hyaenarctos* (now included in *Agriotherium,* a member of the extinct sister group to ursids) (McKenna and Bell 1997; Hunt, chapter 3), also unanimously indicates an ursid origin.

In one of the most thorough and impressive comparative morphological investigations for any species, Davis (1964: 322) declared that "every morphological feature examined indicates that the giant panda is nothing more than a highly specialized bear." However, a few morphological features have been used to argue against such a relationship, even if they do not indicate a relationship with either procyonids or *Ailurus* ("negative evidence"; see below and table 1.1). These include the greater than expected robustness of the skull, skeleton, and dentition of *Ailuropoda* for an animal of its size (Morris and Morris 1981).

Virtually all molecular evidence points to an ursid affinity for *Ailuropoda* (Waits, brief report 1.1). Only the karyotype is equivocal. In its diploid number, *Ailuropoda* ($2N = 42$) resembles procyonids (e.g., $2N = 42$ for *Procyon lotor*) and *Ailurus* ($2N = 44$) more so than ursine bears ($2N = 74$). Although this observation is correctly attributed to Newnham and Davidson (1966), the attendant implication of procyonid ancestry is not. Newnham and Davidson (1966: 161) explicitly pointed out that large differences in karyotypes and diploid numbers occur within such families as Canidae (dogs) and that the evidence merely indicates *Ailuropoda* to be a different species from ursine bears. They added that the number of chromosomal arms ("nombre fondamental") might be a more informative measure in this regard. Along these lines, O'Brien et al. (1985) discovered that, although *Ailuropoda* possesses a procyonid-like diploid number, the banding patterns of its chromosomes are virtually identical to those of ursine bears. Together with other molecular evidence, they persuasively argued that *Ailuropoda* is related to ursids and that its reduced diploid number is a result of extensive chromosomal fusion in the past (see also Nash et al. 1998). It is noteworthy that similar, albeit independent, fusions have been inferred to explain the karyotype of the spectacled bear *(Tremarctos ornatus)* ($2N = 52$) (Nash and O'Brien 1987; Nash et al. 1998).

Finally, behavioral characteristics strongly group the two panda species together. Only life history traits are similar between *Ailuropoda* and ursids (Garshelis, chapter 4). Otherwise, *Ailuropoda* resembles *Ailurus* in its unusual scent-marking behavior and in its mating and feeding behaviors. The latter is the most remarkable, with both pandas being renowned for the ability to manipulate precisely their herbivorous food items, although only *Ailuropoda* possesses an enlarged sesamoid that acts as an analog of an opposable thumb.

Researchers have sought to make sense of the conflicting signals within the phenotypic evidence (i.e., morphological and behavioral data) through one of two evolutionary scenarios. The first holds that *Ailuropoda* is an ursid that has shifted to an almost exclusively herbivorous diet (the so-called "bear school"). This scenario accounts for such features as the more robust dentition as being obvious (convergent) adaptations for herbivory. The second holds that *Ailuropoda* is instead derived from a small herbivore, typically with procyonid affinities, that has converged secondarily on a larger bearlike body plan (the "raccoon school"). The greater than expected robustness of the skull and skeleton has been used as evidence of rapid growth in the lineage leading to *Ailuropoda* (Morris and Morris 1981). The raccoon school in particular has relied on the negative evidence found in table 1.1 in combination with behavioral information.

Proponents of the raccoon school concede that the majority of evidence places *Ailuropoda* with ursids (e.g., Ewer 1973; Morris and Morris 1981). In supporting a nonursid origin, they instead argue that the fewer features that cluster *Ailuropoda* with either procyonids or *Ailurus* represent evolutionary novelties that are more difficult to envisage evolving on multiple occasions (Ewer 1973; Morris and Morris 1981). Similar reasons have been used to cluster megachiropteran bats with primates on the basis of a shared optic network and other neural features, in spite of an overwhelming number of similarities, mostly related to flying, with microchiropteran bats (Pettigrew 1986, 1991). An extension of this general argument is that organisms with a similar body plan are more likely to develop convergently evolved features, given similar selective regimes. This argument was used to explain why the two main groups of pinnipeds (true seals versus sea lions and walruses) were so similar morphologically, despite formerly being believed to have separate ancestors (McLaren 1960; Mitchell 1967; Repenning 1990). This conclusion is now held to be false (Wyss 1987; Vrana et al. 1994; Bininda-Emonds et al. 1999). With respect to *Ailuropoda*, the full argument is that its overall similarity with ursids applies only to features that are phenotypically plastic or that are expected to show a greater degree of convergence in response to similar selection pressures, possibly due to the inheritance of a common, primitive, arctoid body plan (Ewer 1973; Morris and Morris 1981).

The issue is not easily resolved. Molecular evidence has played a valuable role, because convergence at this level is unlikely to mirror that at the morphological level. However, problems remain within a purely morphological domain. Character weighting continues to be a controversial area in phylogenetic systematics, with no clear guidelines. Presumably, the use of as much evidence as possible, both morphological and molecular ("total evidence") (sensu Kluge 1989), is the key to resolving this issue. It is widely held that the best phylogenetic inference is the one supported by the most independent lines of evidence (Mickevich 1978; Farris 1983; Penny and Hendy 1986; Kluge 1989; Novacek 1992; De Jong 1998). So long as homoplasy, of which convergence is one form, remains relatively rare and randomly distributed both among features and the relationships it infers (see Sanderson and Hufford 1996), the true phylogenetic history will be reflected in the majority of features. Thus, surveying as many features as possible in a cladistic framework (to distinguish shared primitive and shared derived features) (Hennig 1966) should be sufficient to overrule any instances of convergence, however improbable they might seem.

METHODS

To examine the affinities of both panda species through time, I surveyed the systematic literature from the description of *Ailuropoda* by David (1869) to the present. In total, 105 studies presented evidence on the position of either *Ailuropoda* or *Ailurus;* this list is not exhaustive. A breakdown of the studies according to data source and whether they provided phylogenetic information about *Ailuropoda, Ailurus,* or both is provided in tables 1.2 and 1.3.

DATA

Information from the literature was analyzed in one of two ways. These methods differ with respect to whether panda relationships were examined in isolation or not. The first method assessed simple statements of phylogenetic affinity of the form "*Ailuropoda* is most closely related to. . . . " To summarize this information quantitatively, I derived a simple affinity metric. Statements advocating an ursid origin were scored arbitrarily as 1, those advocating a procyonid origin as –1. If ursids or procyonids formed the sister group, but not the immediate sister group to either panda species (an "extended" relationship), scores of 0.5 and –0.5 were given, respectively. When neither ursids nor procyonids could be said to be more closely related than the other, a score of 0 was given, regardless of the identity of the inferred sister group. This includes the case for which the pandas were held to be one another's closest relatives. For any set of studies, the average value of the metric varies between –1 and 1, with more positive values indicating increasing ursid affinity and more negative values indicating increasing procyonid affinity. Values tending to zero indicate a relationship to neither group, whether due to conflicting opinions and/or an inferred relationship to another carnivore taxon.

The second method used the supertree construction method of matrix representation with parsimony analysis (MRP) (Baum 1992; Ragan 1992) to maintain the context of all other carnivore taxa mentioned in the source study. In this way, a consensus of carnivore phylogeny at any given time could be obtained, something that is possible only through supertree analysis. Combination of the primary data (total evidence) (sensu Kluge 1989) requires that these data be available and compatible. For many studies, particularly the older ones, the primary data were either not provided or were given simply in the form of a statement of phylogenetic affinity. Data types were also incompatible, meaning that they could not be analyzed simultaneously using a common algorithm. Combination of the source tree topologies using various consensus techniques ("taxonomic congruence") (sensu Mickevich 1978) was also impossible, due to the requirement that all source trees possess the same set of species.

TABLE 1.2

Survey of the Systematic Literature since 1869 Bearing on the Phylogenetic Placement of Ailuropoda and Ailurus

STUDY	SISTER GROUP OF AILUROPODA	SISTER GROUP OF AILURUS	DATA SOURCE	EVIDENCE
David (1869)	Ursidae		(1)	Morphology
Milne-Edwards (1870)*	Procyonidae	Procyonidae	(1)	Osteological characters and dentition similar to lesser panda
Gervais (1870)	Ursidae	Procyonidae	(1)	Intracranial cast; skeletal morphology
Mivart (1885)*	Procyonidae	Procyonidae	(1)	Overall morphology, but primarily skull architecture and dental morphology
Flower and Lydekker (1891)*	Ursidae	Procyonidae	(1)	Review of mammals; similarity to fossil ursid, *Hyaenarctos*
Winge (1895, 1941)*	Ursidae	Procyonidae	(1)	Skeletal morphology; similarity to fossil ursid, *Hyaenarctos*; dentition
Trouessart (1898, 1904)*	*Ailurus* (Ursidae)	*Ailuropoda* (Ursidae)	(1)	Taxonomy of mammals
Schlosser (1899)*	Ursidae	Procyonidae	(1)	Similarity to fossil ursid, *Hyaenarctos*; dentition
Lankester (1901)*	Procyonidae	Procyonidae	(1)	Skull, limb, and dental morphology
Lydekker (1901)	Procyonidae	Procyonidae	(1)	Skull, limb, and dental morphology
Beddard (1902)*	Ursidae	Procyonidae	(1)	Review of mammals
Kidd (1904)*	Felidae	Procyonidae	(1)	Arrangement of hair on nasal region
Weber (1904)	Ursidae	Procyonidae	(1)	Similarity to fossil ursid, *Hyaenarctos*
Bardenfleth (1914)	Ursidae	Procyonidae	(1)	Dental and osteological morphology
Pocock (1921)	Ailuropodidae	Ailuridae	(1)	External morphology (primarily feet, ears, rhinaria, and genitalia)
Matthew and Granger (1923)	Ursidae		(1)	Morphology; similarity to fossil ursid, *Hyaenarctos*

(*continued*)

TABLE 1.2 (continued)

STUDY	SISTER GROUP OF AILUROPODA	SISTER GROUP OF AILURUS	DATA SOURCE	EVIDENCE
Pocock (1928)	Ailuropodidae	Ailuridae	(1)	Soft external features of head and foot
Weber (1928)*	Ursidae	Procyonidae	(1)	Overall morphology; fossil affinities
Matthew (1929)	Ursidae	Ursidae	(1)	Morphological review of ursids
De Carle Sowerby (1932)*	*Ailurus*	*Ailuropoda*	(1)	Gross cranial morphology; dentition; coloration
Boule and Piveteau (1935)*	Ursidae	Procyonidae	(1)	Dental and basicranial morphology; fossil affinities
Gregory (1936)*	*Ailurus* (Procyonidae)	*Ailuropoda* (Procyonidae)	(1)	Skull and dental morphology
Raven (1936)*	*Ailurus* (Procyonidae)	*Ailuropoda* (Procyonidae)	(1)	Visceral and vascular anatomy
Segall (1943)*	*Ailurus*	*Ailuropoda*	(1)	Morphology of auditory region and ossicles
Kretzoi (1945)*	Ailuropodidae	Ailuridae	(1)	Dentition; fossil affinites
Simpson (1945)	*Ailurus* (Procyonidae)	*Ailuropoda* (Procyonidae)	(1)	Review of mammals
Mettler and Goss (1946)	Ursidae	Ursidae	(1)	Gross external brain morphology
Erdbrink (1953)	Ursidae	Ursidae	(1)	Dentition
Colbert (1955)*	*Ailurus* (Procyonidae)	*Ailuropoda* (Procyonidae)	(1)	Review of vertebrates
Leone and Wiens (1956)*	Ursidae	Ursidae	(2)	Precipitin test; serum proteins
Piveteau (1961)*	Procyonidae	Procyonidae	(1)	Dental morphology; affinity with fossil taxa
Davis (1964)	Ursidae	Unresolved	(1)	Comparative anatomy
Walker (1964)	Procyonidae	Procyonidae	(1)	Review of mammals
Wurster and Benirschke (1968), and Wurster (1969)*	Ursidae	Procyonidae	(2)	Karyology
Kretzoi (1971)	Ailuropodidae	Ailuridae	(1)	Dentition
Hendey (1972, 1980)	Ursidae	Ursidae	(1)	Morphology (primarily of dentition); fossil relationships
Sarich (1973, 1975, 1976)*	Ursidae	Ursidae	(2)	Immunological distance

Ewer (1973)*	*Ailurus* (Procyonidae)	*Ailuropoda* (Procyonidae)	(1)	Review of carnivores
Cave (1974)*	Procyonidae		(1)	Morphology of epipharyngeal bursa
Chu (1974)*	Ailuropodidae	Procyonidae	(1)	Morphology; dentition; ethology; neonatal allometry
Pei (1974)	Ursidae		(1)	Fossil evidence; dentition
Wang (1974)	Ursidae		(1)	Fossil evidence; cranial and dental morphology
Bugge (1978)*	Procyonidae		(1)	Morphology of cephalic arterial system
Starck (1978)*	Ursidae	Unresolved	(1)	Morphology, paleontology, and geography
De Ridder (1979)*	Ursidae	Procyonidae	(3)	Review of all available evidence
Thenius (1979)*	Ursidae	Procyonidae	(3)	Review of paleontological, morphological, serological, karyological, and ethological characters
Wurster-Hill and Bush (1980)*	Procyonidae (extended)	Procyonidae (extended)	(2)	Banded karyology
Eisenberg (1981)	Uncertain	Uncertain	(3)	Review of mammals
Morris and Morris (1981)*	*Ailurus* (Procyonidae)	*Ailuropoda* (Procyonidae)	(3)	Summary of available data (morphological, molecular, and ethological)
Pan et al. (1981)*	Ursidae		(2)	Serology
Schmidt-Kittler (1981)*	Procyonidae	Procyonidae	(1)	Morphology of dentition and skull; fossil evidence
Ginsburg (1982)*	Ursidae	Ursidae	(1)	Skull and dental morphology
Honacki et al. (1982)	Ursidae	Procyonidae	(3)	Review of mammals
Peters (1982)*	Procyonidae	Procyonidae	(1)	Vocalization structure
Nowak and Paradiso (1983)	Ursidae	Procyonidae	(3)	Review of mammals
Wozencraft (1984)*	Ursidae	Ursidae	(1)	Morphology
Braunitzer et al. (1985)	Ursidae		(2)	Globin sequence
Couturier and Dutrillaux (1985)*	Procyonidae	Procyonidae	(2)	Karyology
Eisentraut (1985)*	Ursidae	Procyonidae	(1)	Ridges on hard palate
Feng et al. (1985, 1991)*	*Ailurus*	*Ailuropoda*	(2)	Protein electrophoresis

(*continued*)

TABLE 1.2 (*continued*)

STUDY	SISTER GROUP OF AILUROPODA	SISTER GROUP OF AILURUS	DATA SOURCE	EVIDENCE
O'Brien et al. (1985)*	Ursidae	Procyonidae	(2)	DNA hybridization; isozyme genetic distance; immunological distance; karyological evidence
Pirlot et al. (1985)*	Procyonidae		(1)	Relative brain size; external brain morphology
Schaller et al. (1985)	Ursidae (extended)	Ursidae (extended)	(3)	Review of fossil, morphological, and molecular data
Romer and Parsons (1986)*	Ursidae	Procyonidae	(1)	Review of vertebrates
Tagle et al. (1986)*	Procyonidae (extended)	Procyonidae (extended)	(2)	Globin sequence
Braunitzer and Hofmann (1987)*	*Ailurus* (Procyonidae)	*Ailuropoda* (Procyonidae)	(2)	Globin sequence
Hofmann and Braunitzer (1987)*	Ursidae	Ursidae	(2)	Globin sequence
Liang and Zhang (1987)*	Ailuropodidae		(2)	Amino acid composition of LDH-M4 isozymes
Ramsay and Dunsbrack (1987)*	Ursidae		(1)	Life history traits
Flynn et al. (1988)*	Ursidae	Uncertain	(3)	Review of carnivores
Kamiya and Pirlot (1988)*	Ursidae	Unresolved	(1)	Brain size and morphology
Goldman et al. (1989)*	Ursidae	Procyonidae	(2)	Protein electrophoresis
Qiu and Qi (1989)*	Ursidae	Procyonidae	(1)	Dental morphology of new fossil discovery
Wang et al. (1989)*	Ursidae		(2)	Immunology
Wayne et al. (1989)*	Ursidae	Procyonidae	(2)	DNA hybridization
Wozencraft (1989)*	Ursidae	Ursidae	(1)	Overall morphology
Czelusniak et al. (1990)*	Ursidae	Ursidae	(2)	Globin sequences
Taylor (1990)*	*Ailurus* (Ursidae)	*Ailuropoda* (Ursidae)	(3)	Review of fossil, morphological, and molecular data
Czelusniak et al. (1991)*	Ursidae	Ursidae	(2)	Globin sequences
Dziurdzik and Nowogrodzka-Zagórska (1991)*	Ursidae	Procyonidae	(1)	Histological structure of hairs

Study	Ailuropoda	Ailurus	Data source	Data type
Nowak (1991)	Ursidae	Procyonidae	(3)	Review of mammals
Zhang and Shi (1991)*	Ailuropoda (Ursidae)	Ailurus (Ursidae)	(2)	mtDNA restriction site analysis
Baryshnikov and Averianov (1992)*		Procyonidae	(1)	Morphology of deciduous dentition
Hashimoto et al. (1993)*	Ursidae	Ursidae	(2)	Globin sequence
Wolsan (1993)*		Procyonidae (extended)	(1)	Skull and dental morphology
Wozencraft (1993)*	Ailuropoda (Ursidae)	Ailuropoda (Ursidae)	(3)	Review of carnivores
Wyss and Flynn (1993)*		Ursidae (extended)	(1)	Morphology
Zhang and Ryder (1993)*	Ursidae	Unresolved	(2)	mtDNA sequence analysis
Vrana et al. (1994)*	Ursidae	Ursidae (extended)	(3)	Total evidence of mtDNA and morphology
Zhang and Ryder (1994)*	Ursidae	Procyonidae	(2)	mtDNA sequence analysis
Lento et al. (1995)*	Ursidae	Unresolved	(2)	Spectral analysis of mtDNA sequence data
Pecon Slattery and O'Brien (1995)*	Ursidae	Procyonidae	(2)	mtDNA; protein electrophoresis
Ledje and Arnason (1996a)*	Ursidae	Unresolved	(2)	Cytochrome b sequence analysis
Ledje and Arnason (1996b)*	Ursidae	Unresolved	(2)	12S rDNA sequence analysis
Talbot and Shields (1996)*	Ursidae		(2)	mtDNA sequence analysis
Lin et al. (1997)	Ursidae		(2)	RAPD DNA hybridization
Wang (1997)*		Procyonidae	(1)	Skull and dental morphology
Flynn and Nedbal (1998)*	Ursidae	Procyonidae (extended)	(3)	Total evidence of morphology and DNA sequences
Lan and Wang (1998)*	Ursidae	Procyonidae	(2)	RFLP analysis of rDNA
Schreiber et al. (1998)*		Ailuridae	(2)	Comparative determinant analysis (immunology)
Waits et al. (1999)	Ursidae		(2)	mtDNA sequence analysis

NOTES: A statement such as "Ailurus (Ursidae)" means that Ailurus was held to be the sister taxon to Ailuropoda with ursids forming a sister group to both. The term "extended" means that the group listed was the closest sister group between ursids and procyonids to the panda species; however, the panda species had an even closer sister group relationship with some other carnivore taxon. For data source: (1), morphological; (2), molecular; (3), both morphological and molecular (total evidence). Studies marked with an asterisk provided source trees for the supertree analyses.

TABLE 1.3
Number of Studies Providing Statements of Phylogenetic Affinity for Ailuropoda *and* Ailurus, *or Providing Source Trees for the Supertree Analysis*

	PHYLOGENETIC AFFINITY	SUPERTREE ANALYSIS
Ailuropoda	90	71
Ailurus	79	76
Either panda species	105	80
Both panda species	72	64
Total	105	116

SOURCES: Thirty-six source trees that do not mention either *Ailuropoda* or *Ailurus* were included in the supertree analysis to give a better estimate of the family level relationships within Carnivora. Additional source trees were obtained from Gregory and Hellman (1939), Sarich (1969a,b), Seal et al. (1970), Hunt (1974), Radinsky (1975), Tedford (1976), Arnason (1977), Hendey (1978), Ling (1978), Schmidt-Kittler (1981), Dutrillaux et al. (1982), Flynn and Galiano (1982), Goodman et al. (1982), De Jong (1986), Wyss (1987), Holmes (1988), Rodewald et al. (1988), Ahmed et al. (1990), Nojima (1990), McKenna (1991), Janczewski et al. (1992), Arnason and Ledje (1993), Garland et al. (1993), Hunt and Tedford (1993), Veron and Catzeflis (1993), Berta and Wyss (1994), Hunt and Barnes (1994), Masuda and Yoshida (1994), Slade et al. (1994), Arnason et al. (1995), Austin (1996), Bininda-Emonds and Russell (1996), Werdelin (1996), and Ortolani (1999).

In contrast, MRP can combine phylogenetic information from any study, be it in the form of a tree or a simple statement, by coding it as a series of binary elements. These elements are then combined into a single matrix that is analyzed using parsimony to derive a tree that best summarizes the hierarchical information in the set of source trees. Briefly, each node from every source tree is coded in turn as follows: if a given species is descended from that node, it is scored as 1; if it is not, it is scored as 0. Species that are not present in a particular study, but are present in others, are scored as ? for that particular study (figure 1.1) (Sanderson et al. 1998; Bininda-Emonds 2000b). In this way, supertree construction can combine studies examining different sets of species. Simulation studies show that MRP is as accurate as total evidence in cases in which both methods can be applied (Bininda-Emonds and Sanderson 2001).

Matrix representations for all source trees were constructed by eye. Supertree analysis used PAUP* version 4.0b2 (Swofford 1999). Searches always used the exact branch-and-bound algorithm, thereby guaranteeing that all of the most parsimonious solutions for the data were found.

The supertree was the strict consensus of all equally most parsimonious solutions. Differential support for the relationships within a supertree was quantified using the Bremer decay index (Bremer 1988; Källersjö et al. 1992), because the bootstrap is inappropriate due to character nonindependence (Purvis 1995). The Bremer decay index measures the number of additional steps over the most optimal length before a node of interest is contradicted. Nodes that remain in the strict consensus solution of increasingly suboptimal trees are not readily contradicted and are therefore inferred to have more support.

SLIDING WINDOW ANALYSIS

To view changes in phylogenetic opinion over time, I employed a sliding window approach to time series analysis. Specifically, the data sources were arranged in ascending chronological order and secondarily by author name in ascending alphabetical order. Contiguous, overlapping sets of data sources (e.g., sources 1–10, 2–11, 3–12) were then analyzed.

For statements of phylogenetic affinity, I calculated the average value of the affinity metric

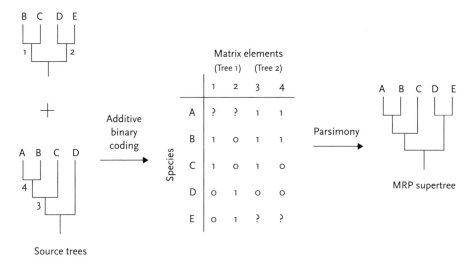

FIGURE 1.1. Basic procedure of supertree construction using matrix representation with parsimony (MRP).

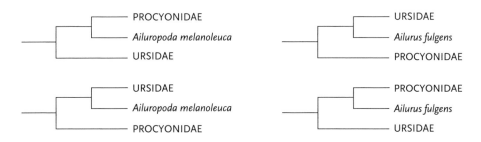

FIGURE 1.2. Backbone constraint trees used for the supertree analyses in PAUP*, forcing *Ailuropoda* or *Ailurus* into a sister group relationship with either ursids or procyonids.

for windows that were either five or ten studies in size. The overall consensus opinion at any given point in time was also obtained from the cumulative average of the affinity metric for all studies thus far included.

For the supertree analyses, the window size was fifteen studies. I used the affinity metric to summarize the placement of both *Ailuropoda* and *Ailurus* in the supertree of each window. I also examined the support for these inferred placements by using backbone constraint trees (see Swofford 1999) to force PAUP* to search only for solutions in which each panda species was more closely related to ursids than to procyonids and vice versa (figure 1.2). Support for these alternative placements was quantified by how much less parsimonious they were than the optimal length for that window. Unlike the Bremer decay index, higher values in this case indicate decreasing support for the constrained placement.

RESULTS

TYPES OF EVIDENCE

Unsurprisingly, morphological evidence dominates until the late 1970s (see table 1.2). Thereafter, molecular data come to bear increasingly on the question of panda relationships, either alone or in concert with morphological evidence (total evidence). From the late 1980s, the phylogenetic placement of either *Ailuropoda* or *Ailurus* has been examined using molecular data almost exclusively. Behavioral information has

TABLE 1.4
Summary of Statements of Phylogenetic Affinity for Ailuropoda *and* Ailurus

	NUMBER OF STUDIES	
SISTER GROUP	AILUROPODA	AILURUS
Ursidae	57	12
Other panda species within Ursidae	4	4
Total	61	16
Procyonidae	11	41
Other panda species within Procyonidae	7	7
Total	18	48
Other panda species	3	3
Other panda species within Ursidae or Procyonidae	11	11
Total	14	14
Unresolved or other	8	4

only been used sporadically throughout the survey period.

STATEMENTS OF PHYLOGENETIC AFFINITY

Taken together, all statements of phylogenetic affinity strongly place *Ailuropoda* and *Ailurus* within separate carnivore families (table 1.4). Roughly two-thirds of the ninety studies mentioning *Ailuropoda* cluster it with ursids, whereas a slightly smaller fraction of the seventy-nine studies for *Ailurus* place it with procyonids. The two panda species were held to be one another's closest relative only fourteen times, and usually within either Ursidae or Procyonidae. These observations are captured by the affinity metric. Over all studies, *Ailuropoda* possesses a value of 0.48, whereas *Ailurus* shows a value of –0.41.

The sliding window analysis demonstrates that these overall opinions are largely reflected in any time window since 1869 (figure 1.3). The trends are roughly identical for windows of either five or ten studies in size, although the former unsurprisingly displayed slightly greater fluctuations. Except for two occasions, *Ailuropoda* is always inferred to be more closely related to ursids on average. This is especially true from the late 1980s on, when all windows unequivocally indicate *Ailuropoda* to be a member of the Ursidae. Procyonid or uncertain affinities for *Ailuropoda* are only obtained sporadically across a relatively broad period from the 1940s to the mid-1970s, and a single instance around the mid-1980s. *Ailurus* is usually firmly held to have procyonid affinities, although the windows tend toward 0 (i.e., unresolved or other affinities) with time. In the 1990s, many windows indicate *Ailurus* to have ursid affinities. However, the most recent windows again cluster *Ailurus* more closely with procyonids.

These same trends are also evident when statements of phylogenetic affinity are viewed cumulatively (figure 1.4). Even during an initial period of uncertainty (marked by large fluctuations), the weight of all opinion up to a given time almost always has *Ailuropoda* more closely related to ursids than to procyonids. Moreover, this opinion is generally strengthening with time, particularly from the mid-1980s, as the line moves to more positive values. In contrast, *Ailurus* is always held as being more closely related

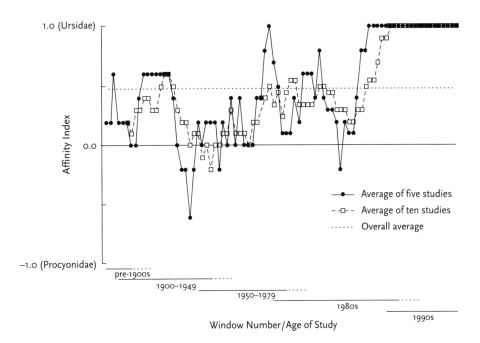

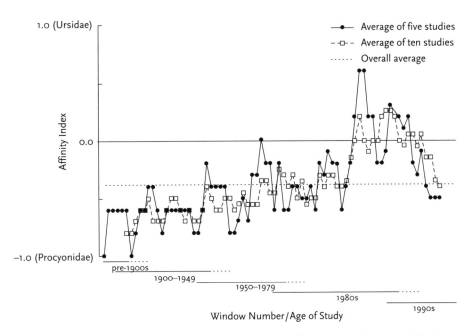

FIGURE 1.3. Sliding window analysis of statements of phylogenetic affinity for (A) *Ailuropoda* and (B) *Ailurus*, using the affinity metric discussed in the text. Approximate time spans of the windows are given on the *x*-axis. The dotted extensions apply to the windows of ten studies only. The overall averages for all studies were 0.48 for *Ailuropoda* and −0.41 for *Ailurus*.

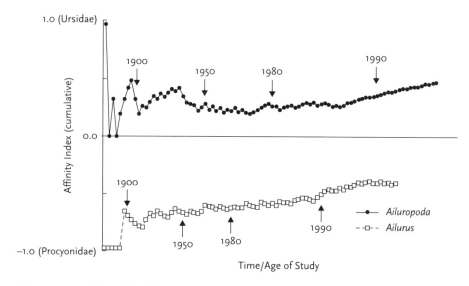

FIGURE 1.4. Cumulative values of statements of phylogenetic affinity for *Ailuropoda* or *Ailurus* using the affinity metric discussed in the text.

to procyonids. Again, this placement is becoming slightly more uncertain with time, as the line tends toward less negative values.

SUPERTREE ANALYSIS

The sliding window analysis of consensus supertrees, in which the positions of the panda species were put in the context of higher-level carnivore relationships, largely identified the same trends as noted above (figure 1.5). The only area of disagreement exists before 1950. During this period, both panda species show identical placements, either (1) not distinctly related to either ursids or procyonids or (2) as the sister taxon to procyonids plus some other carnivore taxon. Although this disagreement mirrors the sliding window analysis of phylogenetic statements for *Ailuropoda*, it weakly contradicts the analogous findings during this time that held *Ailurus* to be more closely related to procyonids (see figure 1.3B; but see below). After the 1950s, supertrees in virtually every window place *Ailuropoda* as the sister group to ursids. *Ailurus* meanwhile is usually clustered with procyonids, except for periods in the 1980s and 1990s, when it clusters distantly with ursids or its placement is equivocal between ursids and procyonids. On the whole, 81.0% of the windows in figure 1.5 placed *Ailuropoda* more closely with ursids, whereas 81.9% placed *Ailurus* more closely with procyonids.

Support for an ursid versus procyonid relationship for each panda species is given in figure 1.6. For *Ailuropoda*, an ursid affinity is usually the more parsimonious solution, particularly from the mid-1980s onward. A sister group relationship with procyonids is more parsimonious only before 1950 and for a brief time during the mid-1980s. The reverse is true for *Ailurus*: a procyonid affinity is usually the more parsimonious. This includes the period before 1950, indicating that *Ailurus* also clusters equally parsimoniously with other nonursid carnivore groups to produce the unresolved result seen in figure 1.5. It is only during the late 1980s to mid-1990s that an *Ailurus*-ursid pairing is the more parsimonious. The placement of *Ailurus* is also generally more uncertain than that of *Ailuropoda*. The difference in the length of the competing topologies for *Ailurus* (maximum = 7.9%) is typically much less than those for *Ailuropoda* (maximum = 14.8%), revealing that placements of *Ailurus* are not as strongly supported. Moreover, whereas

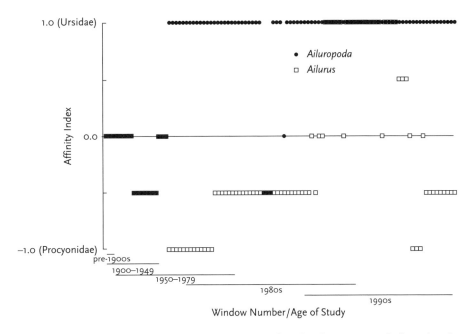

FIGURE 1.5. Sliding window analysis of supertrees to examine the inferred sister groups of *Ailuropoda* and *Ailurus*, as quantified using the affinity metric discussed in the text.

the length difference is increasing for *Ailuropoda* with time (indicating increasing certainty), it is decreasing slightly for *Ailurus*.

The supertree obtained from all 116 source trees is completely resolved (figure 1.7). The high values for the goodness-of-fit measures the consistency index (CI), the retention index (RI), and the rescaled consistency index (RC) (see Farris 1989) indicate generally good agreement among the source trees. There is strong support for a sister group relationship between *Ailuropoda* and ursids. *Ailurus* clusters with procyonids, but this is comparatively weakly supported. Supertrees obtained for each of the major data sources used (morphological, molecular, and total evidence; figure 1.8) also place *Ailuropoda* with ursids. Again, support for this placement is strong, but comparatively higher for the molecular and total evidence supertrees. The different data sources indicate different affinities for *Ailurus*: as the sister group to procyonids (morphological) or musteloids (mustelids plus procyonids; molecular), or unresolved within arctoids (total evidence). Except for the morphological supertree, support for each placement is comparatively weak within each supertree.

DISCUSSION

Despite being one of the most celebrated cases of controversy in mammalian systematics, virtually all lines of evidence hold *Ailuropoda* to be more closely related to ursids than it is to procyonids. Moreover, such a placement is favored relatively consistently through time and by each of morphological and behavioral (i.e., phenotypic), molecular, and total evidence studies. This arrangement enjoys strong support at most times and has not been contradicted since the mid-1980s. Little doubt should now remain that *Ailuropoda* is the sister group to the true bears.

Instead, despite receiving much less attention due to a greater apparent consensus, it is the position of *Ailurus* within carnivores that is much more doubtful. Although it is usually held to have procyonid affinities at any given period since 1869, the strength of this inference is comparatively weak and perhaps decreasing with

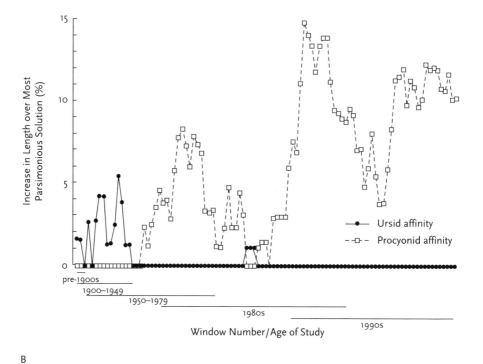

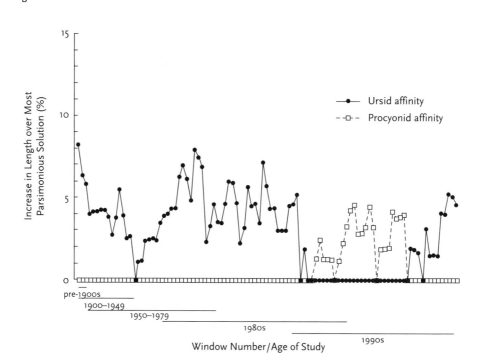

FIGURE 1.6. Sliding window analysis of the relative increase in length over the most parsimonious solution for that window when (A) *Ailuropoda* or (B) *Ailurus* are constrained to have either ursid or procyonid affinities in the supertree analysis (see figure 1.2). Higher values indicate decreasing support for the constrained placement.

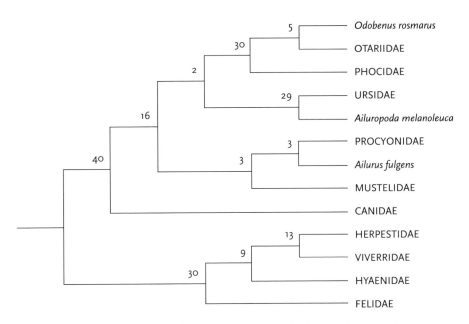

FIGURE 1.7. Overall supertree of family level relationships within Carnivora, as determined from 116 source trees spanning the years 1870 to 1999 inclusive. The single most parsimonious solution had a length of 614 steps, CI = 0.681, RI = 0.749, and RC = 0.510. Support throughout the supertree is given above each branch in the form of Bremer decay indices. Higher values indicate increasing support.

time. Only morphological studies provide reasonable support for this arrangement. Instead, several recent, mostly molecular studies propose an ursid affinity for *Ailurus* and many others are equivocal on the matter (see table 1.2). As such, the position of *Ailurus* is perhaps less clear now than at any time in the past. Much of this may derive from other evidence that indicates *Ailurus* to be the last surviving member of a relatively ancient lineage, one that may extend close to the origins of the major arctoid lineages (Sarich 1976; O'Brien et al. 1985; Bininda-Emonds et al. 1999). Compounded with evidence of a rapid adaptive radiation around this time (Bininda-Emonds et al. 1999), it has proved extremely difficult to resolve the position of *Ailurus* with any certainty or consistency. Much more research effort, using a wide variety of data types, is required.

Of the alternative evolutionary scenarios mentioned earlier, *Ailuropoda* should be viewed as a bear adapted to a herbivorous diet (as are *Helarctos*, *Tremarctos*, and *Ursus thibetanus*) rather than a small herbivore that has converged on a larger ursid body plan (cf. Davis 1964). Features shared with the herbivorous procyonids, such as an enlarged dentition, are therefore instances of convergence, possibly facilitated by both lineages being derived from the same arctoid body plan.

Likewise, the apparent procyonid affinities of *Ailurus* mean that derived features shared by the two panda species (perhaps including the common name "panda") (see Mayr 1986) should be viewed as convergent. However, a definitive statement in this regard is not possible, given the uncertain position of *Ailurus*. If *Ailurus* does have ursid affinities, as suggested by several recent studies, then its similarities with *Ailuropoda* would cease to be convergent, although they might still be primitive.

I refrain from making any taxonomic conclusions in this chapter, even for *Ailuropoda*, for which the phylogenetic position seems reasonably secure. Although conservation decisions and priorities can be based on taxonomic information (e.g., Lockwood 1999; see also May 1990; Vane-Wright et al. 1991), such information is often only a crude approximation to the phylogenetic history of a group. Furthermore, taxonomic

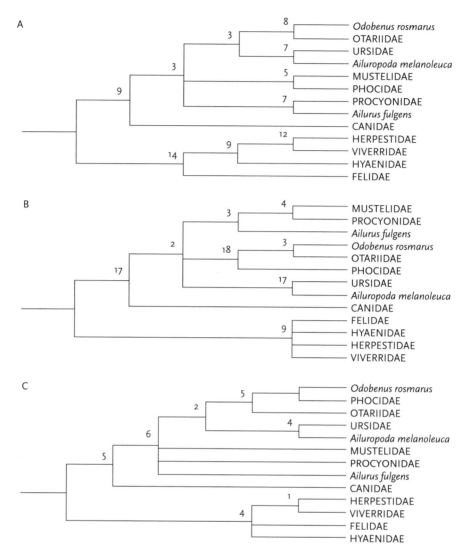

FIGURE 1.8. Supertrees of family level relationships within Carnivora, as determined from source trees derived from (A) morphological ($N = 2$, length = 287, CI = 0.662, RI = 0.734; RC = 0.486), (B) molecular ($N = 6$, length = 241, CI = 0.718, RI = 0.764; RC = 0.548), or (C) total evidence ($N = 4$, length = 74, CI = 0.757, RI = 0.861; RC = 0.651) data. Support throughout each supertree is given above each branch in the form of Bremer decay indices.

assessments are subjective and can frequently obscure or even misrepresent phylogenetic information. For instance, by placing *Ailuropoda* in its own family (Ailuropodidae), we gain the knowledge that it is (subjectively) "distinct" at the cost of realizing its close relationship and therefore similarity with ursids. Although the former piece of information is an important factor in establishing conservation priorities, the latter is critical for conservation practice. In managing *Ailuropoda*, we will likely have greater success by adapting existing ursid conservation programs because of key similarities between all the species (e.g., the slow reproductive rate and associated life history traits) (see Garshelis, chapter 4). Instead, we would be better served by using the more resolved and accurate phylogenetic information whenever possible (Crozier 1997; Nee and May 1997; Vázquez and Gittleman 1998). Conservation priorities can be set using

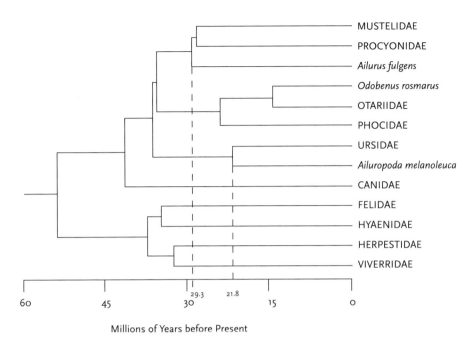

FIGURE 1.9. Best estimate of the relationships of *Ailuropoda* and *Ailurus*, together with times of divergence. Adapted from Bininda-Emonds et al. (1999).

metrics that quantify phylogenetic diversity or distinctiveness from phylogenies with branch length information (e.g., Faith 1994; Crozier 1997). In this regard, both *Ailuropoda* and *Ailurus* would be regarded as worthy of conservation, because they represent the sole surviving members of relatively ancient lineages. This can be clearly seen from figure 1.9, which contains what I think to be the best current estimate of higher-level relationships and divergence times within the carnivores. Based on both fossil and molecular data, the lineage giving rise to *Ailurus* probably diverged from the common ancestor of mustelids and procyonids about 29.3 million years before present, whereas the lineage for *Ailuropoda* separated from that leading to the true ursids about 21.8 million years before present (Bininda-Emonds et al. 1999).

ACKNOWLEDGMENTS

I thank Lee Hagey and Don Lindburg for the opportunity to speak at the Phylogeny and Conservation session of *Panda 2000*, and John Gittleman for his support and encouragement. Steve O'Brien kindly supplied copies of some otherwise difficult to obtain source studies. Financial support for this study was provided by both *Panda 2000* and a Natural Sciences and Engineering Research Council of Canada postdoctoral fellowship.

REFERENCES

Ahmed, A., M. Jahan, and G. Braunitzer. 1990. Carnivora: The primary structure of hemoglobin from adult coati *(Nasua nasua rufa*, Procyonidae). *J Protein Chem* 9:23–29.

Arnason, U. 1977. The relationship between the four principal pinniped karyotypes. *Hereditas* 87: 227–42.

Arnason, U., and C. Ledje. 1993. The use of highly repetitive DNA for resolving cetacean and pinniped phylogenies. In *Mammalian phylogeny: Placentals*, edited by F. S. Szalay, M. J. Novacek, and M. C. McKenna, pp. 74–80. New York: Springer-Verlag.

Arnason, U., K. Bodin, A. Gullberg, C. Ledje, and S. Mouchaty. 1995. A molecular view of pinniped relationships with particular emphasis on the true seals. *J Mol Evol* 40:78–85.

Austin, P. K. 1996. Systematic relationships of the Herpestidae (Mammalia: Carnivora). M.Sc. thesis. University of Illinois, Chicago.

Bardenfleth, K. S. 1914. On the systematic position of *Aeluropus melanoleucus*. *Mindeskrift Japetus Steenstrup Født8sel* 17:1–15.

Baryshnikov, G. F., and A. O. Averianov. 1992. Deciduous teeth of carnivorous mammals (order Carnivora) part III. The family Procyonidae. *Trudy Zool Inst* 246:103–28.

Baum, B. R. 1992. Combining trees as a way of combining data sets for phylogenetic inference, and the desirability of combining gene trees. *Taxon* 41: 3–10.

Beddard, F. E. 1902. *Mammalia*. London: Macmillan and Company.

Berta, A., and A. R. Wyss. 1994. Pinniped phylogeny. In *Contributions in marine mammal paleontology honoring Frank C. Whitmore, Jr.*, edited by A. Berta and T. A. Deméré. *Proc San Diego Soc Nat Hist* 29: 33–56.

Bininda-Emonds, O. R. P. 2000a. Factors influencing phylogenetic inference: A case study using the mammalian carnivores. *Mol Phylogenet Evol* 16: 113–26.

Bininda-Emonds, O. R. P. 2000b. Systematics: Supertree reconstruction. In *McGraw-Hill Yearbook of Science and Technology 2001*. New York: McGraw-Hill.

Bininda-Emonds, O. R. P., and A. P. Russell. 1996. A morphological perspective on the phylogenetic relationships of the extant phocid seals (Mammalia: Carnivora: Phocidae). *Bonn Zool Monogr* 41:1–256.

Bininda-Emonds, O. R. P., and M. J. Sanderson. 2001. Assessment of the accuracy of matrix representation with parsimony supertree construction. *Syst Biol* 50:565–79.

Bininda-Emonds, O. R. P., J. L. Gittleman, and A. Purvis. 1999. Building large trees by combining phylogenetic information: A complete phylogeny of the extant Carnivora (Mammalia). *Biol Rev* 74:143–75.

Boule, M., and J. Piveteau. 1935. *Les fossiles. Éléments de Paléontologie*. Paris: Masson.

Braunitzer, G., and O. Hofmann. 1987. Les hémoglobins des pandas. *C R Séances Soc Biol Fil* 181:116–21.

Braunitzer, G., A. Stangl, and R. Göltenboth. 1985. Preliminary results of a comparative study on the amino acid sequence of the giant panda (*Ailuropoda melanoleuca*, Carnivora). *Bongo (Berlin)* 10:183–84.

Bremer, K. 1988. The limits of amino acid sequence data in angiosperm phylogenetic reconstruction. *Evolution* 42:795–803.

Bugge, J. 1978. The cephalic arterial system in carnivores, with special reference to the systematic classification. *Acta Anat* 101:45–61.

Cave, A. J. E. 1974. The sacculus epipharyngeus in the giant panda, *Ailuropoda melanoleuca*. *J Zool* 172:123–31.

Chu, C. 1974. On the systematic position of the giant panda, *Ailuropoda melanoleuca* (David). *Acta Zool Sin* 20:174–87.

Colbert, E. H. 1955. *Evolution of the vertebrates: A history of the backboned animals through time*. New York: John Wiley and Sons.

Couturier, J., and B. Dutrillaux. 1985. Evolution chromosomique chez les Carnivores. *Mammalia* 50A:124–62.

Crozier, R. H. 1997. Preserving the information content of species: Genetic diversity, phylogeny, and conservation worth. *Annu Rev Ecol Syst* 28: 243–68.

Czelusniak, J., M. Goodman, B. F. Koop, D. A. Tagle, J. Shoshani, G. Braunitzer, T. K. Kleinschmidt, W. W. de Jong, and G. Matsuda. 1991. Perspectives from amino acid and nucleotide sequences on cladistic relationships among higher taxa of Eutheria. In *Current mammalogy*, edited by H. H. Genoways, pp. 545–72. New York: Plenum.

Czelusniak, J., M. Goodman, N. D. Moncrief, and S. M. Kehoe. 1990. Maximum parsimony approach to construction of evolutionary trees from aligned homologous sequences. *Methods Enzymol* 183: 601–15.

David, A. 1869. Extrait d'une lettre du même, datée de la Principauté Thibetaine (independente) de Moupin, le 21 Mars. *Nouv Arch Mus Hist Nat Paris, Bull* 5:12–13.

Davis, D. D. 1964. The giant panda: A morphological study of evolutionary mechanisms. *Fieldiana Zool Mem* 3:1–339.

De Carle Sowerby, A. 1932. The pandas or cat-bears. *China J* 17:296–99.

De Jong, W. W. 1986. Protein sequence evidence for monophyly of the carnivore families Procyonidae and Mustelidae. *Mol Biol Evol* 3:276–81.

De Jong, W. W. 1998. Molecules remodel the mammalian tree. *Trends Ecol Evol* 13:270–75.

De Ridder, M. 1979. De systematische plaats der panda's (Mammalia: Carnivora fissipedia). *Natuurweten Tijdschr* 61:163–73.

Dutrillaux, B., J. Couturier, and G. Chauvier. 1982. Notes et discussions sur "Édentes," Carnivores, "Pinnipèdes" et leurs parasites. 3. Les Pinnipèdes, monophylétiques, sont issus de Procyonidae ancestraux, et non d'Ursidae ni de Mustelidae. *Mém Mus Natl Hist Nat Sér A Zool* 123:141–43.

Dziurdzik, B., and M. Nowogrodzka-Zagórska. 1991. The histological structure of hairs of the giant panda, *Ailuropoda melanoleuca* (David, 1869), and the lesser panda, *Ailurus fulgens* (F. Cuvier, 1825), and the systematic position of these species. *Acta Zool Cracov* 34:463–74.

Eisenberg, J. F. 1981. *The mammalian radiations: An analysis of trends in evolution, adaptation, and behavior.* Chicago: University of Chicago Press.

Eisentraut, M. 1985. The pattern of ridges in the hard palate in procyonids and bears. *Bongo (Berlin)* 10:185–96.

Erdbrink, D. P. 1953. *A review of fossil and recent bears of the Old World, with remarks on their phylogeny, based upon their dentition.* Deventer: Jan de Lange.

Ewer, R. F. 1973. *The carnivores.* Ithaca: Cornell University Press.

Faith, D. P. 1994. Phylogenetic diversity: A general framework for the prediction of feature diversity. In *Systematics and conservation evaluation*, edited by P. L. Forey, C. J. Humphries, and R. I. Vane-Wright, pp. 251–268. Oxford: Oxford University Press.

Farris, J. S. 1983. The logical basis of phylogenetic analysis. In *Advances in cladistics*, edited by N. I. Platnick and V. A. Funk, pp. 7–36. New York: Columbia University Press.

Farris, J. S. 1989. The retention index and the rescaled consistency index. *Cladistics* 5:417–19.

Feng, W., C. Luo, Z. Ye, A. Zhang, and G. He. 1985. The electrophoresis comparison of serum protein and LDH isoenzyme in 5 Carnivora animals—giant panda, red panda, Asiatic black bear, cat and dog. *Acta Theriol Sin* 5:151–56.

Feng, W., C. Luo, Z. Ye, A. Zhang, and G. He. 1991. The electrophoresis comparison of serum protein and LDH isoenzyme in 5 Carnivora animals—giant panda, red panda, Asiatic black bear, cat and dog. *Sichuan Daxue Xuebao* 28:155–60.

Flower, W. H., and R. Lydekker. 1891. *An introduction to the study of mammals living and extinct.* London: Adam and Charles Black.

Flynn, J. J., and H. Galiano. 1982. Phylogeny of the early Tertiary Carnivora, with a description of a new species of *Protictis* from the middle Eocene of northwestern Wyoming. *Am Mus Novit* 2725:1–64.

Flynn, J. J., and M. A. Nedbal. 1998. Phylogeny of the Carnivora (Mammalia): Congruence vs incompatibility among multiple data sets. *Mol Phylogenet Evol* 9:414–26.

Flynn, J. J., N. A. Neff, and R. H. Tedford. 1988. Phylogeny of the Carnivora. In *The phylogeny and classification of the Tetrapods*, edited by M. J. Benton, pp. 73–116. Oxford: Clarendon.

Garland, T., Jr., A. W. Dickerman, C. M. Janis, and J. A. Jones. 1993. Phylogenetic analysis of covariance by computer simulation. *Syst Biol* 42:265–92.

Gervais, P. 1870. Mémoire sur les formes cérébrales propres aus carnivores vivants et fossiles suivi de remarques sur la classification de ces animaux. *Nouv Arch Mus Hist Nat Paris* 1:103–62.

Ginsburg, L. 1982. Sur la position systématique du petit panda, *Ailurus fulgens* (Carnivora, Mammalia). *Géobios Mém Spéc* 6:247–58.

Goldman, D., P. R. Giri, and S. J. O'Brien. 1989. Molecular genetic-distance estimates among the Ursidae as indicated by one- and two-dimensional protein electrophoresis. *Evolution* 43:282–95.

Goodman, M., A. E. Romero-Herrera, H. Dene, J. Czelusniak, and R. E. Tashian. 1982. Amino acid sequence evidence on the phylogeny of primates and other eutherians. In *Macromolecular sequences in systematics and evolutionary biology*, edited by M. Goodman, pp. 115–191. New York: Plenum.

Gregory, W. K. 1936. On the phylogenetic relationships of the giant panda *(Ailuropoda)*. *Am Mus Novit* 878:1–29.

Gregory, W. K., and M. Hellman. 1939. On the evolution and major classification of the civets (Viverridae) and allied fossil and recent Carnivora: A phylogenetic study of the skull and dentition. *Proc Am Philos Soc* 81:309–92.

Hashimoto, T., E. Otaka, J. Adachi, K. Mizuta, and M. Hasegawa. 1993. The giant panda is closer to a bear, judged by α- and β-hemoglobin sequences. *J Mol Evol* 36:282–89.

Hendey, Q. B. 1972. A Pliocene ursid from South Africa. *Ann S Afr Mus* 59:115–32.

Hendey, Q. B. 1978. Late Tertiary Hyaenidae from Langebaanweg, South Africa, and their relevance to the phylogeny of the family. *Ann S Afr Mus* 76:265–97.

Hendey, Q. B. 1980. Origin of the giant panda. *S Afr J Sci* 76:179–80.

Hennig, W. 1966. *Phylogenetic systematics.* Urbana: University of Illinois Press.

Hofmann, O., and G. Braunitzer. 1987. The primary structure of the hemoglobin of spectacled bear (*Tremarctos ornatus*, Carnivora). *Biol Chem Hoppe-Seyler* 368:949–54.

Holmes, T., Jr. 1988. Sexual dimorphism in North American weasels with a phylogeny of the Mustelidae. Ph.D. dissertation. University of Kansas, Lawrence.

Honacki, J. H., K. E. Kinman, and J. W. Koeppl, eds. 1982. *Mammal species of the world: A taxonomic and geographic reference.* Lawrence: Allen Press and the Association of Systematics Collections.

Hunt, R. M., Jr. 1974. The auditory bulla in Carnivora: An anatomical basis for reappraisal of carnivore evolution. *J Morphol* 143:21–76.

Hunt, R. M., Jr., and L. G. Barnes. 1994. Basicranial evidence for ursid affinity of the oldest pinnipeds. In *Contributions in marine mammal paleontology honoring Frank C. Whitmore, Jr.*, edited by A. Berta and T. A. Deméré. *Proc San Diego Soc Nat Hist* 29:57–67.

Hunt, R. M., Jr., and R. H. Tedford. 1993. Phylogenetic relationships within the aeluroid Carnivora and implications of their temporal and geographic distribution. In *Mammalian phylogeny: Placentals*, edited by F. S. Szalay, M. J. Novacek, and M. C. McKenna, pp. 53–73. New York: Springer-Verlag.

Janczewski, D. N., N. Yuhki, D. A. Gilbert, G. T. Jefferson, and S. J. O'Brien. 1992. Molecular phylogenetic inference from saber-toothed cat fossils of Rancho La Brea. *Proc Natl Acad Sci USA* 89:9769–73.

Källersjö, M., J. S. Farris, A. G. Kluge, and C. Bult. 1992. Skewness and permutation. *Cladistics* 8:275–87.

Kamiya, T., and P. Pirlot. 1988. The brain of the lesser panda *Ailurus fulgens*: A quantitative approach. *Z Zool Syst Evolutionsforsch* 26:65–72.

Kidd, W. 1904. On the arrangement of the hair on the nasal region of the parti-coloured bear *(Aeluropus melanoleucus). Proc Zool Soc Lond* 1904:373.

Kluge, A. G. 1989. A concern for evidence and a phylogenetic hypothesis of relationships among *Epicrates* (Boidae, Serpentes). *Syst Zool* 38:7–25.

Kretzoi, M. 1945. Bemerkungen über das Raubtiersystem. *Ann Hist Natur Musei Natl Hungar* 38:59–83.

Kretzoi, N. 1971. Kritische Bemerkungen zur Abstammung der Ursiden. *Vertebrata Hungar* 12:123–32.

Lan, H., and W. Wang. 1998. Phylogenetic relationships among giant panda and related species based on restriction site variations in rDNA spacers. *Zool Res* 19:337–43.

Lankester, E. R. 1901. On the affinities of *Aeluropus melanoleucus*, A. Milne-Edwards. *Trans Linn Soc Lond, Zool* 8:163–65.

Ledje, C., and U. Arnason. 1996a. Phylogenetic analyses of complete cytochrome *b* genes of the order Carnivora with particular emphasis on the Caniformia. *J Mol Evol* 42:135–44.

Ledje, C., and U. Arnason. 1996b. Phylogenetic relationships within caniform carnivores based on analyses of the mitochondrial 12S rRNA gene. *J Mol Evol* 43:641–49.

Lento, G. M., R. E. Hickson, G. K. Chambers, and D. Penny. 1995. Use of spectral analysis to test hypotheses on the origin of pinnipeds. *Mol Biol Evol* 12:28–52.

Leone, C. A., and A. L. Wiens. 1956. Comparative serology of carnivores. *J Mammal* 37:11–23.

Liang, S., and L. Zhang. 1987. A comparison of the primary structures of lactate dehydrogenase isozymes M_4 from giant panda, red panda, black bear and dog. *Sci Sin B (Chem, Biol, Agricult, Med, Earth Sci)* 30:270–82.

Lin, F., Y. Yang, Y. Zhang, H. Chen, L. Fei, Y. F. Song, G. He, and A. Zhang. 1997. A preliminary study on the taxonomy position of giant panda using RAPD. *Acta Theriol Sin* 17:161–64.

Ling, J. K. 1978. Pelage characteristics and systematic relationships in the Pinnipedia. *Mammalia* 42:305–13.

Lockwood, J. L. 1999. Using taxonomy to predict success among introduced avifauna: Relative importance of transport and establishment. *Conserv Biol* 13:560–67.

Lydekker, R. 1901. Detailed description of the skull and limb-bones of *Ailuropoda melanoleucus*. *Trans Linn Soc Lond, Zool* 8:166–71.

Masuda, R., and M. C. Yoshida. 1994. A molecular phylogeny of the family Mustelidae (Mammalia, Carnivora), based on comparison of mitochondrial cytochrome *b* nucleotide sequences. *Zool Sci* 11:605–12.

Matthew, W. D. 1929. Critical observations upon Siwalik mammals (exclusive of Proboscidea). *Bull Am Mus Nat Hist* 56:437–560.

Matthew, W. D., and W. Granger. 1923. New fossil mammals from the Pliocene of Sze-Chuan, China. *Bull Am Mus Nat Hist* 48:563–98.

May, R. M. 1990. Taxonomy as destiny. *Nature* 347:129–30.

Mayr, E. 1986. Uncertainty in science: Is the giant panda a bear or a raccoon? *Nature* 323:769–71.

McKenna, M. C. 1991. The alpha crystallin A chain of the eye lens and mammalian phylogeny. *Ann Zool Fenn* 28:349–60.

McKenna, M. C., and S. K. Bell. 1997. *Classification of mammals above the species level*. New York: Columbia University Press.

McLaren, I. A. 1960. Are the Pinnipedia biphyletic? *Syst Zool* 9:18–28.

Mettler, F. A., and L. J. Goss. 1946. The brain of the giant panda *(Ailuropoda melanoleuca)*. *J Comp Neurol* 84:1–9.

Mickevich, M. F. 1978. Taxonomic congruence. *Syst Zool* 27:143–58.

Milne-Edwards, A. 1870. Note sur quelques Mammifères du Thibet oriental. *Ann Sci Nat Cinq* 13:1.

Mitchell, E. D. 1967. Controversy over diphyly in pinnipeds. *Syst Zool* 16:350–51.

Mivart, St. G. 1885. On the anatomy, classification, and distribution of the Arctoidea. *Proc Zool Soc Lond* 1885:340–404.

Morris, R., and D. Morris. 1981. *The giant panda*. London: Macmillan.

Nash, W. G., and S. J. O'Brien. 1987. A comparative chromosome banding analysis of the Ursidae and their relationship to other carnivores. *Cytogenet Cell Genet* 45:206–212.

Nash, W. G., J. Wienberg, M. A. Ferguson-Smith, J. C. Menninger, and S. J. O'Brien. 1998. Comparative genomics: Tracking chromosome evolution in the family Ursidae using reciprocal chromosome painting. *Cytogenet Cell Genet* 83:182–92.

Nee, S., and R. M. May. 1997. Extinction and the loss of evolutionary history. *Science* 278:692–95.

Newnham, R. E., and W. M. Davidson. 1966. Comparative study of the karyotypes of several species in Carnivora including the giant panda *(Ailuropoda melanoleuca)*. *Cytogenetics* 5:152–63.

Nojima, T. 1990. A morphological consideration of the relationships of Pinnipedia to other carnivorans based on the bony tentorium and bony falx. *Mar Mamm Sci* 6:54–74.

Novacek, M. J. 1992. Mammalian phylogeny: Shaking the tree. *Nature* 356:121–25.

Nowak, R. M. 1991. *Walker's mammals of the world*. Baltimore: Johns Hopkins University Press.

Nowak, R. M., and J. L. Paradiso. 1983. *Walker's mammals of the world*. Baltimore: Johns Hopkins University Press.

O'Brien, S. J., W. G. Nash, D. E. Wildt, M. E. Bush, and R. E. Benveniste. 1985. A molecular solution to the riddle of the giant panda's phylogeny. *Nature* 317:140–44.

O'Brien, S. J., R. E. Beauveniste, W. G. Nash, J. S. Martenson, M. A. Eichelberger, D. E. Wildt, M. Bush, R. K. Wayne, and D. Goldman. 1991. Molecular biology and evolutionary theory: The giant panda's closest relatives. In *New perspectives on evolution*, edited by L. Warren and H. Koprowski, pp. 225–80. New York: John Wiley and Sons.

Ortolani, A. 1999. Spots, stripes, tail tips and dark eyes: Predicting the function of carnivore colour patterns using the comparative method. *Biol J Linn Soc* 67:433–76.

Pan, W., L. Chen, and N. Xiao. 1981. Serological study of giant panda and various mammalians. *Acta Scient Natur Univ* 1:79–88.

Pecon Slattery, J., and S. J. O'Brien. 1995. Molecular phylogeny of the red panda *(Ailurus fulgens)*. *J Hered* 86:413–22.

Pei, W. C. 1974. A brief evolutionary history of the giant panda. *Acta Zool Sin* 20:188–90.

Penny, D., and M. D. Hendy. 1986. Estimating the reliability of evolutionary trees. *Mol Biol Evol* 3:403–17.

Peters, G. 1982. A note on the vocal behaviour of the giant panda, *Ailuropoda melanoleuca* (David, 1869). *Z Säugetierkd* 47:236–46.

Pettigrew, J. D. 1986. Flying primates? Megabats have the advanced pathway from eye to midbrain. *Science* 231:1304–6.

Pettigrew, J. D. 1991. Wings or brain: Convergent evolution in the origins of bats. *Syst Zool* 40:199–216.

Pirlot, P., S. S. Jiao, and J. Q. Xie. 1985. Quantitative morphology of the panda bear in comparison with the brains of the raccoon and the bear. *J Hirnforsch* 26:17–22.

Piveteau, J. 1961. *Carnivora. Traité de Paléontologie*. Paris: Masson et Cie.

Pocock, R. I. 1921. The external characters and classification of the Procyonidae. *Proc Zool Soc Lond* 1921:389–422.

Pocock, R. I. 1928. Some external characters of the giant panda *(Ailuropoda melanoleuca)*. *Proc Zool Soc Lond* 1928:975–81.

Purvis, A. 1995. A modification to Baum and Ragan's method for combining phylogenetic trees. *Syst Biol* 44:251–55.

Qiu, Z., and G. Qi. 1989. Ailuropod found from the Late Miocene deposits in Lufeng, Yunnan. *Vertebrata Palasiatica* 27:153–69.

Radinsky, L. 1975. Viverrid neuroanatomy: Phylogenetic and behavioral implications. *J Mammal* 56:130–50.

Ragan, M. A. 1992. Phylogenetic inference based on matrix representation of trees. *Mol Phylogenet Evol* 1:53–58.

Ramsay, M. A., and R. L. Dunbrack. 1987. Is the giant panda a bear? *Oikos* 50:267.

Raven, H. C. 1936. Notes on the anatomy of the viscera of the giant panda *(Ailuropoda melanoleuca)*. *Am Mus Novit* 877:1–23.

Repenning, C. A. 1990. Oldest pinniped. *Science* 248:499.

Rodewald, K., G. Braunitzer, and R. Göltenboth. 1988. Carnivora: Primary structure of the hemoglobins from Ratel *(Mellivora capensis)*. *Biol Chem Hoppe-Seyler* 369:1137–42.

Romer, A. S., and T. S. Parsons. 1986. *The vertebrate body*. Philadelphia: Saunders College.

Sanderson, M. J., and L. Hufford, eds. 1996. *Homoplasy: The recurrence of similarity in evolution*. San Diego: Academic.

Sanderson, M. J., A. Purvis, and C. Henze. 1998. Phylogenetic supertrees: Assembling the trees of life. *Trends Ecol Evol* 13:105–9.

Sarich, V. M. 1969a. Pinniped origins and the rate of evolution of carnivore albumins. *Syst Zool* 18:286–95.

Sarich, V. M. 1969b. Pinniped phylogeny. *Syst Zool* 18:416–22.

Sarich, V. M. 1973. The giant panda is a bear. *Nature* 245:218–20.

Sarich, V. M. 1975. Pinniped systematics: Immunological comparisons of their albumins and transferrins. *Am Zool* 15:826.

Sarich, V. M. 1976. Transferrin. *Trans Zool Soc Lond* 33:165–71.

Schaller, G. B., J. Hu, W. Pan, and J. Zhu. 1985. *The giant pandas of Wolong*. Chicago: University of Chicago Press.

Schlosser, M. 1899. Über die Bären und bärenähnlichen des europäischen Teriärs. *Palaeontographica* 46:95–147.

Schmidt-Kittler, V. 1981. Zur Stammesgeschichte der marderverwandten Raubtiergruppen (Musteloidea, Carnivora). *Eclogae Geol Helv* 74:753–801.

Schreiber, A., K. Eulenberger, and K. Bauer. 1998. Immunogenetic evidence for the phylogenetic sister group relationship of dogs and bears (Mammalia, Carnivora: Canidae and Ursidae): A comparative determinant analysis of carnivoran albumin, C3 complement and immunoglobulin µ-chain. *Exp Clin Immunogenet* 15:154–70.

Seal, U. S., N. I. Phillips, and A. W. Erickson. 1970. Carnivora systematics: Immunological relationships of bear serum albumin. *Comp Biochem Physiol* 32:33–48.

Segall, W. 1943. The auditory region of the arctoid carnivores. *Zool Ser Field Mus Nat Hist* 29:33–59.

Simpson, G. G. 1945. The principles of classification and a classification of mammals. *Bull Am Mus Nat Hist* 85:1–350.

Slade, R. W., C. Moritz, and A. Heideman. 1994. Multiple nuclear-gene phylogenies: Application to pinnipeds and comparison with a mitochondrial DNA gene phylogeny. *Mol Biol Evol* 11:341–56.

Starck, D. 1978. *Vergleichende Anatomie der Wirbeltiere auf evolutionsbiologischer Grundlage*. Berlin: Springer-Verlag.

Swofford, D. L. 1999. PAUP*. *Phylogenetic analysis using parsimony (*and other methods)*. Version 4. Sunderland: Sinauer Associates.

Tagle, D. A., M. M. Miyamoto, M. Goodman, O. Hofmann, G. Braunitzer, R. Göltenboth, and H. Jalanka. 1986. Hemoglobin of pandas: Phylogenetic relationships of carnivores as ascertained with protein sequence data. *Naturwissenschaften* 73:512–14.

Talbot, S. L., and G. F. Shields. 1996. A phylogeny of the bears (Ursidae) inferred from complete sequences of three mitochondrial genes. *Mol Phylogenet Evol* 5:567–75.

Taylor, D. 1990. *The giant panda*. London: Boxtree.

Tedford, R. H. 1976. Relationships of pinnipeds to other carnivores (Mammalia). *Syst Zool* 25:363–74.

Thenius, E. 1979. Zur systematischen und phylogenetischen Stellung des Bambusbären: *Ailuropoda melanoleuca* David (Carnivora, Mammalia). *Z Säugetierkd* 44:286–305.

Trouessart, É.-L. 1898. *Catalogus mammalium tam viventium quam fossilium*. Berolini: R. Friedländer.

Trouessart, É.-L. 1904. *Catalogus mammalium tam viventium quam fossilium. Quinquennale supplementum*. Berolini: R. Friedländer und Sohn.

Vane-Wright, R. I., C. J. Humphries, and P. H. Williams. 1991. What to protect? Systematics and the agony of choice. *Biol Conserv* 55:235–54.

Vázquez, D. P., and J. L. Gittleman. 1998. Biodiversity conservation: does phylogeny matter? *Curr Biol* 8:R379–81.

Veron, G., and F. M. Catzeflis. 1993. Phylogenetic relationships of the endemic Malagasy carnivore *Cryptoprocta ferox* (Aeluroideae): DNA/DNA hybridization experiments. *J Mamm Evol* 1:169–85.

Vrana, P. B., M. C. Milinkovitch, J. R. Powell, and W. C. Wheeler. 1994. Higher level relationships of the arctoid Carnivora based on sequence data and "total evidence." *Mol Phylogenet Evol* 3:47–58.

Waits, L. P., J. Sullivan, S. J. O'Brien, and R. H. Ward. 1999. Rapid radiation events in the family Ursidae indicated by likelihood phylogenetic estimation from multiple fragments of mtDNA. *Mol Phylogenet Evol* 13:82–92.

Walker, E. P. 1964. *Mammals of the world*. Baltimore: Johns Hopkins University Press.

Wang, T. K. 1974. Taxonomic status of the species, geological distribution and evolutionary history of *Ailuropoda*. *Acta Zool Sin* 20:191–201.

Wang, X. 1997. New cranial material of *Simocyon* from China, and its implications for phylogenetic relationship to the red panda *(Ailurus)*. *J Vertebr Paleontol* 17:184–98.

Wang, X., X. Chen, W. Jiang, C. Zheng, J. Zheng, and J. Ye. 1989. Use of rabbit antisera to IgG of giant panda in defining taxonomic position of giant panda. *Acta Theriol Sin* 9:94–97.

Wayne, R. K., R. E. Benveniste, D. N. Janczewski, and S. J. O'Brien. 1989. Molecular and biochemical evolution of the Carnivora. In *Carnivore behavior, ecology, and evolution*, edited by J. L. Gittleman, pp. 465–94. Ithaca: Cornell University Press.

Weber, M. 1904. *Die Säugetiere. Einführung in die Anatomie und Systematik der recenten und fossilen Mammalia*. Jena: G. Fischer.

Weber, M. 1928. *Die Säugetiere. Einführung in die Anatomie und Systematik der recenten und fossilen Mammalia*. Jena: G. Fischer.

Werdelin, L. 1996. Carnivoran ecomorphology: A phylogenetic perspective. In *Carnivore behavior, ecology, and evolution,* edited by J. L. Gittleman, pp. 582–624. Ithaca: Cornell University Press.

Winge, H. 1895. *Jordfundne og nulevende Rovdyr (Carnivora) fra Lagoa Santa, Minas Geraes, Brasilien. Med Udsigt over Rovdyrenes indbyrdes Slaegtskab [Fossil and living carnivores (Carnivora) from Lagoa Santa, Minas Geraes, Brazil. With a review of the interrelationships of the Carnivores].* Copenhagen: F. Dreyer.

Winge, H. 1941. *The interrelationships of the mammalian genera.* Copenhagen: C. A. Reitzels.

Wolsan, M. 1993. Phylogeny and classification of early European Mustelida (Mammalia: Carnivora). *Acta Theriol* 38:345–84.

Wozencraft, W. C. 1984. A phylogenetic reappraisal of the Viverridae and its relationship to other Carnivora. Ph.D. dissertation. University of Kansas, Lawrence.

Wozencraft, W. C. 1989. The phylogeny of the Recent Carnivora. In *Carnivore behavior, ecology, and evolution,* edited by J. L. Gittleman, pp. 495–535. Ithaca: Cornell University Press.

Wozencraft, W. C. 1993. Order Carnivora. In *Mammal species of the world: A taxonomic and geographic reference,* edited by D. E. Wilson, and D. A. Reeder, pp. 279–348. Washington, D.C.: Smithsonian Institution Press.

Wurster, D. H. 1969. Cytogenetic and phylogenetic studies in Carnivora. In *Comparative mammalian cytogenetics,* edited by K. Benirschke, pp. 310–29. New York: Springer-Verlag.

Wurster, D. H., and K. Benirschke. 1968. Comparative cytogenetic studies in the order Carnivora. *Chromosoma* 24:336–82.

Wurster-Hill, D. H., and M. Bush. 1980. The interrelationships of chromosome banding patterns in the giant panda *(Ailuropoda melanoleuca)*, hybrid bear *(Ursus middendorfi × Thalarctos maritimus)* and other carnivores. *Cytogenet Cell Genet* 27:147–54.

Wyss, A. R. 1987. The walrus auditory region and the monophyly of pinnipeds. *Am Mus Novit* 2871:1–31.

Wyss, A. R., and J. J. Flynn. 1993. A phylogenetic analysis and definition of the Carnivora. In *Mammalian phylogeny: Placentals,* edited by F. S. Szalay, M. J. Novacek, and M. C. McKenna, pp. 32–52. New York: Springer-Verlag.

Zhang, Y. P., and O. A. Ryder. 1993. Mitochondrial DNA sequence evolution in the Arctoidea. *Proc Natl Acad Sci USA* 90:9557–61.

Zhang, Y. P., and O. A. Ryder. 1994. Phylogenetic relationships of bears (the Ursidae) inferred from mitochondrial DNA sequences. *Mol Phylogenet Evol* 3:351–59.

Zhang, Y. P., and L. M. Shi. 1991. Riddle of the giant panda. *Nature* 352:573.

BRIEF REPORT 1.1

Pylogenetic Placement of the Giant Panda Based on Molecular Data

Lisette P. Waits

The use of paleontological and morphological data to reconstruct the genealogical history and taxonomic placement of the giant panda *(Ailuropoda melanoleuca)* has produced conflicting results. The species has been placed with the Ursidae (bear family), Procyonidae (raccoon family), or in a separate family (Ailuropodidae). An alternative method for defining the evolutionary history and taxonomic placement of the giant panda is molecular phylogenetics. The results and conclusions from twelve different molecular studies of the phylogenetic relationship of the giant panda indicate that ten place the giant panda with the Ursidae (table BR.1.1). The analyses used in the two studies positing a placement with the Ailuridae have been challenged by other scholars.

Our recent examination of 1916 base pairs from six different mtDNA regions provides the largest available DNA sequence dataset for the giant panda and other bear species and demonstrates the close relationship between them. We also generated 678 base pairs of mtDNA sequence data for the raccoon *(Procyon lotor)*, seal *(Phoca vitulina)* and domestic dog *(Canis familiaris)*. Based on results from these studies, we find strong molecular support for the conclusion that the giant panda belongs with the Ursidae.

An optimum phylogenetic tree (figure BR.1.1) was generated using maximum likelihood analysis of 678 base pairs of mitochondrial DNA sequence data from the 16S rRNA and cytochrome *b* gene regions. The same optimal topology was also generated using maximum parsimony and maximum likelihood distance methods. The domestic dog was used as an outgroup. The alternate hypothesis of phylogenetic placement with the raccoon was rejected using the Kishino-Hasegawa test.

REFERENCES

Goldman, D., P. R. Giri, and S. J. O'Brien. 1989. Molecular genetic-distance estimates among the Ursidae as indicated by one- and two-dimensional protein electrophoresis. *Evolution* 43:282–95.

Hashimoto, T., E. Otaka, J. Adachi, K. Mizuta, and M. Hasegawa. 1993. The giant panda is closer to a bear, judged by α- and β-hemoglobin sequences. *J Mol Evol* 36:282–89.

Nash, W. G., and S. J. O'Brien. 1987. A comparative chromosome banding analysis of the Ursidae and their relationships to other carnivores. *Cytogenet Cell Genet* 45:206–12.

O'Brien, S. J., W. G. Nash, D. E. Wildt, M. E. Bush, and R. E. Benveniste. 1985. A molecular solution to the riddle of the giant panda's phylogeny. *Nature* 317:140–44.

TABLE BR.1.1
Phylogenetic Placement of the Giant Panda Based on Molecular Genetic Analyses

STUDY	DATA TYPE	PLACEMENT
Sarich (1973)	Immunological distance	Ursidae
O'Brien et al. (1985)	DNA/DNA hybridization and isozyme distance	Ursidae
Tagle et al. (1986)	Amino acid sequences of A and B hemoglobin	Ailuridae
Nash and O'Brien (1987)	Cytogenetic banding patterns	Ursidae
Wayne et al. (1989)	Consensus tree of previous molecular analyses	Ursidae
Goldman et al. (1989)	One- and two-dimensional protein electrophoresis	Ursidae
Zhang and Shi (1991)	mtDNA restriction enzyme analysis	Ailuridae
Hashimoto et al. (1993)	Amino acid sequences of A and B hemoglobin	Ursidae
Zhang and Ryder (1993)	mtDNA sequence data	Ursidae
Vrana et al. (1994)	mtDNA sequence data	Ursidae
Talbot and Shields (1996)	mtDNA sequence data	Ursidae
Waits et al. (1999)	mtDNA sequence data	Ursidae

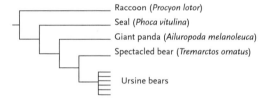

FIGURE BR.1.1. Optimal phylogenetic tree generated using maximum likelihood analysis of 678 base pairs of mtDNA sequence data from the 16S rRNA and cytochrome *b* gene regions. The same optimal topology was also generated using maximum parsimony and maximum likelihood distance methods. The domestic dog *(Canis familiaris)* was used as an outgroup. The alternate hypothesis of phylogenetic placement with the raccoon was rejected using the Kishino-Hasegawa test (Waits, unpubl. data).

Sarich, V. M. 1973. The giant panda is a bear. *Nature* 245:218–20.

Tagle, D. A., M. M. Miyamoto, M. Goodman, O. Hofmann, G. Braunitzer, G. R. Goeltenboth, and H. Jalanke. 1986. Hemoglobins of the pandas: Phylogenetic relationships of carnivores as ascertained by protein sequence data. *Naturwissenschaften* 73:512–14.

Talbot, S. L., and G. F. Shields. 1996. A phylogeny of the bears (Ursidae) inferred from complete sequences of three mitochondrial DNA genes. *Mol Phylogenet Evol* 5:567–75.

Vrana, P. B., M. C. Milinkovitch, J. R. Powell, and W. C. Wheeler. 1994. Higher level relationships of the Arctoid Carnivora based on sequence data and "total evidence." *Mol Phylogenet Evol* 3:47–58.

Waits, L. P., J. Sulivan, J. O'Brien, and R. H. Ward. 1999. Rapid radiation events in the family Ursidae indicated by likelihood phylogentic estimation from multiple fragments of mtDNA. *Mol Phylogenet Evol* 13:82–92.

Wayne, R. K., R. E. Benveniste, D. N. Janczewski, and S. J. O'Brien. 1989. Molecular and biochemical evolution of the Carnivora. In *Carnivore behavior, ecology, and evolution*, edited by J. L. Gittleman, pp. 465–93. Ithaca: Cornell University Press.

Zhang, Y. P., and O. A. Ryder. 1993. Mitochondrial DNA sequence evolution in the Arctoidea. *Proc Natl Acad Sci USA* 90:9557–61

Zhang, Y. P., and L. M. Shi. 1991. Riddle of the giant panda. *Nature* 352:73.

2

What Is a Giant Panda?

A QUERY ABOUT ITS PLACE AMONG THE URSIDAE

Lee R. Hagey and Edith A. MacDonald

THE PHYLOGENY of giant pandas (*Ailuropoda melanoleuca*) has long been a puzzle, particularly as this herbivorous animal has the intestine of a carnivore, a specialized physiology, a unique life history, and a pattern of facial markings that has seemingly evolved to elicit emotion in humans. To take a new approach to this puzzle, we examined the profile of bile salts (breakdown products of cholesterol metabolism) of the giant panda and potentially related carnivores.

Substantial evidence indicates that the bile salts of ursids (bears) are in some way different from those of other carnivores. For example, traditional Oriental medicine has taught the usefulness of "bear bile" for dozens of centuries. In Western science, a novel bile salt, tauroursodeoxycholic acid (see structure in figure 2.1) was isolated from the bile of polar bears collected during an expedition to Greenland (Hammarsten 1901). Although trace proportions of this unusual bile acid can be detected in the bile of different species, in the family Ursidae, it is one of the dominant bile acids. Haslewood (1978) speculated that the structure and form of different bile salts found in species could be a trait useful in determining evolutionary relationships. In an effort to map out the distribution of tauroursodeoxycholic acid in ursids, we reexamined the biliary bile salts of the giant panda, extended the survey of the biliary bile salts in different carnivore families, as listed in a previous report (Hagey et al. 1993), and examined the relationship of this novel bile acid to ursid phylogeny.

SAMPLE COLLECTION AND TREATMENT

Gallbladder bile samples from ursids and other carnivore families were obtained from deceased animals housed at the San Diego Zoo under an approved protocol of the Zoological Society of San Diego. Bile from the sloth bear (*Ursus ursinus*) was provided by the Philadelphia Zoo (Penrose Research Laboratory). Bile from the Asiatic black bear (*U. thibetanus*) was provided by Edgard Espinosa, National Fish and Wildlife Forensics Laboratory, Ashland, Oregon. Giant panda bile was obtained from the Fuzhow Zoo, China.

Samples were dispersed in several volumes of reagent-grade isopropanol immediately after collection to prevent bacterial degradation. Bear,

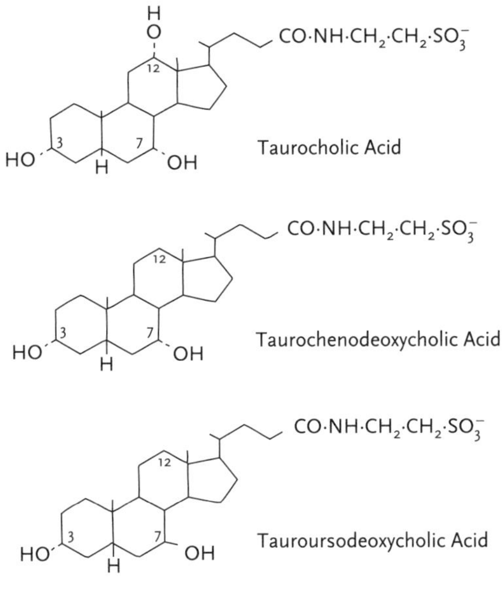

FIGURE 2.1. Structures of the three major primary bile acids found in carnivores. Carbon positions 3, 7, and 12 are labeled on each of the bile acid skeletons.

red panda *(Ailurus fulgens)*, and giant panda fecal samples were obtained from animals housed at the San Diego Zoo. Fecal samples were frozen immediately after collection and kept at −20°C until analysis.

We analyzed bile samples for bile acid conjugates using a reverse-phase high-performance liquid chromatograph (HPLC) and a methanol:water buffer. An octadecylsilane column (RP C-18) was used, with elution at 0.75 ml/min of an isocratic 50 mM KH_2PO_4 buffer (apparent pH 5.25) in methanol:water 68:32 v/v, prepared as described in Rossi et al. (1987). The effluent was monitored at 205 nm (for the amide bond) to quantify the conjugated bile acids. Peaks were assigned by comparison of the relative retention times with those of known standards. The identification of these peaks, and of unconjugated bile acids in fecal samples, was confirmed using gas chromatograph mass spectrometry (GC-MS). A Hewlett-Packard 5890 Gas Chromatograph-5970 MSD, controlled by a Hewlett-Packard Unix Chem Station program, was used with a 30-m capillary column (Supelco SPB35, 0.25-mm inner diameter, SPB 35% phenyl methyl silicone) operated at 275°C (isothermal). A splitless injection was used with an injection temperature of 295°C and interface temperature of 290°C. Helium was used as the carrier gas with a column head pressure of 7 psi.

Fecal bile samples were prepared for GC-MS by refluxing for 1 h in methanolic sodium hydroxide solution (85 ml methanol, 10 ml 10N sodium hydroxide), and after extraction of neutral sterols with hexanes, the remaining solution was acidified (pH < 2) with hydrochloric acid and the unconjugated bile acids extracted into ethyl acetate. Conjugated bile acids, from feces, bile, and collected peaks of the HPLC effluent, were deconjugated by alkaline hydrolysis (1.0 N sodium hydroxide, 130°C, 4 h). Unconjugated and deconjugated bile acids were esterified with methanol (diazomethane) and acetylated by the perchloric acid–catalyzed method of Roovers et al. (1968). Bile acids were incubated for 1.5 h in 2 ml of acetylation fluid, extracted in ethyl acetate, and dried before injection into the GC-MS.

Electrospray mass spectrometry (ESI-MS) was performed on a Hewlett-Packard HP 1100 ESI-MS instrument. Bile samples were first extracted to remove salts, using a Zip-tip$_{C-18}$ (Millipore Corporation, Bedford, Mass.) and 5-µl aliquots were injected into the instrument. The HPLC system was set for loop injection only (no column) using a 90:10 methanol:water buffer and a flow rate of 0.35 ml/min. The ESI-MS was operated in the negative mode with an orifice voltage of 200 V (fragmenter) and an applied capillary voltage of 5000 V.

WHAT DOES BILE ANALYSIS TELL US ABOUT GIANT PANDA PHYLOGENY?

Table 2.1 shows the bile salt composition of a number of species from the six nonursid families of the order Carnivora, as determined using HPLC. All are characterized by the presence of two primary bile acids, taurochenodeoxycholic acid and taurocholic acid (structures shown in figure 2.1), and one secondary bile acid (formed by intestinal bacteria), taurodeoxycholic acid. For most of these species, the proportion of

TABLE 2.1
HPLC Analysis of the Gallbladder Bile Acids from the Carnivora

COMMON NAME	LATIN NAME	BILE ACID COMPOSITION (%)[1]			
		3α7α	3α7β	3α7α12α	3α12α
Canidae					
Coyote	*Canis latrans*	6.1	0.0	84.5	9.4
Crab-eating fox	*Cerdocyon thous*	7.9	0.0	82.0	10.1
Felidae					
Kenya serval	*Felis serval*	1.3	0.0	97.5	1.2
Cheetah	*Acinonyx jubatus*	5.8	0.0	81.0	13.3
Hyaenidae					
Brown hyena	*Hyaena brunnea*	14.5	0.0	57.2	28.3
Cape aardwolf	*Proteles cristatus*	6.5	0.0	82.7	10.8
Mustelidae					
Siberian weasel	*Mustela sibirica*	4.9	0.0	95.1	0.0
Canadian otter	*Lutra canadensis*	6.5	0.0	93.5	0.0
Procyonidae					
Red panda	*Ailurus fulgens*	5.1	0.0	94.9	0.0
Ring-tail coati	*Nasua nasua*	6.7	0.0	92.5	0.8
Raccoon	*Procyon lotor*	16.7	0.0	79.5	3.8
Kinkajou	*Potos flavus*	57.7	0.0	42.3	0.0
Viverridae					
Suricat	*Suricatta suricatta*	5.2	0.0	85.0	9.8
Dwarf mongoose	*Helogale parvala*	13.6	0.0	85.1	1.3

[1]Bile acids are abbreviated according to the position of their hydroxyl groups. 3α7α, taurochenodeoxycholic acid; 3α7β, tauroursodeoxycholic acid; 3α7α12α, taurocholic acid; 3α12α, taurodeoxycholic acid.

taurocholic acid was greater than 80%. The unusual bile acid, tauroursodeoxycholic acid, was not detected in any of these families. The bile acids of the Ursidae (table 2.2), obtained in a manner identical to that used to compile the data in table 2.1, also contained mixtures of taurocholic acid and taurochenodeoxycholic acid. In contrast to the species of table 2.1, the proportion of taurocholic acid in bears was always less than 70% and tauroursodeoxycholic acid was present. The amount detected ranged from a low of 0% in the giant panda to a high of 45% in the Asiatic black bear.

The bile salts of the giant panda (table 2.2) were examined in further detail using ESI-MS. The negative-mode ESI-MS profile is shown in figure 2.2. The overall pattern (upper panel) is dominated by two peaks, one of mass/charge (m/z) 498, a value that corresponds to dihydroxy C_{24} taurine-conjugated bile acids, and the other at m/z 514, for trihydroxy C_{24} taurine-conjugated bile acids. Note the lack of smaller peaks commonly seen at m/z 496 and m/z 512, which are present in the bile of carnivores in general, and are representative of the presence of secondary (microbially produced) bile acids. In the magnified view of the baseline (lower panel), additional trace levels of bile salts are present. These include the taurine conjugates of monohydroxy C_{24} bile acids (m/z 482) and tetrahydroxy C_{24} bile acids (m/z 530), a pentahydroxy C_{27} bile alcohol sulfate (m/z 531), and taurine-conjugated

TABLE 2.2
Analysis of the Bile Acids of the Ursidae and the Giant Panda

		BILE ACID COMPOSITION (%)[1]			
COMMON NAME	LATIN NAME	$3\alpha 7\alpha$	$3\alpha 7\beta$	$3\alpha 7\alpha 12\alpha$	$3\alpha 12\alpha$
Asiatic black bear[2]	*Ursus thibetanus*	55.0	45.0	0.0	0.0
Polar bear[2]	*U. maritimus*	41.7	38.0	19.8	0.5
Black bear[2]	*U. americanus*	29.6	38.8	31.1	0.5
Brown bear[2]	*U. arctos*	28.2	18.3	53.5	0.0
Sun bear[2]	*U. malayanus*	63.6	8.6	27.8	0.0
Sloth bear[2]	*U. ursinus*	31.3	1.4	67.3	0.0
Spectacled bear[2]	*Tremarctos ornatus*	23.6	6.3	70.1	0.0
Giant panda[2]	*Ailuropoda melanoleuca*	61.7	0.0	38.3	0.0
Giant panda[3]	*Ailuropoda melanoleuca*	70.0	0.0	30.0	0.0

[1] Abbreviations as in table 2.1.
[2] HPLC analysis of the gallbladder bile acids.
[3] GC-MS analysis of the fecal bile acids.

di-, tri-, and tetrahydroxylated C_{27} bile acids (at m/z 540, 556, and 572, respectively). The peak at m/z 465 is very likely due to cholesterol sulfate.

The HPLC profile of giant panda biliary bile is shown in figure 2.3 and contains three major peaks. The large peak A (at 15 min) is taurine-conjugated cholic acid (found at m/z 514 in the ESI-MS profile) and the large peak B (at 27 min) is taurine-conjugated chenodeoxycholic acid (m/z 498 in the ESI-MS). The absence of a peak at 32 minutes indicates that none of the secondary bile acid, taurine-conjugated taurodeoxycholic acid, is present. Also not seen was tauroursodeoxycholic acid, the taurine conjugate of which would have appeared at 12 min. The third peak, C (at 34 min), is a plant-derived pigment.

After secretion into the intestine, the structure of bile acids is altered by intestinal bacteria. In carnivores, the effects of bacterial degradation include loss of the taurine conjugate and loss of the hydroxyl group at position 7 on the bile acid nucleus. Because none of these characteristic microbially altered bile acids was observed in the giant panda's profile of biliary bile acids using either ESI-MS or HPLC, its fecal bile acids were further examined for the presence of these metabolites. Initial examination of the fecal bile acids of the giant panda found that its bile acids were surprisingly still conjugated with taurine. After deconjugation, GC-MS analysis found no secondary bile acids, and, as shown in table 2.2, the relative percentages of the two major bile acids, taurochenodeoxycholic acid and taurocholic acid, were nearly identical in bile and feces, and their 7-hydroxyl groups were still intact.

GIANT PANDAS ARE THE OLDEST EXTANT BEARS

Figure 2.4 is a recent phylogeny of ursids, showing the percentage of tauroursodeoxycholic acid found in the bile of each species. The proportion of tauroursodeoxycholic acid in bile parallels the phylogeny of the bears, being lowest in the older extant bears and highest in the more recent bears. The results show that a new, unique bile acid, tauroursodeoxycholic acid, not present in

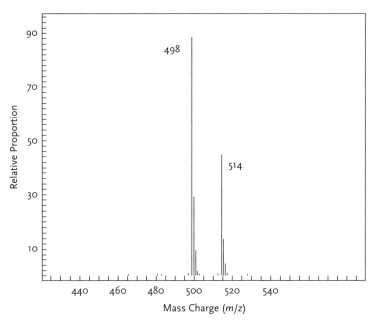

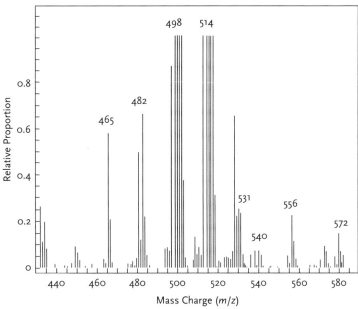

FIGURE 2.2. (Top) ESI-MS profile of bile from a representative giant panda. The two major peaks are m/z 498 (taurine-conjugated dihydroxy C_{24} bile acids), and m/z 514 (taurine-conjugated trihydroxy C_{24} bile acids). (Bottom) A 100× enlargement of the baseline of the top panel. Trace compounds present include m/z 465 (cholesterol sulfate); m/z 482 and 530 (taurine-conjugated mono- and tetrahydroxy-C_{24} bile acids, respectively); m/z 531 (pentahydroxy-C_{27} bile alcohol sulfates); m/z 540, 556, and 572 (taurine-conjugated di-, tri-, and tetrahydroxy-C_{27} bile acids, respectively).

significant proportions in the bile of other carnivores, has recently appeared and is becoming dominant in the bile of bears. We speculate that the forced adaptation of a carnivore to a herbivore-based diet has separated the surviving members of the family Ursidae from the main body of carnivores and has, in turn, led to the induction of this specialized bile acid.

Although it is apparent that a strong evolutionary selection process for tauroursodeoxycholic acid is ongoing in the speciation of bears, the functional advantage of tauroursodeoxycholic acid in bears, if any, is unknown. Bears are characterized as carnivores that have no prey species (with the exception of the recent adaptation of polar bears to seals of the Arctic) and subsist on

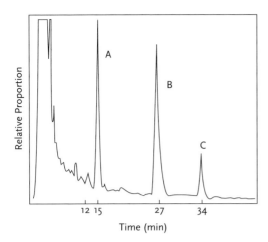

FIGURE 2.3. Reversed phase HPLC profile of the biliary bile acids of a representative giant panda. The relative retention times (to glycine-conjugated chenodeoxycholic acid) are given in parentheses, and the percentage composition follows the semisystematic name of the compound. (A) Taurine-conjugated cholic acid, 27.8% (0.37); (B) taurine-conjugated chenodeoxycholic acid, 62.1% (0.67); (C) a biliary pigment.

an omnivore diet. We can only speculate that as late arrivals to the constant struggle of plant-animal warfare, bears use tauroursodeoxycholic acid to promote their adaptation to a plant-based diet, perhaps as an aid in detoxifying ingested plant xenobiotics. Tauroursodeoxycholic acid is widely used both in Western medicine, to decrease the cytotoxicity of the bile acid pool in a number of chronic cholestatic diseases, such as primary biliary cirrhosis (Poupon et al. 1991) and primary sclerosing cholangitis (O'Brien et al. 1991), and in Oriental medicine, in which (as bear bile) it is the treatment of choice for jaundice, abdominal distension, and upper abdominal pain (Lindley and Carey 1987).

Assuming that bile salts are phylogenetically significant, the progressive increase in the proportion of tauroursodeoxycholic acid in bile can be exploited to predict branch points in ursid phylogeny. To give but one example, the spectacled bear *(Tremarctos ornatus)* is found in the high mountains of South America, a range that it apparently acquired sometime in the past three million years, after continental drift joined the North and South American continents at the isthmus of Panama. Because South America did not have bears prior to the linkage of these two continents (Simpson 1980), the spectacled bear must have originated in North America. At first glance, one could hypothesize that the spectacled bear is a black bear that had simply extended its range to the South. Yet the proportion of tauroursodeoxycholic acid is much lower in the spectacled bear (6%) than in the black bear (39%), indicating that the spectacled bear is not a black bear, but instead is the living descendant of an older, now extinct, North American bear species.

Compared with members of other carnivore families, the biliary bile acid composition of ursids is exceptionally rich in dihydroxy bile acids, both as taurochenodeoxycholic acid in older bears and as taurochenodeoxycholic acid–tauroursodeoxycholic acid mixtures in later bears. The giant panda (with 62% taurochenodeoxycholic acid) shares this characteristic with other

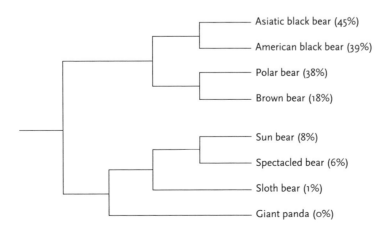

FIGURE 2.4. Phylogeny of the Ursidae with the percentage of ursodeoxycholic acid added after the common name of the bears. Ursodeoxycholic acid proportions in bile greater than 6% are found only in the most recently derived bear. Adapted from Waits et al. (1999).

ursids. However, contrary to findings in all other bears, no tauroursodeoxycholic acid was detected in bile and feces from the giant panda. The lack of tauroursodeoxycholic acid suggests that this species, like most nonursid carnivores, does not have the enzymes necessary to synthesize this bile acid. Examination of bile acid biosynthetic routes (Bjorkhem 1992) indicates that tauroursodeoxycholic acid can be formed in one of two ways. One possibility is to first convert the side chain of cholesterol into that of a bile acid, and then to directly 7β-hydroxylate its steroid nucleus, forming the bile acid structural intermediate 3β-hydroxy-Δ5-cholesten-24-oic acid. The alternative possibility is to interconvert a bile acid normally present in carnivores, taurochenodeoxycholic acid, to tauroursodeoxycholic acid. This process would require two additional enzymatic steps, one to oxidize the 7α-functional group of taurochenodeoxycholic acid and one to reduce this keto group back into the 7β-configuration found in tauroursodeoxycholic acid. Other possibilities for explaining the absence of tauroursodeoxycholic acid in the giant panda are that the bile acids have no functional advantage in the essentially germ-free environment of the giant panda intestine (Hirayama et al. 1990), or that bamboo is less toxic as a dietary food item when compared with the omnivore diets of other bears. As shown in table 2.1, the red panda, which, like the giant panda, consumes bamboo as a major food item, also does not synthesize tauroursodeoxycholic acid, and retains a bile acid profile similar to those of other procyonids. Both of these observations, a high proportion of taurochenodeoxycholic acid and the absence of tauroursodeoxycholic acid, place the giant panda at the base of ursid phylogeny, making it the oldest extant bear.

ACKNOWLEDGMENTS

We thank Gary Suizdak and the staff at the Scripps Research Institute Mass Spectrometry Laboratory, La Jolla, California, for the use of the Hewlett-Packard HP 1100 ESI-MS instrument. We thank Dr. Alan Hofmann for critically reading the manuscript and for his helpful suggestions.

REFERENCES

Björkhem, I. 1992. Mechanism of degregation of the steroid side chain in the formation of bile acids. *J Lipid Res* 33:455–72.

Hagey L. R., D. L. Crombie, E. Espinosa, M. C. Carey, H. Igimi, and A. F. Hofmann. 1993. Ursodeoxycholic acid in the Ursidae: Biliary bile acids of bears, pandas, and related carnivores. *J Lipid Res* 34:1911–17.

Hammarsten O. 1901. Untersuchungen über die Gallen einiger Polarthiere. *Hoppe-Seylers Z Physiol Chem* 32:435–66.

Haslewood G. A. D. 1978. *The biological importance of bile salts.* Amsterdam: North-Holland.

Hirayama K., S. Kawamura, T. Mitsuoka, and K. Tashiro. 1990. The fecal flora of the giant panda. In *Giant panda: Proceedings of the second international symposium on giant pandas, Tokyo, November 10–13, 1987,* edited by S. Asakura and S. Nakagawa, pp. 153–58. Tokyo: Tokyo Zoological Park Society.

Lindley P. F., and M. C. Carey. 1987. Molecular packing of bile acids: Structure of ursodeoxycholic acid. *J Crystallogr Spectrosc Res* 17:231–49.

O'Brien C. B., J. R. Senior, R. Aroramirchandani, and A. K. Batta. 1991. Ursodeoxycholic acid for the treatment of primary sclerosing cholangitis—a 30-month pilot study. *Hepatology* 14:838–47.

Poupon R. E., B. Balkau, E. Eschwege, R. Poupon, and the UDCA-PBC Study Group. 1991. A multicenter, controlled trial of ursodiol for the treatment of primary biliary cirrhosis. *New Engl J Med* 324:1548–54.

Roovers J., E. Evrard, and H. Vanderhaeghe. 1968. An improved method for measuring human blood bile acids. *Clin Chim Acta* 19:449–57.

Rossi S. S., J. L. Converse, and A. F. Hofmann. 1987. High pressure liquid chromatographic analysis of conjugated bile acids in human bile: Simultaneous resolution of sulfated and unsulfated lithocholyl amidates and the common conjugated bile acids. *J Lipid Res* 28:589–95.

Simpson G. G. 1980. *Splendid isolation: The curious history of South American mammals.* New Haven: Yale University Press.

Waits L. P., J. Sullivan, S. J. O'Brien. 1999. Rapid radiation events in the family Ursidae indicated by likelihood phylogenetic estimation from multiple fragments of mtDNA. *Mol Phylogenet Evol* 13:82–92.

3

A Paleontologist's Perspective on the Origin and Relationships of the Giant Panda

Robert M. Hunt, Jr.

FIRST DESCRIBED in 1869 by the Abbé David from a young individual killed by hunters in western Sichuan (Fox 1949), the giant panda *(Ailuropoda melanoleuca)* was placed by David with other living bears in the genus *Ursus*. A year later, the zoologist Milne-Edwards (1870), impressed by skeletal and dental differences from living bears, created the genus *Ailuropoda* (αιλουρος, –πομς, *Ailurus*, -foot; having a foot like the Asian lesser panda, *Ailurus*). The unique ecology of the giant panda, namely, its restricted geographic range and habitat, dependence on bamboo, slow reproductive rate, prolonged infancy, and small populations warrant grave concern for the survival of the species. Recently acquired fossil specimens of the giant panda from southern China, unavailable to zoologists and ecologists even a few years ago, clarify its ancestry and emphasize its importance as a relict carnivoran species of ancient lineage. Here I discuss a revised concept of the bear family, Ursidae, which has emerged from the discovery of fossils over the past half century, and place the giant panda within that context. *Ailuropoda* is a bear, but evidently one of distinctive and singular pedigree.

URSID FOSSILS AND THE ANCESTRY OF LIVING PANDAS

Biologists who study living bears place almost all species in the genus *Ursus* or else recognize the distinctiveness of some species by allocating them to unique genera. Those bears most often segregated from *Ursus* are the Malayan sun bear *(Helarctos)*, Tibetan black bear *(Selenarctos)*, sloth bear *(Melursus)*, and polar bear *(Thalarctos)*. However, regardless of the approach adopted in classification, paleontologists and mammalogists acknowledge a marked similarity in skeletal and dental features among all living species of *Ursus*. Most striking are the anatomical specializations of the feet, the skull, and the teeth. Living *Ursus* toe-in with the forefeet (Kurtén 1966), retain a uniform and specialized fore- and hindfoot anatomy, and, perhaps most importantly, share a highly distinctive dentition.

The living spectacled bear *(Tremarctos)* of South America, and the giant panda *(Ailuropoda)* are the two living ursids that are generally regarded as distinct genera, and not referrable to *Ursus* (Nowak 1991). Today we know that both *Tremarctos* and *Ailuropoda* separated from *Ursus* at least by the late Miocene ~7 million years ago, and have since evolved independently. Whereas the panda has remained a conservative lineage since that time, *Tremarctos* belongs to a group of short-faced ursine bears that evolved a modest species diversity. Short-faced bears are placed in the tribe Tremarctini (Hunt 1998), the Tremarctinae of Kurtén (1966), which comprises the late Miocene-Pliocene *Plionarctos*, and the descendant genera *Tremarctos* and *Arctodus*, the latter including the huge Pleistocene arctodonts (Kurtén 1967; Richards et al. 1996), the largest terrestrial carnivorans that ever lived.

The cheek teeth of living bears (*Ursus*, *Tremarctos*) are unique among all extant carnivorans in the form and size of their cutting (carnassial) teeth and molars. In Carnivora, the carnassial teeth are abbreviated by the symbols P4 (fourth upper premolar) and m1 (first lower molar). Members of the Carnivora are identified by P4/m1 acting as carnassial teeth, often the locus of the greatest bite force. In this chapter the term "carnivoran" refers to species of the Order Carnivora; carnivorous animals in general are termed "carnivores." In most fossil and living carnivorans with teeth adapted to eating flesh, the upper and lower carnassial teeth are prominent, developed cutting blades (e.g., in sabertooth cats, wolves, mountain lions). But in living bears, these teeth are much reduced in size and shearing ability. The vestigial form of the carnassials, however, betrays the carnivorous habits of their earliest ancestors.

Living bears compensate for these small, poorly developed cutting teeth by enlarging the molars behind them. Bear molars are broad enamel platforms that effectively crush and grind tough food, either meat or plant material. Despite the preference of some living ursids (e.g., the polar bear) for meat, they are essentially omnivorous, opportunistic feeders.

So similar are the teeth and skeletal anatomy of living bears that paleontologists do not question their origin from a common ancestral species. *Ursus* can be traced—primarily through teeth, jaws, and occasional skulls—to the early Pliocene (~4–5 million years [Ma, from Megaannum]) in Europe and North America (Hunt 1998). To paleontologists, this is not a very ancient pedigree. The radiation of species included today in *Ursus* must have developed shortly after this time, near the beginning of intensification of global cooling that led to the major glaciations of the northern hemisphere.

The ancestor of *Ursus* almost certainly belonged to one of several species of *Ursavus*, a Miocene ursine genus mostly known from teeth (often in jaw fragments) recovered both in Eurasia and North America. Species of *Ursavus* range in size from that of a domestic cat to a small black bear. The genus first appears in the early Miocene of Europe at about 20 Ma, persisting to about 9 Ma. Two sites in Asia have produced *Ursavus*: Shanwang in eastern China, which dates to 16–17 Ma (Qiu and Qiu 1990; Yang and Yang 1994), and the much younger (~7–8 Ma) Lufeng locality in south China, discussed below (Qi 1979; Flynn and Qi 1982; Badgley et al. 1988; Qiu and Storch 1990). Multiple localities in North America place *Ursavus* within a temporal range spanning about 12.5 to 18 Ma (Hunt 1998). The only complete skeleton of *Ursavus* was found in eastern China at the Shanwang diatomite site in 1981, and at first was not recognized as an ursid (Qiu et al. 1985). It is the size of a large domestic cat, well-furred, and clearly adapted to both a terrestrial and arboreal lifestyle. Exactly which species of *Ursavus* was ancestral to modern bears remains uncertain, in part due to the time gap occurring between the last record of *Ursavus* (~7–8 Ma) and the earliest *Ursus* (~5 Ma). But the close correspondence in the form of the teeth of these two genera leaves little doubt that they are closely related.

THE URSID RADIATIONS

The many *Ursus* species we see today are the result of a "radiation," or splitting off of lineages, that began as early as 4–5 Ma, in the Pliocene. These modern bears, formally classified as a subfamily (the Ursinae), are the last of three such radiations that occurred in ursid evolution. The radiations each involved a particular subfamily

TABLE 3.1
A Classification of the Family Ursidae

Order Carnivora Bowdich 1821
    Infraorder (Division) Arctoidea Flower 1869
    Family Ursidae Fischer de Waldheim 1817
        Subfamily Amphicynodontinae Simpson 1945
        Subfamily Hemicyoninae Frick 1926
        Subfamily Ursinae Fischer de Waldheim 1817
            Tribe Ursavini *(Ursavus, Indarctos)*—early to late Miocene
            Tribe Ursini *(Ursus)*—Pliocene to Recent
            Tribe Tremarctini *(Tremarctos, Arctodus, Plionarctos)*—late Miocene to Recent
            Tribe Ailuropodini *(Ailuropoda, Ailurarctos)*—late Miocene to Recent

of ursids and were consecutive: first, the Amphicynodontinae, next the Hemicyoninae, and finally, the Ursinae (table 3.1). Such speciation events were largely confined to the northern continents (Eurasia and North America), and only late in time did ursids migrate into the southern continents of Africa and South America. Both hemicyonines and ursines reached Africa in the Miocene (documented by very few fossils); only the short-faced ursine bears (Tremarctini: *Arctodus, Tremarctos*) entered South America, some time after 2 Ma.

The first of the three radiations involved the amphicynodonts, very small to mid-sized ancestral ursids, most weighing only about 1–10 kg. They would appear to the casual observer to have been similar to small dogs or foxes. Their fossil remains have chiefly been found in fissure deposits in the Quercy district of southern France, Nei-Mongol in northern China, and a few localities in North America. Skeletons are infrequently found, but those that exist indicate small terrestrial mammals able to climb and run along uneven surfaces. But it is their teeth, especially the carnassials and molars, as well as cranial anatomy (particularly their hearing apparatus, which is enclosed in a bony capsule at the base of the skull) that attest to their ursid affinity.

If we follow the evolution of these amphicynodont lineages in Europe, we find that a second ursid radiation, the hemicyonines, emerges from the amphicynodont stem. Hemicyonine ursids are eventually distributed across Eurasia and North America. Their teeth are instrumental in their identification as members of the ursid family, linking them to the amphicynodonts and to the later ursine bears. The detailed anatomy of the back of the skull also corresponds to that of amphicynodonts and ursines, confirming this relationship. However, hemicyonines share one unique trait that almost all amphicynodont and ursine bears lack: they walked with a digitigrade gait, much like a living wolf, and so have evolved beyond the plantigrade stance found in all amphicynodonts and living ursines.

The three ursid radiations succeeded each other in time. Amphicynodonts are known from about 37 to 29 Ma (late Eocene-Oligocene); hemicyonines existed from about 32 Ma (early Oligocene) to about 9–10 Ma, becoming extinct worldwide in the late Miocene. Ursines appear at about 20 Ma in the early Miocene and diversify into the modern species in the Plio-Pleistocene.

ORIGIN OF THE GIANT PANDA

For most of the past century, fossils of the giant panda have been limited to individual teeth, a few jaws, and rare skulls from Pleistocene cave

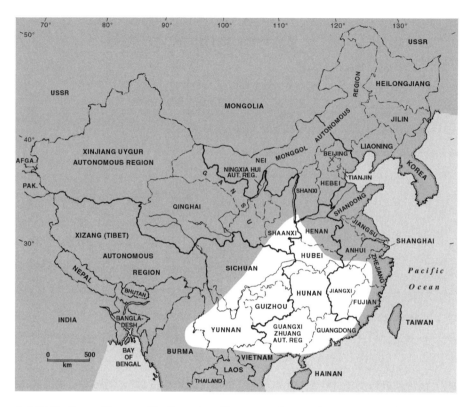

FIGURE 3.1. Geographic range of fossil remains of the giant panda in southeast Asia. Known localities are all of Pleistocene age.

and fissure deposits in south China and Burma (figure 3.1) (Granger 1938; Colbert and Hooijer 1953; Wang 1974). All were assigned either to the living *Ailuropoda melanoleuca* or to closely related subspecies; new names were applied when some Pleistocene populations averaged slightly larger in size than the living giant panda. Several lower jaws and a number of isolated teeth of a smaller form of the giant panda (*Ailuropoda microta*) from the earlier Pleistocene of Guangxi (Kwangsi) Province (Pei 1987) are considered distinct, but neither *A. microta* nor other Pleistocene fossils clarified the ancestry of *Ailuropoda*.

However, because of the remarkable discovery of uniquely important fossils in southern China in the 1980s, the ancestry of the giant panda can now be identified with some certainty. At the late Miocene (~7–8 Ma) site of Lufeng (figure 3.2), nestled in the mountains of the province of Yunnan, Chinese paleontologists found skulls, mandibles, and teeth of rare hominoid fossils. Together with the hominoid remains were the bones and teeth of over fifty species of mammals that lived in the company of these early man-apes (Badgley et al. 1988). Some fossils were initially attributed to *Ursavus* in 1984–85 (Qi 1984), and indeed this taxon exists in the fauna (Qui and Qi 1990). But among the scarce fossils of these ursids were a number of teeth that were later recognized as those of an early predecessor of the giant panda (Qiu and Qi 1989). Named *Ailurarctos lufengensis* in 1989, these teeth establish not only the ancestry of the giant panda but also demonstrate its evolution from a species of *Ursavus* at some time in the Miocene epoch.

Were living bears and giant pandas not so specialized in terms of tooth size and the cusp pattern of the cheek teeth, it would be difficult

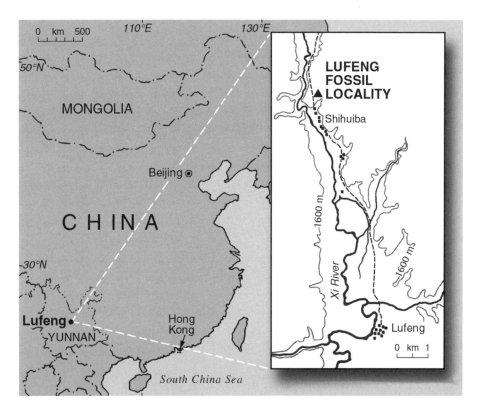

FIGURE 3.2. Fossil teeth of the ursid *Ailurarctos lufengensis*, a plausible ancestor of the living giant panda, were found north of the village of Lufeng, Yunnan, China, in sediments of late Miocene age (~7–8 Ma); the significance of the teeth was first recognized in the 1980s. From Badgley et al. (1988), with permission.

to uncover their point of origin among earlier carnivorans. But the teeth of living ursines, the giant panda, and the Lufeng *Ailurarctos* exhibit a unique cachet, indicating their relatedness by the enamel pattern of the molars, premolars, and carnassials (figure 3.3). There are five upper teeth of *Ailurarctos* that can be assembled into the upper tooth row (Qiu and Qi 1989). An ursid affinity is demonstrated by the form of the upper carnassial (P4), with its retracted internal cusp (protocone); by the quadrate first upper molar (M1) in which the metaconule has migrated to the posterointernal corner of the tooth; and by the second upper molar (M2), where an enamel platform or "talon" has been added to its posterior border. Two lower teeth, the last two molars (m2-3), also show ursid traits: they have broad, expanded enamel crowns that are typical of ursine bears. All these features are seen in Miocene *Ursavus* and suggest derivation from a species of the subfamily Ursinae.

Not only are the Lufeng teeth clearly ursine, they also share specialized characteristics with the living and Pleistocene giant panda *Ailuropoda*. Although a small animal, the Lufeng *Ailurarctos* has already expanded its upper and lower molars (M1-2 and m2-3) in the fashion of the giant panda. Moreover, the enamel of the molar crowns from Lufeng is strongly "wrinkled," an evident match with the specialized molar enamel of *Ailuropoda*. The Lufeng panda also has enlarged and added cusps to its premolars, evidently the initial stage preceding the enlarged and multicusped premolars of the living animal. Taken together, the Lufeng teeth not only link the fossil species with the living giant panda, but

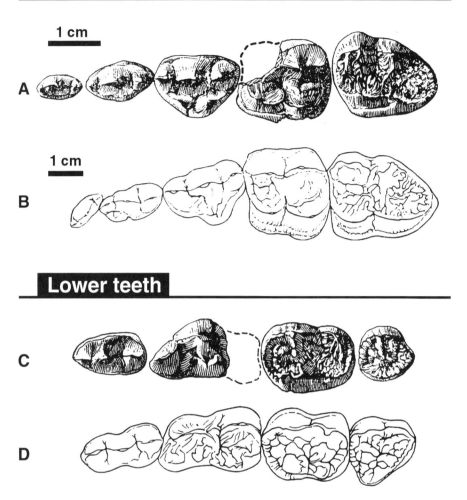

FIGURE 3.3. Comparison of the cheek teeth of the late Miocene Lufeng panda *(Ailurarctos lufengensis)* and the living giant panda. Upper teeth (P2-M2) of (A) *Ailurarctos;* (B) *Ailuropoda.* Lower teeth of (C) *Ailurarctos* (p4-m3, posterior part of m1 missing); (D) *Ailuropoda* (p4-m3). Teeth of the smaller Lufeng panda are scaled to the size of those of the giant panda for comparison. Modified from Qiu and Qi (1989).

clearly show its derivation from the ursine bears, and most probably from a species of Miocene *Ursavus* at some time prior to 7 Ma.

CONCLUSIONS

Fossil teeth from near the village of Lufeng in south China provide the first firm evidence of the ancestry of the giant panda. The panda teeth were found with a late Miocene fauna, including skulls and jaws of hominoid primates, and are 7–8 million years old. Enamel crown patterns of upper and lower cheek teeth of the Lufeng panda *(Ailurarctos lufengensis)* ally it with both the living giant panda and with ancestral ursine bears of the genus *Ursavus* (figure 3.4). The origin of the giant panda lies within the subfamily Ursinae, one of the three great evolutionary radiations of the family Ursidae. Ursine bears arise from species of small amphicynodont ursids,

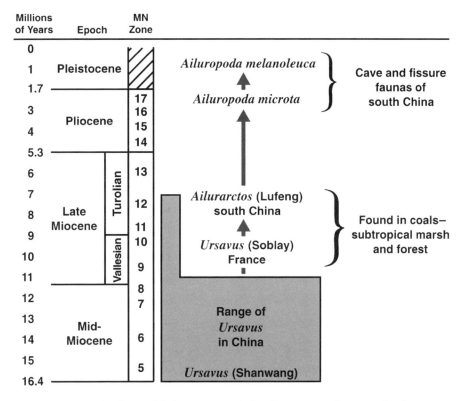

FIGURE 3.4. Suggested evolution of the living giant panda from late Miocene *Ailurarctos* and earlier Miocene *Ursavus* in Eurasia. Whether rare teeth of *Ursavus* from Soblay, France, are involved in the ancestry of the Lufeng panda remains undecided. Modified from Qiu and Qi (1989).

who also gave rise to the larger digitigrade hemicyonines, an important group of Miocene predators on the northern continents.

ACKNOWLEDGMENTS

This chapter could not have been written without access to the publications of the primary authors of research articles on living and fossil pandas in China. I am pleased to acknowledge the friendly cooperation and goodwill of paleontologists at the Institute of Vertebrate Paleontology and Paleoanthropolgy at Beijing, who allowed examination of Miocene fossils from Lufeng and Shanwang during my visits to the institute in 1984 and 1998. I appreciate the help and advice of Qiu Zhanxiang, Qiu Zhuding, Wang Banyue, and Li Chuankui during my research in China. My friend Xue Xiang-xu of Northwestern University, Xian (China), kindly arranged field trips to the Miocene sites at Lufeng and Shanwang that were of great value in understanding the geological setting of these localities. My thanks to University of Nebraska State Museum illustrator Angie Fox for adapting the illustrations for this chapter.

REFERENCES

Badgley, C., G. Qi, W. Chen, and D. Han. 1988. Paleoecology of a Miocene tropical upland fauna: Lufeng, China. *Nat Geogr Res* 4:178–95.

Colbert, E. H., and D. A. Hooijer. 1953. Pleistocene mammals from the limestone fissures of Szechwan, China. *Bull Am Mus Nat Hist* 102:1–134.

Flynn, L. J., and G. Qi. 1982. Age of the Lufeng, China, hominoid locality. *Nature* 298:746–47.

Fox, H. M., ed. 1949. *Abbé David's diary, being an account of the French naturalist's journeys and observations in China in the years 1866 to 1869 (translated from the French)*. Cambridge: Harvard University Press.

Granger, W. 1938. Medicine bones. *Nat Hist* 42:264–71.

Hunt, R. M., Jr. 1998. Evolution of North American Tertiary Ursidae (Mammalia, Carnivora). In *Tertiary mammals of North America*, edited by C. Janis, K. Scott, and L. Jacobs, pp. 174–95. London: Cambridge University Press.

Kurtén, B. 1966. Pleistocene bears of North America: Genus *Tremarctos*, spectacled bears. *Acta Zool Fenn* 115:1–120.

Kurtén, B. 1967. Pleistocene bears of North America: Genus *Arctodus*, short-faced bears. *Acta Zool Fenn* 117:1–60.

Milne-Edwards, A. 1870. Note sur quelques mammifères du Thibet oriental. *Ann Sci Nat Série V* 13(10):1.

Nowak, R. M. 1991. *Walker's mammals of the world*, 5th ed. Vol. II. Baltimore: Johns Hopkins University Press.

Pei, W. 1987. Carnivora, Proboscidea, and Rodentia from the Liucheng *Gigantopithecus* cave and other caves in Guangxi. *Mem Inst Vertebrate Palaeontol Palaeoanthropol (Acad Sin)* 18:1–134.

Qi, G. 1979. Pliocene mammalian fauna of Lufeng, Yunnan. *Vertebrata PalAsiatica* 17:14–22.

Qi, G. 1984. First discovery of *Ursavus* in China and note on other Ursidae specimens from the *Ramapithecus* fossil site of Lufeng. *Acta Anthropol Sin* 3:53–61.

Qiu, Z.-D., and G. Storch. 1990. New murids (Mammalia: Rodentia) from the Lufeng hominoid locality, late Miocene of China. *J Vertebrate Paleontol* 10:467–72.

Qiu, Z.-X., and G. Qi. 1989. Ailuropod found from the late Miocene deposits in Lufeng, Yunnan. *Vertebrata PalAsiatica* 27:153–69.

Qiu, Z.-X., and G. Qi. 1990. Restudy of mammalian fossils referred to Ursinae indet. from *Lufengpithecus* locality. *Vertebrata PalAsiatica* 28:270–83.

Qiu, Z.-X., and Z.-D. Qiu. 1990. Late Tertiary mammalian local faunas of China. *J Strat* 14:241–60.

Qiu, Z.-X., D. Han, H. Jia, and B.-Z. Wang. 1985. Dentition of the *Ursavus* skeleton from Shanwang, Shandong Province. *Vertebrata PalAsiatica* 23:264–75.

Richards, R. L., C. S. Churcher, and W. D. Turnbull. 1996. Distribution and size variation in North American short-faced bears, *Arctodus simus*. In *Palaeoecology and palaeoenvironments of late Cenozoic mammals*, edited by K. M. Stewart and K. L. Seymour, pp. 191–246. Toronto: University of Toronto Press.

Wang, T. 1974. On the taxonomic status of the species, geological distribution, and evolutionary history of *Ailuropoda*. *Acta Zool Sin* 20:191–201.

Yang, H., and S. Yang. 1994. The Shanwang fossil biota in eastern China: A Miocene Konservat-Lagerstätte in lacustrine deposits. *Lethaia* 27:345–54.

4

Variation in Ursid Life Histories

IS THERE AN OUTLIER?

David L. Garshelis

THE URSIDAE, comprising only eight extant species of bears, is the second-smallest family in the order Carnivora. Given the small number of species, less variation might be expected in life histories of bears than among species of other carnivore families. However, sizes of bears vary greatly, from 30-kg sun bears (*Helarctos malayanus*) to polar bears (*Ursus maritimus*) and brown bears (*U. arctos*) exceeding 500 kg, and they inhabit a vast area, from the tropics to the arctic in both the Old and New Worlds. The potential for considerable variation in life histories therefore exists. Most of the individual species also span a broad geographic area, so intraspecific variation in life histories could also be significant. This chapter investigates variation in the life histories of bears, both within and across species.

A key objective is to address a related question: is there a life history outlier among the ursids? For a long time, giant pandas (*Ailuropoda melanoleuca*) were considered taxonomically enigmatic, falling somewhere between bears and raccoons (Procyonidae), but not clearly within either family (Schaller et al. 1985; O'Brien 1993; Schaller 1993). There is a growing consensus, however, that the giant panda is a true ursid, although this species appears to have diverged very early from the ancestral lineage, based on its genetic and morphological differentiation from other bears (O'Brien et al. 1985; Hunt 1996, chapter 3; Bininda-Emonds et al., 1999; Bininda-Emonds, chapter 1; Waits, brief report 1.1). But how similar or different are the panda's life history traits? Gittleman (1994) suggested that some aspects of the life history of the giant panda distinctly separate it from patterns observed in other members of the Carnivora.

This question is important because it sets a framework for conservation strategies. All species of bears have at some time experienced a severe population decline from overexploitation and habitat alteration (Servheen 1989; Servheen et al. 1999). In some cases, successful management has reversed the decline. Can the same management tactics that succeeded in these cases be employed across the family, or are life histories sufficiently diverse to hamper application of similar approaches across taxa?

Before proceeding further, it is important to define exactly what life history strategies are. Many authors writing on the subject have restricted their discussion to reproductive strategies, including seasonality of breeding and birthing; age of maturity; reproductive lifespan; litter size; litter frequency; birth weight; birth sex ratio; age of weaning, independence, and dispersal of offspring; and survival of offspring (Stearns 1976, 1992; Horn 1978; Boyce 1988; Partridge and Harvey 1988; Roff 1992; Gittleman 1993). I take a broader approach that encompasses a variety of biological and behavioral strategies related both to survival and reproduction. This corresponds more closely with the treatment by Southwood (1988), who categorized life history tactics into five areas: adaptations to harsh weather or other physical conditions; defense against predation; foraging; migration and other escapes from altered habitat; and reproduction. In this chapter, I condense the first four of these into strategies for survival, and then consider reproductive strategies separately.

My purpose is mainly to compare the life history strategies used by the eight species of bears and explain some of the differences. Much of the variation among and within species is explained by geography, habitat, and—related to these two—food habits; thus, I begin by discussing these factors. However, I do not attempt to quantitatively relate ecological variables and life history tactics. Such an effort is probably unwarranted at this time, as the data are quite sparse for at least half the species. Moreover, quantitative comparisons involving life history traits often necessitate condensing wide variation into mean values (Gittleman 1986; Ferguson and Larivière 2002), whereas a focal point of this chapter is on the amount of variation and the relationship between known variation and number of studies that have been conducted.

Some species of bears have been studied extensively. Although I attempted a comprehensive review of this literature, I cite only key references regarding each species on each topic, as well as references for extreme values (minimums and maximums) shown in tables. I also cite material from unpublished reports and correspondence related to studies in progress on some of the lesser-known bear species. My aim is to compare the available data, and in the process, highlight areas where more research is needed.

GEOGRAPHIC RANGES, HABITAT, AND FOOD HABITS

Bears presently occupy all three northern continents, as well as South America. Bears once occupied Africa, but never reached Australia (Hunt 1996). Only brown bears and polar bears range across more than one continent. The American black bear *(U. americanus)* is endemic to North America, the Andean (spectacled) bear *(Tremarctos ornatus)* is endemic to South America, and the sun bear, sloth bear *(Melursus ursinus)*, Asiatic black bear *(U. thibetanus)* and giant panda are all endemic to Asia (although historically the Asiatic black bear ranged into Europe [Kurtén 1976]). Only two countries, China and India, have four species of bears, and there is only one small place in eastern India where three bear species may be sympatric (Choudhury 1993).

Four bears live in the tropics; three, the sloth, sun, and Andean, are mainly tropical, and the latter two live on both sides of the equator. The brown bear, American black bear, and giant panda are temperate species. However, brown bears, and to a slightly lesser extent, American black bears, occupy a diverse array of habitats, ranging from coastal rainforest to barren-ground tundra and even desert.

Six of the eight species of bears span a latitudinal range of at least 30° (figure 4.1). The giant panda has by far the most restricted range. Four species occupy an elevational gradient exceeding 3,000 m. Polar bears, which live on the frozen Arctic Ocean and only occasionally come ashore, have the narrowest elevational range, followed by lowland-dwelling sloth bears. The polar bear is the only species that does not inhabit forested environments, although brown bears (Vaisfeld and Chestin 1993), American black bears (Veitch and Harrington 1996), and Andean bears (Peyton 1980; Brown and Rumiz 1989) sometimes live latitudinally or elevationally above treeline.

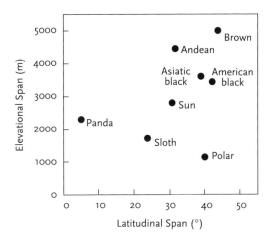

FIGURE 4.1. Current latitudinal and elevational span (i.e., lowest to highest occupancy) of each of the eight species of bears. Most latitudinal data were obtained from range maps in Servheen et al. (1999). Other references for latitudinal and elevational data: polar (Schweinsburg 1979; Amstrup 2000), brown (Schaller 1979; Vaisfeld and Chestin 1993); American black (Beck 1991); Asiatic black (Prater 1971; Schaller 1977; M. Shrestha, pers. comm.); sloth (Seidensticker 1993); sun (Choudhury 1993; Santiapillai and Santiapillai 1996); Andean (Brown and Rumiz 1989); giant panda (Schaller et al. 1985).

Polar bears and giant pandas occupy the least diverse habitats and have the most restricted food habits. At one dietary extreme, polar bears are almost strictly carnivorous, preying mainly on one species of seal *(Phoca hispida);* they consume vegetative matter only in southern areas, where they are forced ashore during summer by melted sea ice (Derocher et al. 1993; Hobson and Stirling 1997). At the other extreme, the diet of giant pandas is limited almost entirely to bamboo (Schaller et al. 1985, 1989; Pan et al. 2001). When preferred species of bamboo become less available, as after a mass flowering and die off, they shift to alternate species of bamboo, but not to other types of food (Reid et al. 1989). The other species of bears are much more omnivorous and opportunistic; consequently, their diets vary spatially and temporally. For example, although fruits tend to be the mainstay for brown bears throughout their range, roots and vertebrates are large dietary components in alpine and tundra habitats, whereas insects tend to be important in broadleaf forests (Mattson 1998). However, different populations, even in similar habitats, can have appreciably different food habits. For instance, ungulates are a key food in some populations of brown (grizzly) bears (Jacoby et al. 1999) and Asiatic black bears (Hwang et al. 2002). Fish (primarily salmon) often constitute an important part of the diet in Pacific coastal brown bear populations. Coastal American black bears also take advantage of plentiful supplies of spawning fish, but only when brown bears are absent. In some years and in some places, both brown bears and American black bears consume large quantities of insects (French et al. 1994; Noyce et al. 1997; Swenson et al. 1999; Mattson 2001). Sloth bears are noted insect specialists, with particular morphological adaptations for myrmecophagy, but their diets vary from predominantly ants and termites to predominantly fruits, depending on season and geographic location (Joshi et al. 1997). Conversely, sun bears, which are regarded as primarily frugivorous, can switch to an entirely insect diet and sustain this for more than a year when fruiting fails (Wong et al. 2002; G. Fredriksson, pers. comm.). In short, excluding giant pandas and polar bears, the main generalization that can be made about the feeding habits of bears is that they are highly adaptable and variable.

SURVIVAL STRATEGIES

Other than humans, the most likely predator on bears is another bear. Larger brown bears may displace or kill black bears, and bears also kill members of their own species. Young bears are the most common victims, and adult males the most common killers (Mattson et al. 1992b; Garshelis 1994; McLellan 1994; Derocher and Wiig 1999; Swenson et al. 2001a). Young males (1–4 years old), the dispersing cohort, are particularly prone to encountering mortality risks, especially in human-dominated landscapes (Schwartz and Franzmann 1992; Garshelis 1994; Swenson et al. 1998). In the absence of humans (especially hunting pressure), adult bears have very high survivorship (>95%) (Taylor

1994; Amstrup and Durner 1995; Sorensen and Powell 1998; McLellan et al. 1999; Pan et al. 2001).

Survival strategies of bears have been affected more by availability of food than by predators or competitors. In both tropical and temperate environments, fruit availability can vary markedly by habitat, elevation, season, and year. Dietary flexibility is one adaptation to this problem. However, when food entirely disappears, as it does in northern latitudes during winter, bears must undergo a prolonged period of fasting. They endure these long winter fasts in a den, employing a complex and as yet not fully understood physiological adaptation known as hibernation (Hellgren 1998). The process requires, and moreover, seems instigated by, the accumulation of vast quantities of fat. Notably, only three of the eight bear species hibernate due to seasonal disappearance of their foods, and even among these three, none is an obligate hibernator. If adequate food supplies persist during the winter, they remain active (brown bears and American black bears: Linnell et al. [2000]; Asiatic black bears: Hwang [2003]). South-latitude polar bears, fattened from months of foraging on seals, are able to live off this fat during the summer, when melted sea ice forces them on shore. It is unclear whether this summer fasting involves the same physiological processes as winter hibernation, but the effect is the same (Atkinson et al. 1996).

The large size of bears enables them not only to carry enough fat to fast when necessary, but also enhances their mobility, and hence capacity to locate seasonally abundant food sources. Bears of several species have been known to make large seasonal movements, either laterally or elevationally (polar: Amstrup et al. [2000]; brown: LeFranc et al. [1987]; American black: Garshelis and Pelton [1981], Beck [1991]; Asiatic black: Hwang [2003]; sloth: Joshi et al. [1995]). Bears also have large home ranges that correspond with their large body mass and hence large food requirements (Swihart et al. 1988).

Home range sizes of bears seem to relate to the distribution and abundance of food—the more abundant the food, the smaller the area needed for food gathering (brown: McLoughlin et al. [1999, 2000]; American black: Gompper and Gittleman [1991]). Bears have enormously variable home range sizes, both across and within species (table 4.1). Despite this variability, some species-specific differences seem evident. The polar bear is distinctive in having the largest home ranges. Even the smallest measured mean range size for a population of polar bears is more than double the largest population mean for any other species. Brown bear home ranges are next largest, but some brown bears have smaller ranges than those of American and Asiatic black bears. American black bears and brown bears have been studied the most, and consequently exhibit the greatest variability. Both of these species show a two-order-of-magnitude spread among population means and even greater variation among individuals. Seemingly smaller intraspecific differences for other bear species are likely due, at least in part, to the limited number of populations and individuals that have been studied (table 4.1). However, from this limited number of studies, it appears that giant pandas have smaller home ranges than most other bears.

Most bears also utilize three-dimensional space by climbing trees. Polar bears do not climb trees, and neither do brown bears in North America, but brown bears climb readily in Europe and Asia (Vaisfeld and Chestin 1993). All other species are able tree climbers. However, the degree to which bears use trees varies across and within species. Sun bears and Andean bears are generally regarded as the most arboreal, a belief that probably stems from their obvious agility in trees and their use of tree "nests." However, use of tree nests by these two species, although not quantified, appears to vary significantly among populations (Meijaard 1999; Goldstein 2002; R. Steinmetz, pers. comm.). Both species of black bears, and sometimes even Asian brown bears (Vaisfeld and Chestin 1993: 189) also build tree nests of sorts, but these seem to result mainly from bears pulling (and thereby breaking) thinner branches in toward

TABLE 4.1
Home Range Sizes Observed in Radiotelemetry Studies of the Eight Species of Bears

	VARIATION IN HOME RANGE AREA (KM^2)[1]		
BEAR SPECIES	AMONG INDIVIDUALS	AMONG SEX-SPECIFIC POPULATION MEANS[2]	NUMBER OF POPULATIONS STUDIED[3]
Polar	900–600,000	20,000–250,000	10
Brown	7–30,000	50–8000 (20)[4]	>30
American black	3–1100	5–500 (2–7000)[4]	>30
Asiatic black	2–200		3
Sloth	4–120		3
Sun	4–20		2
Andean	>10		1
Giant panda	1–60	5–15	3

SOURCES: References are listed for the extreme values shown in the table. Polar (Garner et al. 1990; Ferguson et al. 1999; Amstrup et al. 2000); brown (LeFranc et al. 1987; McLoughlin et al. 1999, unpubl. data; Huber and Roth 1993); American black (Quigley 1982; Pacas and Paquet 1994); Asiatic black (Hwang 2003; T. Hazumi and M. Koyama, unpubl. rep.; A. Katayama, unpubl. rep.); sloth (Joshi et al. 1995; K. Yoganand, unpubl. data); sun (Wong 2002; G. Fredriksson, unpubl. data); Andean (Paisley 2001); giant panda (Pan et al. 2001; Schaller et al. 1985, 1989 [only one panda of each sex]); Yong et al., chapter 10).

[1]Values are rounded off for clarity and because they vary with methods and number of telemetry locations.

[2]Females generally have smaller home ranges than males, so except for polar bears (see next note), the range in area represents the smallest average female home range to the largest average male range.

[3]Studied populations refer to cases where at least two individuals of the same sex were radiotracked for an entire year. Note strong association between amount of variability in home range sizes and number of populations studied. For polar bears, only adult females have been studied, because the necks of males are larger than their head, and hence cannot be fitted with a radiocollar.

[4]Parenthetical values represent the minimum for island populations of brown bears (Schoen et al. 1986; Smith and Van Daele 1990) and American black bears (Lindzey and Meslow 1977), and maximum for black bears in an unusual tundra environment that resembled brown bear habitat (Veitch and Harrington 1996).

the trunk to more easily acquire fruits or nuts, especially acorns (Domico 1988; Schaller et al. 1989; Hwang et al. 2002; pers. observ.). Although acorns are a favorite food and oak trees occur widely across the ranges of these species, nests are prevalent in only a few populations. Black bears sometimes rest in trees, and their cubs routinely climb trees to escape such threats as another bear or a person; cubs may remain in refuge trees for several hours while their mother goes off foraging. Likewise, giant panda cubs often wait in a tree, typically for 4–8 h (in the extreme case, more than 2 full days), while their mother is away foraging (Lü et al. 1994). Adult pandas, however, do not regularly climb trees (Schaller et al. 1985). Sloth bears occasionally climb trees to obtain honey from beehives (Joshi et al. 1997), but they do not use trees for refuge. Instead, possibly because they evolved in a largely treeless environment with such dangerous predators as tigers *(Panthera tigris)* and tree-climbing leopards *(Panthera pardis)*, sloth bear cubs are carried on the mother's back, and climb aboard whenever threatened (Joshi et al. 1999). Polar bears and brown bears also regularly inhabit areas with few or no trees, and frightened cubs occasionally climb on their mother's back (Rosing 2000). However, these two species of bears typically rely on their large size and aggressiveness to ward off potential danger.

The large size of bears in comparison with the small, scattered, and/or nonnutritive food that they consume (Dierenfeld et al. 1982; Welch et al. 1997; Rode et al. 2001) generally limits them to a solitary existence. Forested habitats (restricting visibility) and potential intraspecific predation with little interspecific predation also select for solitariness. In fact, bears are the least social of the carnivores (Gittleman 1989). However, at concentrated food sources like garbage dumps, salmon-spawning streams, insect aggregations, large animal carcasses, cornfields, or even rich berry patches, bears congregate peaceably (polar: Latour [1981]; brown: French et al. [1994]; Craighead et al. [1995]; American black: Herrero [1983]; sloth: Brander [1982]). Congregations of denning polar bears (Hansson and Thomassen 1983; Ovsyanikov 1998) and sloth bears (N. Akhtar, pers. comm.) also have been observed in places where prime den sites are restricted to small, select areas. These two species, although normally active year-round (except parturient females), use dens as shelter from harsh weather conditions—polar bears as protection from winter storms on open Arctic sea ice (Ferguson et al. 2000), and sloth bears to escape intense heat in tropical habitats with little shade (K. Yoganand, pers. comm.; N. Akhtar, pers. comm.).

Despite their generally solitary nature, bears are rarely territorial. Home ranges, even core areas, tend to overlap. Tree marking (biting or clawing) and scent marking (rubbing) have been observed in all but polar bears (brown: Tschanz et al. [1970], Vaisfeld and Chestin [1993: 245]; American black: Burst and Pelton [1983]; Asiatic black: M. Hwang [pers. comm.]; sloth: Laurie and Seidensticker [1977]; sun: G. Fredriksson [pers. comm.]; Andean: B. Peyton [pers. comm.]; panda: Schaller et al. [1985]), and intraspecific avoidance by females of males has been documented or surmised for at least five of the species (polar: Derocher and Stirling [1990]; brown: Wielgus and Bunnell [1994]; American black: Garshelis and Pelton [1981]; Asiatic black: Huygens and Hayashi [2001]; sloth: Joshi et al. [1995]). However, few populations, and in those, only females show indications of territoriality (exclusive use of parts of their home range) (brown: McLoughlin et al. [2000]; American black: Rogers [1987]; Asiatic black: T. Hazumi and M. Koyama [unpubl. ms.]). Some variation in results among studies may be due to different definitions and methodologies for assessing the degree of territoriality (Samson and Huot 2001). Nevertheless, the amount of home range overlap does seem to vary with food conditions and possibly bear density (Powell et al. 1997; McLoughlin et al. 2000).

All told, it is difficult to make species-specific generalizations about many of these life history characteristics, because traits not observed in a given species may be manifested under circumstances that have not yet been studied. Species that have not been well studied are likely to exhibit far more variation than is currently recognized.

REPRODUCTIVE STRATEGIES

Information is available on the mating systems of five of the eight bears (polar, brown, American black, sloth, and panda), all of which can be characterized as promiscuous. Males, depending on their size/social status, may mate with multiple females (or may be entirely precluded from mating). Likewise, females frequently mate with more than one male (often three or more) (polar: Ramsay and Stirling [1986]; brown: Craighead et al. [1995]; American black: Barber and Lindzey [1986]; sloth: Joshi et al. [1999]; panda: Zhu et al. [2001]). They copulate numerous times, presumably to induce ovulation (Boone et al. 1998). In contrast, in some low-density populations of polar and brown bears, males may sequester a lone female for long enough to mate and possibly preclude matings by other males (Herrero and Hamer 1977; Ramsay and Stirling 1986; Hamer and Herrero 1990).

Mating and birthing seasons have been documented in the wild for six of the species (figure 4.2). The brown bear and two black bear species are similar, with mating generally occurring during May–July, delayed implantation of the blasto-

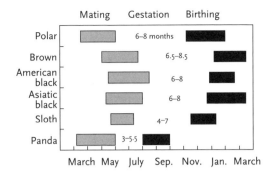

FIGURE 4.2. Seasonality of reproduction in bears outside the tropics. Data for Asiatic black bears and sloth bears living in tropical areas are unavailable. Mating and birthing periods (obtained mainly from references in table 4.2) represent the general norm rather than the full range observed. Indicated gestation includes several months of delayed implantation, followed by a period of fetal development (true gestation) lasting about 2 months.

cyst for several months, about 2 months gestation postimplantation, and birth in winter dens, mainly during January–February (sometimes later for Asiatic black bears [Bromlei 1973]). The mating season appears to be more latitudinally variable than the birthing season (Garshelis and Hellgren 1994). The relatively fixed birthing season but more variable mating season is made possible by flexibility in the period of delayed implantation, and indeed this flexibility is believed to be the primary advantage of delayed implantation, which commonly occurs among carnivores living in highly seasonal environments (Ferguson et al. 1996). The same basic brown–black bear pattern also occurs in polar bears, although it is shifted about a month and a half earlier (figure 4.2). The cycle is also shifted earlier in giant pandas, and the period between mating and birthing is shortened, so births occur during August–September (Wang et al., brief report 4.1). For sloth bears in the northern part of their range, mating coincides with the brown–black bear pattern, but births occur a month or more earlier. In contrast, in the southernmost part of their range (Sri Lanka), sloth bears are reported to have less-defined mating and birthing seasons (Phillips 1984; de Silva and de Silva 1996). Similarly, the tropical sun bear also shows no clear birthing peak (figure 4.3), with the caveat that the only available information on this species are anecdotal reports from the wild and records of births in northern zoos. Zoo records indicate a period of 3–3.5 months (Dathe 1970) or 6–8 months (McCusker 1974) between mating and birthing, the former being more common and indicative of a greatly shortened period of delayed implantation.

Zoo data may or may not accurately reflect the situation in the wild. In giant pandas, matings in Chinese zoos occurred over a wider span of months than was observed in the wild, although birthing dates were similar (Wang et al., brief report 4.1). In Andean bears, reproductive data are not available from the wild, but extensive data are available from captivity. The period between mating and birthing (including apparent delayed implantation) is similar to northern bears (figure 4.2), ranging from 5.5 to 8.5 months (Rosenthal 1989). In northern zoos (North America and Europe) nearly 90% of Andean bear births occurred during December–February, matching the birthing peak in north

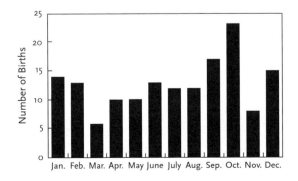

FIGURE 4.3. Birthing dates of sun bears in North American and European zoos. Data (1938–1997) were obtained from North American studbooks (Frederick 1998) and a European summary (Kolter 1995).

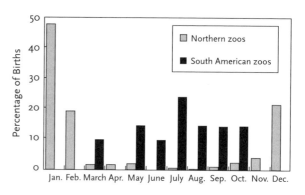

FIGURE 4.4. Birthing dates of Andean (spectacled) bears in northern (North American and European) zoos versus South American zoos. Data (1960–1999) were obtained from international studbooks for the spectacled bear (Rosenthal, 1986–1999). Captive data may not be representative of the situation in the wild.

temperate brown and black bears. However, in zoos on their home continent in the southern hemisphere, the birthing season appears shifted about six months (figure 4.4). This result may be an artifact of small sample size, as some anecdotal evidence from the wild suggests that birthing can vary from year to year with changes in rainfall, but generally peaks during December–January (B. Peyton, pers. comm.).

Bear cubs are born small and helpless (altricial); they are much smaller, in fact, than what would be expected based on the size of their mothers. One explanation is that mother bears, in a state of hibernation without access to food, cannot adequately support fetal development through the placenta, so gestation (after implantation) is shortened and nourishment of cubs is transferred to the mammae (Ramsay and Dunbrock 1986). Although, as previously discussed, most species of bears do not hibernate, there seems to be a long history of denning (presumably hibernation) among ursid ancestors (Hunt et al. 1983), so perhaps due to phylogenetic constraints, all bears still have shortened gestation and give birth to small cubs. As a result, bears of all species are born in secluded dens.

Types of dens used by bears include natural rock caves, hollow trees, beneath the roots of trees, fallen trees, brush piles, aboveground nests, or excavations in the ground or snow (polar bears). American black bears seem to use the most variable types of dens (Linnell et al. 2000), but this may be an artifact of the relatively large number of denning studies that have been conducted on this species. Asiatic black bears also use a variety of den types (Xu et al. 1994). Both species of black bear seem to prefer to den in large, hollow trees, where these are available (unlogged areas) (Johnson and Pelton 1981; Huygens et al. 2001). Giant pandas also may prefer hollow tree dens (Schaller et al. 1985; Durnin et al. 2001, pers. comm.), but readily use rock caves where tree dens are not available (Zhu et al. 2001). Denning information for wild sun and Andean bears is unavailable, but for the other species, cubs stay in dens until 2–4 months old and mothers remain in dens with their newborn cubs for weeks to months without eating or drinking.

For most populations of brown and American black bears, and northern populations of Asiatic black bears, all sex-age classes hibernate and are thus fasting anyway during the winter birthing period. Conversely, parturient polar bears are forced to fast, and hence hibernate while caring for newborn cubs during the winter, even though food is readily available during this season (Watts and Hansen 1987). It may be that parturient mothers of south-latitude species employ this physiological adaptation to enable them to fast while protecting their newborns, despite food being available. For example, sloth bear mothers in maternity dens fast for about two months both in the wild (Joshi et al. 1999) and in captivity (Jacobi 1975).

Giant panda mothers fast for only 2–3 weeks, and unlike other bears, they regularly change dens (Schaller et al. 1985; Lü et al. 1994; Zhu et al. 2001). This behavior is as yet unexplained, but may be because newborn pandas are much smaller than other bears (0.1% of mother's weight, one-third to one-quarter that of other bears) (Lü 1993; Zhu et al. 2001) and are easily transported in the mother's mouth (Schaller et al. 1985: 195). Another possibility is that den switching is at least partly a result of disturbance by researchers. Even hibernating bears, particularly those in warmer climes, sometimes switch dens after visits by researchers or disturbances from other people (polar, brown, and American black: Linnell et al. [2000]; Asiatic black: Huygens et al. [2001]; sun: pers. observ.). Very little is known about the denning habits of parturient females in tropical areas. Possibly, without such limiting factors as snow or food scarcity, tropical bears, like pandas, switch dens and come out to feed on a more regular basis. In captivity, sun bears have been observed to transport young cubs in their mouth or cradled in their front paws (Poglayen-Neuwall 1986).

Reproductive output in bears is strongly linked to maternal condition; because maternal condition varies spatially and temporally, reproduction does as well. Within a population, litter size appears to be less variable than other reproductive parameters (Noyce and Garshelis 1994; McDonald and Fuller 2001), but there is a great deal of variation among populations within species (Taylor 1994; Derocher 1999; Coy 1999; Hilderbrand et al. 1999). Despite this variation, there are also clear differences in litter sizes across species (table 4.2). American black bears have the most variable but also the largest maximum litter sizes, with some populations averaging three or more cubs per litter (Alt 1989; Bittner 2001; Doan-Crider 2003) and particularly heavy individuals producing as many as five cubs (or in at least one case, six [Rowan 1974]). Brown bears rarely produce five cubs (although again, there was at least one case of six [Wilk et al. 1988]); the maximum is typically four, and the highest average litter size for a population is only 2.5 (LeFranc et al. 1987; McLellan 1994).

Polar bears may produce three cubs, with rare instances of four (Larsen 1985; Ramsay and Stirling 1988), although population means almost never exceed two (Derocher 1999). Litters of three cubs have been observed on occasion for sloth bears, both in the wild and in captivity, but because mothers carry their cubs, they have difficulty raising three (Brander 1982; Heath and Mellen 1983); in the wild, births of two-cub litters are most common (Joshi et al. 1999; K. Yoganand, pers. comm.). Limited information indicates that two-cub litters are also most common in Asiatic black bears (Prater 1971; Bromlei 1973; Lekagul and McNeely 1977; Yudin 1993). Giant pandas have the smallest litters, with rare cases of three in captivity, and only one case in captivity where a mother successfully raised twins without human assistance (Zhu et al. 2001). In the wild, pandas that give birth to twins normally raise only one (Schaller et al. 1985; Zhu et al. 2001), although mothers with twins outside the den have been observed in some areas (Schaller et al. 1985; Pan and Lü 1993).

Zoo data may be an unreliable indicator of litter size in the wild. For example, litter size for captive sun bears averages only 1.02 (Kolter 1995), whereas observational data from the wild indicate a prevalence of twins (Fetherstonhaugh 1940; Erdbrink 1953; Domico 1988; G. Fredriksson, pers. comm.). An average litter size of only 1.5 for captive Andean bears (calculated from data in Rosenthal [1986–1999]) also may be less than typical in the wild (B. Peyton, pers. comm.). This may seem counterintuitive, in that captive bears should be in superior health to individuals in the wild, but captivity entails other stresses and nonoptimal conditions.

Number of mammae should be a reflection of maximum litter size, but the trend in bears is equivocal (table 4.2). Female brown bears and black bears (both species) may produce litters of four and sometimes five, so accordingly they have six mammae. However, neither sloth bears nor Andean bears have been observed with four cubs and both rarely have three, yet they also

TABLE 4.2
Reproductive Data for Eight Species of Bears

	FEMALE		LITTER SIZE			LITTER INTERVAL (YEARS)	CUB SURVIVAL (%)	
BEAR SPECIES	MEAN AGE OF FIRST BIRTHING (YEARS)	NUMBER OF MAMMAE	MAXIMUM (RARE)		POPULATION MEAN	POPULATION MEAN	POPULATION MEAN	OVERALL AVERAGE[1]
Polar	4.6–7.2	4	3 (4)		1.5–2.0	2.1–3.6	44–88	60
Brown	4.4–9.6	6	4 (5)		1.6–2.5	2.4–5.2	34–85	75
American black	3.2–6.5	6	5		1.5–3.0	2.0–3.2	27–91	65
Asiatic black		6	4		Mode 2			
Sloth		6	2 (3)		1.6–2.0	≥2.5[2]		70[2]
Sun		4	2					
Andean		6	2 (3)[3]					
Giant panda	5–7[4]	4	2 (3)		1.1–1.7[4]	2.2[2]		60[2]

SOURCES: Polar (Lentfer et al. 1980; Derocher et al. 1992; Derocher and Taylor 1994; Derocher and Sterling 1995; Wiig 1998; Derocher 1999); brown (Bromlei 1973; LeFranc et al. 1987; Vaisfeld and Chestin 1993; McLellan 1994; Case and Buckland 1998; Mace and Waller 1998; Sellers et al. 1999; Ferguson and McLoughlin 2000; Swenson et al. 2001b); American black (Alt 1989; Beecham and Rohlman 1994; Garshelis 1994; Coy 1999); Asiatic black (Prater 1971; Bromlei 1973; Lekagul and McNeely 1977; Yudin 1993); sloth (Laurie and Seidensticker 1977; Joshi et al. 1999; K. Yoganand, pers. comm.), sun (Fetherstonhaugh 1940; Erdbrink 1953; Domico 1988; G. Fredriksson, pers. comm.), Andean (Rosenthal 1986–1999), giant panda (Schaller et al. 1985; Wei et al. 1997; Pan et al. 2001; Zhu et al. 2001; Wang et al., brief report 4.1).

NOTES: Data are from wild populations except where noted. Blank entries in the table represent research needs.

[1] Average among population means.

[2] Data from a single wild study (sloth bear: Joshi et al. [1999]; giant panda: Pan et al. [2001]). The estimate of cub survival for wild pandas is based on observations of litter sizes outside the natal den compared with an assumed average litter size at birth of 1.4 (D. Wang, pers. comm.).

[3] Captive data only.

[4] "Typical" for wild and captive animals.

have six mammae. Conversely, polar bears, which recently evolved from brown bears (Talbot and Shields 1996), have lost a pair of mammae, and yet produce litters as large or larger than sloth bears and Andean bears. Sun bears and giant pandas, with litters of one or two, also have only four mammae.

The sex ratio of litters in wild bears is typically close to 50M:50F. However, sample sizes in most wild studies are too small to detect statistically significant deviations from equity. The general trend among wild studies with large samples of cubs is a bias toward males (51–71%), but adequately large samples are available for only three species: polar, brown, and American black. In captivity, only sloth bears exhibit a consistently male-biased sex ratio at birth (Faust and Thompson 2000). Andean bears in zoos produced strongly male-biased litters during the 1990s, but not previously (based on studbook data since the 1960s: Rosenthal [1986–1999]), suggesting some recent change in captive conditions. Captive sun bears show an unusual bias toward female births (Faust and Thompson 2000). Sex allocation theory predicts that in a polygynous or promiscuous mating system, like that in bears, mothers in good physical condition should favor male offspring (Trivers and Willard 1973); hence zoo bears, which are typically well fed, would be expected to produce male-biased litters. However, stress may cause disproportionately higher mortality of male fetuses (Clutton-Brock and Iason 1986), so captivity-induced stress might be responsible for the abnormally small, female-biased litters in sun bears.

Survival of cubs is related to nutrition as well as other factors, such as bear density (propensity for intraspecific cub-killing), presence of other predators, weather, and hazards (e.g., terrain, roads). For five species of bears studied in the wild, cub survival centers around 60–75%, but this is misleading, as the range among populations within species is quite large (table 4.2). Also, the estimate for giant pandas is derived from sightings of cubs outside the den, presuming an average litter size at birth (based on captive data) of 1.4 (Wang et al., brief report 4.1). If litter size at birth is really close to 1.0, then first-year cub survival would approach 90% (Pan et al. 2001).

Tremendous intraspecific variation also exists in the age that bears of each species produce their first cubs and the interval between litters (table 4.2). Adequate data are available only for polar, brown, and American black bears. Among these, American black bears tend to mature the earliest (some populations having an average age of first reproduction under four years) and produce cubs most frequently (every other year in some populations). However, there are brown and polar bear populations that are more prolific than some populations of American black bears. Intraspecific variation in reproductive age and litter interval are strongly linked to nutrition (polar: Derocher and Stirling [1995]; brown: Ferguson and McLoughlin [2000]; American black: Schwartz and Franzmann [1991]; Noyce and Garshelis [1994]). In some cases, though, late ages of maturity and extended litter intervals may be less a result of poor nutrition than a consequence of females attempting to synchronize their reproduction with sporadically favorable food conditions (Ferguson and McLoughlin 2000). This life history strategy, often referred to as "bet-hedging," appears to be a characteristic of species that live in relatively unproductive but highly seasonal environments; this fits the situation for some ursids (Ferguson and Larivière 2002), but not all (pandas being a notable exception).

Limited information indicates that reproductive rates of bears may decline in old age (polar: Derocher and Stirling [1994]; brown: McLoughlin et al. [2003]; Schwartz et al. [2003]; panda: Zhu et al. [2001]); nevertheless, they continue to produce cubs well into their twenties. All three species for which good data exist on longevity in the wild have been known to live at least 30 years (recorded maximums: polar, 32 years; brown, 36–39 years; American black, 35 years). The oldest known skull of a wild panda was reportedly 26 years old (Wei et al., cited in Reid and Gong [1999]). All species of bears have lived to at least 30 years old in captivity (Jones 1982, 1992, pers. comm.; Schaller et al. 1985; Frederick 1998; L. Wachsberg, pers. comm.).

SHARED TRAITS, VARIATION, AND OUTLIERS

Key life history traits shared among the eight species of bears include solitariness, except gatherings at concentrated food sources; lack of territoriality among males and most females; promiscuous mating; ability for delayed implantation; births of altricial cubs in secluded dens, necessitating fasting by mothers; extended maternal care of cubs, resulting in intervals of at least two years between litters (unless a litter is completely lost); high survivorship (in the absence of human-induced mortality), and long lifespans. Considerable inter- and intraspecific variation has been observed in diet, habitat use, home range size, arboreality, den selection, hibernation, age of sexual maturity, litter interval, litter size, and cub survival.

Probably each of the species is distinctive in some respect. Behaviorally, the most deviant may be the polar bear, because of the radically different conditions in which it lives. It is the only ursid that prefers a nonforested environment (it is, in fact, a marine mammal), and hence is apparently unable to climb trees. It is also the only obligate carnivore among the ursids. Its home range size is larger than that of other bears, and larger than expected for terrestrial carnivores of this size (Ferguson et al. 1999). Polar bear mothers give birth and thus are forced to hibernate during a season of readily accessible food. In some southerly populations, the maternal denning season follows a summer of on-shore fasting imposed by melting of the sea ice, so females may go without food for as long as eight months, during which they give birth to and nurse cubs (Atkinson and Ramsay 1995). Warming trends, which lengthen the period that the bears are without access to ice, and hence food, have negatively affected body condition and reproductive output (Stirling et al. 1999).

Major environmental events have affected other species of bears as well. Giant pandas starved in some areas after widespread flowering and die back of bamboo (Schaller et al. 1985; Reid and Gong 1999). Sun bears starved after fruiting failed following an El Niño (Wong 2002; G. Fredriksson, pers. comm.). American black bears failed to produce cubs, or cubs died or were abandoned after fruiting failures (Fair 1978; Eiler et al. 1989; M. Pelton, pers. comm.). Adult female grizzly (brown) bears suffered high human-caused mortality when mast crops failed and they were lured toward areas of human habitation (Mattson et al. 1992a).

American black bears and brown bears show tremendous plasticity in life history traits; however, it is unclear whether such aspects as their home range size and reproductive output are really more variable than in other bears or if this is simply a reflection of the large number of long-term studies conducted on these two species. Asiatic black bears and Andean bears occupy a climatic gradient related to latitudinal and elevational variation, respectively, that matches or exceeds that of brown and American black bears (see figure 4.1), yet the life histories of the South American and Asian species are not nearly as well known.

Sloth bears are distinctive in carrying their cubs on their back. Sun bears are distinctive in their apparent lack of a birthing season, bimodal gestation period (shorter periods possibly not involving delayed implantation), and female-biased births in captivity. The giant panda gives birth to the smallest cubs, and is the only species that frequently gives birth to more cubs than it can care for. However, it may not be the only bear that reduces litter size after birth. American black bears may consume cubs after birth (M. Vaughan, pers. comm.), brown bears may abandon single cubs (Tait 1980), and sloth bears may be forced to abandon a third cub that cannot fit on the mother's back (Brander 1982; Heath and Mellen 1983).

Giant pandas appear to be an outlier in terms of litter size, but not necessarily in terms of total reproductive output. An "average" female panda—with first cub production at six years old, litters of single cubs upon emergence from the natal den, cub survival (post den emergence) of 70–90%, and an interval of 2.2 years between litters (see table 4.2)—would wean five or six cubs by age 20. A typical female brown bear in northern Alaska or Canada, with a litter size of two and cub survival of 60%, but first reproduction delayed until 8 years and a litter interval of 4 years, would similarly wean five cubs by age 20 (data from McLellan 1994; Ferguson and McLoughlin 2000). Brown bears in other areas are more prolific, but at the periphery of their range, reproductive output tends to be quite low. Like many endangered species (Lomolino and Channell 1995), giant pandas persist only at the periphery of their historic range (Schaller et al. 1985); the reproductive output in these remnant populations may be considerably lower than the former norm for the species. Some have posited that the birth of a second panda cub, which is typically not raised, is insurance in case one dies (Schaller et al. 1985) or, alternately, a nonadaptive trait inherited from their ursid ancestry (Zhu et al. 2001). Another explanation, in accordance with the extraordinary variation in life history traits found among ursids, is that ancient pandas, living in the center of their historic range, were more apt to raise twin cubs. That is, the life history strategies of today's pandas may be an outlier from the past. A similar idea was advanced by Kleiman (1983) who, ironically, was studying reproduction in healthy captive pandas that nevertheless raised only single cubs. As noted previously, however, well-fed bears in captivity may produce smaller litters than those observed in the wild; indeed, captive pandas often have trouble raising even one cub (Zhang et al. 2000).

CONSERVATION IMPLICATIONS

Good life history data are available for few populations of bears. For some species, there are very little data from the wild. Despite this, it is imperative to continue to attempt to assess the status of bears and promote their conservation, not just as eight individual species, but as innumerable disjunct populations.

Population viability analyses have been conducted on various populations of brown (grizzly) bears (Shaffer 1983; Suchy et al. 1985; Sæther et al. 1998; Wiegand et al. 1998; Herrero et al. 2000; Boyce et al. 2001), Asiatic black bears (Wang et al. 1994; Horino and Miura 2000; Park 2001), and giant pandas (Carter and Wang 1993; Wei et al. 1997; Carter et al. 1999). To perform such analyses, estimates of vital rates (not only survival and reproduction, but immigration and emigration) are needed. A key problem is that viability analyses are typically prompted by small population size, and small populations yield small numbers of study subjects and hence poor estimates of relevant variables. From the data reviewed in this chapter, it is apparent that variability in life history parameters within and between populations is characteristic of the Ursidae. Hence, population projections based on limited data from the target population, or data extrapolated from elsewhere, are likely to be quite imprecise.

The most endangered bear in the world, and thus the one deserving particular conservation attention, is the giant panda. Great concern has centered around its low reproductive rate. However, current population models indicate that, except during catastrophic food failures, this reproductive rate should be sufficient to produce a positive population growth rate (Wei et al. 1997; Carter et al. 1999; Pan et al., chapter 5). Population growth rates for pandas, however, have been negatively affected by the impacts of humans, particularly habitat destruction and poaching (Schaller 1993; Reid 1994), the two major threats to other species of bears as well (Servheen et al. 1999).

Little can be done about the slow reproductive rates of some species and populations of bears. Limited improvements in reproduction may be achieved through certain habitat enhancements (e.g., for pandas, saving den trees, creating corridors between reserves, manipulating forests to promote growth of certain species and ages of bamboo and cover trees) (Taylor et al. 1991; Reid and Gong 1999). Conversely, survivorship of adult bears, which is very high in the absence of human intervention, is effectively under our control. Heightened protection, particularly of adult females, from human sources of mortality has enabled some once-troubled bear populations to recover (Swenson et al. 1995; Derocher et al. 1998; Servheen 1999). With no further knowledge about specific life history traits of the lesser-known species, this strategy, if successfully implemented, would dramatically improve the status of bears around the world. Notably, human-caused bear mortality is closely linked to habitat fragmentation and access (Mace et al. 1996; Huber et al. 1998; Garshelis 2002), so habitat conditions, which may affect both reproduction and survival, deserve special attention. The long-term conservation of small, isolated, and increasingly human-impacted bear populations will require innovative, pragmatic, and site-specific approaches (Primm 1996). The situation is especially critical and challenging for the giant panda, because of its precariously low numbers and restricted geographic range, juxtaposed against a complex array of human-related issues that impinge upon many aspects of its life history (Johnson et al. 1996; Liu et al. 1999).

ACKNOWLEDGMENTS

I appreciate Devra Kleiman's invitation to present the paper on which this chapter is based at the *Panda 2000* conference. The Zoological Society of San Diego kindly supported my attendance at the conference, and the Minnesota Department of Natural Resources supported my writing of the paper. I thank the many people who graciously provided summaries of unpublished data, unpublished manuscripts, or insights from their experiences, especially Naim Akhtar, Gabriella Fredriksson, Isaac Goldstein, Toshihiro Hazumi, Mei-hsui Hwang, Lydia Kolter, Phil McLoughlin, Susy Paisley, Bernie Peyton, Mahendra Shrestha, Dajun Wang, and K. Yoganand. Mike Pelton provided an encouraging review of an earlier draft.

REFERENCES

Alt, G. L. 1989. Reproductive biology of female black bears and early growth and development of cubs in northeastern Pennsylvania. Ph.D. dissertation. West Virginia University, Morgantown.

Amstrup, S. C. 2000. Polar bear. In *The natural history of an arctic oil field,* edited by J. C. Truett and S. R. Johnson, pp. 133–57. San Diego: Academic.

Amstrup, S. C., and G. M. Durner. 1995. Survival rates of radio-collared female polar bears and their dependent young. *Can J Zool* 73:1312–22.

Amstrup, S. C., G. M. Durner, I. Stirling, N. J. Lunn, and F. Messier. 2000. Movements and distribution of polar bears in the Beaufort Sea. *Can J Zool* 78:948–66.

Atkinson, S. N., and M. A. Ramsay. 1995. The effects of prolonged fasting on the body composition and reproductive success of female polar bears *(Ursus maritimus)*. *Funct Ecol* 9:559–67.

Atkinson, S. N., R. A. Nelson, and M. A. Ramsay. 1996. Changes in the body composition of fasting polar bears *(Ursus maritimus):* The effect of relative fatness on protein conservation. *Physiol Zool* 69:304–16.

Barber, K. R., and F. G. Lindzey. 1986. Breeding behavior of black bears. *Int Conf Bear Res Manag* 6: 129–36.

Beck, T. D. I. 1991. Black bears of west-central Colorado. *Colo Div Wildl Tech Publ* 39.

Beecham, J. J., and J. Rohlman. 1994. *A shadow in the forest. Idaho's black bear.* Moscow, Idaho: University of Idaho Press.

Bininda-Emonds, O. R. P., J. L. Gittleman, and A. Purvis. 1999. Building large trees by combining phylogenetic information: A complete phylogeny of the extant Carnivora (Mammalia). *Biol Rev* 74:143–75.

Bittner, S. L. 2001. Maryland status report. *East Black Bear Workshop* 16:36–38.

Boone, W. R., J. C. Catlin, K. J. Casey, E. T. Boone, P. S. Dye, R. J. Schuett , J. O. Rosenberg, T. Tsubota, and J. M. Bahr. 1998. Bears as induced ovulators—A preliminary study. *Ursus* 10:503–5.

Boyce, M. S., ed. 1988. *Evolution of life histories of mammals: Theory and pattern.* New Haven: Yale University Press.

Boyce M. S., B. M. Blanchard, R. R. Knight, and C. Servheen. 2001. Population viability for grizzly bears: A critical review. *Int Assoc Bear Res Manag Monogr Ser* 4.

Brander, A. A. D. 1982. *Wild animals in central India.* Dehra Dun: Natraj.

Bromlei, F. G. 1973. *Bears of the south far-eastern USSR.* Translated from Russian. New Delhi: Indian National Scientific Documentation Center.

Brown, A. D., and D. I. Rumiz. 1989. Habitat and distribution of the spectacled bear *(Tremarctos ornatus)* in the southern limit of its range. In *Proceedings of the first international symposium on the spectacled bear,* edited by M. Rosenthal, pp. 93–103. Chicago: Lincoln Park Zoo.

Burst, T. L., and M. R. Pelton. 1983. Black bear mark trees in the Smoky Mountains. *Int Conf Bear Res Manag* 5:45–53.

Carter, J., and H. Wang. 1993. A model linking the population dynamics of the giant panda *Ailuropoda melanoleuca* with bamboo life history dynamics. *Proc Int Union Game Biol* 21(1):299–309.

Carter, J., A. S. Ackleh, B. P. Leonard, and H. Wang. 1999. Giant panda *(Ailuropoda melanoleuca)* population dynamics and bamboo (subfamily Bambusoideae) life history: A structured population approach to examining carrying capacity when the prey are semelparous. *Ecol Model* 123:207–23.

Case, R. L., and L. Buckland. 1998. Reproductive characteristics of grizzly bears in the Kugluktuk area, Northwest Territories, Canada. *Ursus* 10:41–47.

Choudhury, A. 1993. Potential biosphere reserves in Assam (India). *Tiger Paper* 20(1):2–8.

Clutton-Brock, T. H., and G. R. Iason. 1986. Sex ratio variation in mammals. *Q Rev Biol* 61:339–74.

Coy, P. L. 1999. Geographic variation in reproduction of Minnesota black bears. M.Sc. thesis. University of Minnesota, Minneapolis.

Craighead, J. J., J. S. Sumner, and J. A. Mitchell. 1995. *The grizzly bears of Yellowstone. Their ecology in the Yellowstone ecosystem, 1959–1992.* Washington, D.C.: Island.

Dathe, H. 1970. A second generation birth of captive sun bears at East Berlin zoo. *Int Zoo Yrbk* 10:79.

Derocher, A. E. 1999. Latitudinal variation in litter size of polar bears: Ecology or methodology? *Polar Biol* 22:350–56.

Derocher, A. E., and I. Stirling. 1990. Distribution of polar bears *(Ursus maritimus)* during the ice-free period in western Hudson Bay. *Can J Zool* 68:1395–1403.

Derocher, A. E., and I. Stirling. 1994. Age-specific reproductive performance of female polar bears *(Ursus maritimus). J Zool Lond* 234:527–36.

Derocher, A. E., and I. Stirling. 1995. Temporal variation in reproduction and body mass of polar bears in western Hudson Bay. *Can J Zool* 73:1657–65.

Derocher, A. E., and M. T. Taylor. 1994. Density-dependent population regulation of polar bears. In *Density-dependent population regulation of black, brown, and polar bears,* edited by M. Taylor, pp. 25–30. *Int Conf Bear Res Manag Monogr Ser* 3.

Derocher, A. E., and Ø. Wiig. 1999. Infanticide and cannibalism of juvenile polar bears *(Ursus maritimus)* in Svalbard. *Arctic* 52:307–10.

Derocher, A. E., I. Stirling, and D. Andriashek. 1992. Pregnancy rates and serum progesterone levels of polar bears in western Hudson Bay. *Can J Zool* 70:561–66.

Derocher, A. E., D. Andriashek, and I. Stirling. 1993. Terrestrial foraging by polar bears during the ice-free period in western Hudson Bay. *Arctic* 46:251–54.

Derocher, A. E., G. W. Garner, N. J. Lunn, and Ø. Wiig, eds. 1998. *Polar bears: Proceedings of the Twelfth Working Meeting of the IUCN/SSC Polar Bear Specialist Group.* Occasional Paper of the IUCN Species Survival Commission no. 19. Gland: IUCN.

de Silva, M., and P. K. de Silva. 1996. The sloth bear *(Melursus ursinus* Shaw) and its feeding habits in the Ruhuna National Park, Sri Lanka. In *Abstracts of the 2nd International Symposium on the Coexistence of Large Carnivores with Man,* p. 84. Saitama, Japan, November 19–23.

Dierenfeld, E. S., H. F. Hintz, J. B. Robertson, P. J. Van Soest, and O. T. Oftedal. 1982. Utilization of bamboo by the giant panda. *J Nutrition* 112:636–41.

Doan-Crider, D. L. 2003. Movements and spatio-temporal variation in relation to food productivity and distribution, and population dynamics of the Mexican black bear in the Serranias del Burro, Coahuila, Mexico. Ph.D. thesis. Texas A&M University, Kingsville.

Domico, T. 1988. *Bears of the world.* New York: Facts on File.

Durnin, M. E., D. R. McCullough, J. Huang, and H. Zhang. 2001. Estimation and characterization of potential giant panda *(Ailuropoda melanoleuca)* dens in the Wolong Nature Reserve, Sichuan, China. In *Abstracts of the 13th International Conference on Bear Research and Management,* p. 60. Jackson, Wyoming, May 20–26.

Eiler, J. H., W. G. Wathen, and M. R. Pelton. 1989. Reproduction in black bears in the southern Appalachian Mountains. *J Wildl Manag* 53:353–60.

Erdbrink, D. P. 1953. *A review of fossil and recent bears of the old world with remarks on their phylogeny based upon their dentition.* Deventer: Drukkerij Jan de Lange.

Fair, J. S. 1978. Unusual dispersal of black bear cubs in Utah. *J Wildl Manag* 42:642–44.

Faust, L. J., and S. D. Thompson. 2000. Birth sex ratio in captive mammals: Patterns, biases, and the implications for management and conservation. *Zoo Biol* 19:11–25.

Ferguson, S. H., and S. Larivière. 2002. Can comparing life histories help conserve carnivores? *Anim Conserv* 5:1–12.

Ferguson, S. H., and P. D. McLoughlin. 2000. Effect of energy availability, seasonality, and geographic range on brown bear life history. *Ecography* 23:193–200.

Ferguson, S. H., J. A. Virgil, and S. Larivière. 1996. Evolution of delayed implantation and associated grade shifts in life history traits of North American carnivores. *Ecoscience* 3:7–17.

Ferguson, S. H., M. K. Taylor, E. W. Born, A. Rosing-Asvid, and F. Messier. 1999. Determinants of home range size for polar bears *(Ursus maritimus)*. *Ecol Letters* 2:311–18.

Ferguson, S. H., M. K. Taylor, A. Rosing-Asvid, E. W. Born, and F. Messier. 2000. Relationships between denning of polar bears and conditions of sea ice. *J Mammal* 81:1118–27.

Fetherstonhaugh, A. H. 1940. Some notes on Malayan bears. *Malayan Nat J* 1:15–22.

Frederick, C. 1998. *North American regional sun and Asiatic black bear studbook*. Seattle: Woodland Park Zoo.

French, S. P., M. G. French, and R. R. Knight. 1994. Grizzly bear use of army cutworm moths in the Yellowstone ecosystem. *Int Conf Bear Res Manag* 9(1):389–99.

Garner, G. W., S. T. Knick, and D. C. Douglas. 1990. Seasonal movements of adult female polar bears in the Bering and Chukchi seas. *Int Conf Bear Res Manag* 8:219–26.

Garshelis, D. L. 1994. Density-dependent population regulation of black bears. In *Density-dependent population regulation of black, brown, and polar bears*, edited by M. Taylor, pp. 3–14. *Int Conf Bear Res Manag Monogr Ser* 3.

Garshelis, D. L. 2002. Misconceptions, ironies, and uncertainties regarding trends in bear populations. *Ursus* 13:321–34.

Garshelis, D. L., and E. C. Hellgren. 1994. Variation in reproductive biology of male black bears. *J Mammal* 75:175–88.

Garshelis, D. L., and M. R. Pelton. 1981. Movements of black bears in the Great Smoky Mountains National Park. *J Wildl Manag* 45:912–25.

Gittleman, J. L. 1986. Carnivore life history patterns: Allometric, phylogenetic, and ecological associations. *Am Nat* 127:744–71.

Gittleman, J. L. 1989. Carnivore group living: Comparative trends. In *Carnivore behavior, ecology, and evolution*, edited by J. L. Gittleman, pp. 183–207. Vol. 1. Ithaca: Cornell University Press.

Gittleman, J. L. 1993. Carnivore life histories: A reanalysis in light of new models. *Symp Zool Soc Lond* 65:65–86.

Gittleman, J. L. 1994. Are the pandas successful specialists or evolutionary failures? *BioScience* 44:456–64.

Goldstein, I. 2002. Andean bear–cattle interactions and tree nest use in Bolivia and Venezuela. *Ursus* 13:369–72.

Gompper, M. E., and J. L. Gittleman. 1991. Home range scaling: Intraspecific and comparative trends. *Oecologia* 87:343–48.

Hamer, D., and S. Herrero. 1990. Courtship and use of mating areas by grizzly bears in the front ranges of Banff National Park, Alberta. *Can J Zool* 68:2695–97.

Hansson, R., and J. Thomassen. 1983. Behavior of polar bears with cubs in the denning area. *Int Conf Bear Res Manag* 5:246–54.

Heath, D. C., and J. D. Mellen. 1983. Development of maternally reared sloth bear cubs in captivity. Presented at the 6th International Conference on Bear Research and Management, Grand Canyon, Arizona, February 18–22.

Hellgren, E. C. 1998. Physiology of hibernation in bears. *Ursus* 10:467–77.

Herrero, S. 1983. Social behaviour of black bears at a garbage dump in Jasper National Park. *Int Conf Bear Res Manag* 5:54–70.

Herrero, S., and D. Hamer. 1977. Courtship and copulation of a pair of grizzly bears, with comments on reproductive plasticity and strategy. *J Mammal* 58:441–44.

Herrero, S., P. S. Miller, and U. S. Seal, eds. 2000. Population and habitat viability assessment for the grizzly bear of the central Rockies ecosystem. Eastern slopes grizzly bear project of the University of Calgary, Alberta, and the Conservation Breeding Specialist Group. Apple Valley, Minnesota: Conservation Breeding Specialist Group.

Hilderbrand, G. V., C. C. Schwartz, C. T. Robbins, M. E. Jacoby, T. A. Hanley, S. M. Arthur, and C. Servheen. 1999. The importance of meat, particularly salmon, to body size, population productivity, and conservation of North American brown bears. *Can J Zool* 77:132–38.

Hobson, K. A., and I. Stirling. 1997. Low variation in blood $\delta^{13}C$ among Hudson Bay polar bears: Implications for metabolism and tracing terrestrial foraging. *Mar Mammal Sci* 13:359–67.

Horino, S., and S. Miura. 2000. Population viability analysis of a Japanese black bear population. *Pop Ecol* 42:37–44.

Horn, H. S. 1978. Optimal tactics of reproduction and life history. In *Behavioural ecology: An evolutionary approach,* edited by J. R. Krebs and N. B. Davies, pp. 411–29. Sunderland: Sinauer.

Huber, D., and H. U. Roth. 1993. Movements of European brown bears in Croatia. *Acta Theriol* 38: 151–59.

Huber, D., J. Kusak, and A. Frkovic. 1998. Traffic kills of brown bears in Gorski Kotar, Croatia. *Ursus* 10:167–71.

Hunt, R. M., Jr. 1996. Biogeography of the order Carnivora. In *Carnivore behavior, ecology, and evolution,* edited by J. L. Gittleman, pp. 485–541. Vol. 2. Ithaca: Cornell University Press.

Hunt, R. M., Jr., X. Xiang-Xu, and J. Kaufman. 1983. Miocene burrows of extinct bear dogs: Indication of early denning behavior of large mammalian carnivores. *Science* 221:364–66.

Huygens, O. C., and H. Hayashi. 2001. Use of stone pine seeds and oak acorns by Asiatic black bears in central Japan. *Ursus* 12:47–50.

Huygens, O. C., M. Goto, S. Izumiyama, H. Hayashi, and T. Yoshida. 2001. Denning ecology of two populations of Asiatic black bears in Nagano prefecture, Japan. *Mammalia* 65:417–28.

Hwang M.-H. 2003. Ecology of Asiatic black bears and people–bear interactions in Yushan National Park, Taiwan. Ph.D. thesis. University of Minnesota, Minneapolis.

Hwang, M-H., D. L. Garshelis, and Y. Wang. 2002. Diets of Asiatic black bears in Taiwan, with methodological and geographical comparisons. *Ursus* 13:111–25.

Jacobi, E. F. 1975. Breeding sloth bears in Amsterdam Zoo. In *Breeding endangered species in captivity,* edited by R. D. Martin, pp. 351–56. London: Academic.

Jacoby, M. E., G. V. Hilderbrand, C. Servheen, C. C. Schwartz, S. M. Arthur, T. A. Hanley, C. T. Robbins, and R. Michener. 1999. Trophic relations of brown and black bears in several western North American ecosystems. *J Wildl Manag* 63:921–29.

Johnson, K. G., and M. R. Pelton. 1981. Selection and availability of dens for black bears in Tennessee. *J Wildl Manag* 45:111–19.

Johnson, K. G., Y. Yao, C. You, S. Yang, and Z. Shen. 1996. Human/carnivore interactions: Conservation and management implications from China. In *Carnivore behavior, ecology, and evolution,* edited by J. L. Gittleman, pp. 337–70. Vol. 2. Ithaca: Cornell University Press.

Jones, M. L. 1982. Longevity of captive mammals. *Der Zool Garten NF Jena* 52:113–28.

Jones, M. L. 1992. Longevity of mammals in captivity —An update. *In Vivo* 6:363–66.

Joshi, A. R., D. L. Garshelis, and J. L. D. Smith. 1995. Home ranges of sloth bears in Nepal: Implications for conservation. *J Wildl Manag* 59:204–14.

Joshi, A. R., D. L. Garshelis, and J. L. D. Smith. 1997. Seasonal and habitat-related diets of sloth bears in Nepal. *J Mammal* 78:584–97.

Joshi, A. R., J. L. D. Smith, and D. L. Garshelis. 1999. Sociobiology of the myrmecophagous sloth bear in Nepal. *Can J Zool* 77:1690–704.

Kleiman, D. 1983. Ethology and reproduction of captive giant pandas *(Ailuropoda melanoleuca). Z Tierpsychol* 62:1–46.

Kolter, L. 1995. *European regional studbook of the sun bear (Helarctos malayanus).* Cologne: Zoologischer Garten.

Kurtén, B. 1976. *The cave bear story. Life and death of a vanished animal.* New York: Columbia University Press.

Larsen, T. 1985. Polar bear denning and cub production in Svalbard, Norway. *J Wildl Manag* 49: 320–26.

Latour, P. B. 1981. Spatial relationships and behavior of polar bears *(Ursus maritimus* Phipps) concentrated on land during the ice-free season of Hudson Bay. *Can J Zool* 59:1763–74.

Laurie, A., and J. Seidensticker. 1977. Behavioural ecology of the sloth bear *(Melursus ursinus). J Zool Lond* 182:187–204.

LeFranc, M. N., Jr., M. B. Moss, K. A. Patnode, and W. C. Sugg III. 1987. *Grizzly bear compendium.* Washington, D.C.: Interagency Grizzly Bear Committee.

Lekagul, B., and J. A. McNeely. 1977. *Mammals of Thailand.* Bangkok: Sahakarnbhat.

Lentfer, J. W., R. J. Hensel, J. R. Gilbert, and F. E. Sorensen. 1980. Population characteristics of Alaskan polar bears. *Int Conf Bear Res Manag* 4:109–15.

Lindzey, F. G., and E. C. Meslow. 1977. Home range and habitat use by black bears in southwestern Washington. *J Wildl Manag* 41:413–25.

Linnell, J. D. C., J. E. Swenson, R. Andersen, and B. Barnes. 2000. How vulnerable are denning bears to disturbance? *Wildl Soc Bull* 28:400–413.

Liu, J., Z. Ouyang, W. W. Taylor, R. Groop, Y. Tan, and H. Zhang. 1999. A framework for evaluating the effects of human factors on wildlife habitat: The case of giant pandas. *Conserv Biol* 13:1360–70.

Lomolino, M. V., and R. Channell. 1995. Splendid isolation: Patterns of geographic range collapse in endangered mammals. *J Mammal* 76:335–47.

Lü, Z. 1993. Newborn panda in the wild. *Nat Geogr* 183(2):60–65.

Lü, Z., W. Pan, and J. Harkness. 1994. Mother-cub relationships in giant pandas in the Qinling

Mountains, China, with comment on rescuing abandoned cubs. *Zoo Biol* 13:567–68.

Mace, R., and J. S. Waller. 1998. Demography and population trend of grizzly bears in the Swan Mountains, Montana. *Conserv Biol* 12:1005–16.

Mace, R., J. S. Waller, T. J. Manley, L. J. Lyon, and H. Zuuring. 1996. Relationships among grizzly bears, roads, and habitat in the Swan Mountains, Montana. *J Appl Ecol* 33:1395–1404.

Mattson, D. J. 1998. Diet and morphology of extant and recently extinct northern bears. *Ursus* 10: 479–96.

Mattson, D. J. 2001. Myrmecophagy by Yellowstone grizzly bears. *Can J Zool* 79:779–93.

Mattson, D. J., B. M. Blanchard, and R. R. Knight. 1992a. Yellowstone grizzly bear mortality, human habituation, and whitebark pine seed crops. *J Wildl Manag* 56:432–42.

Mattson, D. J., R. R. Knight, and B. M. Blanchard. 1992b. Cannibalism and predation on black bears by grizzly bears in the Yellowstone ecosystem, 1975–1990. *J Mammal* 73:422–25.

McCusker, J. S. 1974. Breeding Malayan sun bears at Fort Worth Zoo. *Int Zoo Yrbk* 15:118–19.

McDonald, J. E., and T. K. Fuller. 2001. Prediction of litter size in American black bears. *Ursus* 12:93–102.

McLellan, B. 1994. Density-dependent population regulation of brown bears. In *Density-dependent population regulation of black, brown, and polar bears*, edited by M. Taylor, pp. 15–24. *Int Conf Bear Res Manag Monogr Ser* 3.

McLellan, B. N., F. W. Hovey, R. D. Mace, J. G. Woods, D. W. Carney, M. L. Gibeau, W. L. Wakkinen, and W. F. Kasworm. 1999. Rates and causes of grizzly bear mortality in the interior mountains of British Columbia, Alberta, Montana, Washington, and Idaho. *J Wildl Manag* 63:911–20.

McLoughlin, P. D., R. L. Case, R. J. Gau, S. H. Ferguson, and F. Messier. 1999. Annual and seasonal movement patterns of barren-ground grizzly bears in the central Northwest Territories. *Ursus* 11:79–86.

McLoughlin, P. D., S. H. Ferguson, and F. Messier. 2000. Intraspecific variation in home range overlap with habitat quality: A comparison among brown bear populations. *Evol Ecol* 14:39–60.

McLoughlin, P. D., M. K. Taylor, H. D. Cluff, R. J. Gau, R. Mulders, R. L. Case, S. Boutin, and F. Messier. 2003. Demography of barren-ground grizzly bears. *Can J Zool* 81:294–301.

Meijaard, E. 1999. *Ursus (Helarctos) malayanus*, the neglected Malayan sun bear. Netherlands Commission for International Nature Protection. Leiden: Backhuys.

Noyce, K. V., and D. L. Garshelis. 1994. Body size and blood characteristics as indicators of condition and reproductive performance in black bears. *Int Conf Bear Res Manag* 9(1):481–96.

Noyce, K. V., P. B. Kannowski, and M. R. Riggs. 1997. Black bears as ant-eaters: Seasonal associations between bear myrmecophagy and ant ecology in north-central Minnesota. *Can J Zool* 75:1671–86.

O'Brien, S. J. 1993. Fuzzy thinking about the giant panda's ancestry. In *Bears. Majestic creatures of the wild*, edited by I. Stirling, pp. 34–35. Emmaus, Penn.: Rodale.

O'Brien, S. J., W. G. Nash, D. E. Wildt, M. E. Bush, and R. E. Benveniste. 1985. A molecular solution to the riddle of the giant panda's phylogeny. *Nature* 317:140–44.

Ovsyanikov, N. 1998. Den use and social interactions of polar bears during spring in a dense denning area on Herald Island, Russia. *Ursus* 10:251–58.

Pacas, C. J., and P. C. Paquet. 1994. Analysis of black bear home range using a geographic information system. *Int Conf Bear Res Manag* 9(1):419–25.

Paisley, S. L. 2001. Andean bears and people in Apolobamba, Bolivia: Culture, conflict and conservation. Ph.D. thesis, University of Kent, Canterbury.

Pan, W., and Z. Lü. 1993. The giant panda. In *Bears. Majestic creatures of the wild*, edited by I. Stirling, pp. 140–49. Emmaus, Penn.: Rodale.

Pan, W., Z. Lü, X. Zhu, D. Wang, H. Wang, Y. Long, D. L. Fu, and X. Zhou. 2001. *A chance for lasting survival*. (In Chinese.) Beijing: Peking University Press.

Park, S. 2001. Habitat-based population viability analysis for the Asiatic black bear in Mt. Chiri National Park, Korea. *Cent Biol Mångfald (CBM:s) Skriftserie (Uppsala)* 3:149–65.

Partridge, L., and P. H. Harvey. 1988. The ecological context of life history evolution. *Science* 241:1449–55.

Peyton, B. 1980. Ecology, distribution, and food habits of spectacled bears, *Tremarctos ornatus*, in Peru. *J Mammal* 61:639–52.

Phillips, W. W. A. 1984. *Manual of the mammals of Sri Lanka*. Second edition. Part III. Colombo: Wildlife and Nature Protection Society of Sri Lanka.

Poglayen-Neuwall, I. 1986. An unusual method of transport of young sun bears by their mothers. *Der Zool Garten NF Jena* 56:437–38.

Powell, R. A., J. W. Zimmerman, and D. E. Seaman. 1997. *Ecology and behaviour of North American black bears: Home ranges, habitat and social organization*. London: Chapman and Hall.

Prater, S. H. 1971. *The book of Indian animals*. Bombay: Oxford University Press.

Primm, S. A. 1996. A pragmatic approach to grizzly bear conservation. *Cons Biol* 10:1026–35.

Quigley, H. B. 1982. Activity patterns, movement ecology, and habitat utilization of black bears in the

Great Smoky Mountains National Park, Tennessee. M.Sc. thesis. University of Tennessee, Knoxville.

Ramsay, M. A., and R. L. Dunbrock. 1986. Physiological constraints on life history phenomena: The example of small bear cubs at birth. *Am Nat* 127: 735–43.

Ramsay, M. A., and I. Stirling. 1986. On the mating system of polar bears. *Can J Zool* 64:2142–51.

Ramsay, M. A., and I. Stirling. 1988. Reproductive biology and ecology of female polar bears *(Ursus maritimus)*. *J Zool Lond* 214:601–34.

Reid, D. G. 1994. The focus and role of biological research in giant panda conservation. *Int Conf Bear Res Manag* 9(1):23–33.

Reid, D. G., and J. Gong. 1999. Giant panda conservation action plan *(Ailuropoda melanoleuca)*. In *Bears. Status survey and conservation action plan*, edited by C. Servheen, S. Herrero, and B. Peyton, pp. 241–54. Gland: IUCN.

Reid, D. G., J. Hu, S. Dong, W. Wang, and Y. Huang. 1989. Giant panda *Ailuropoda melanoleuca* behaviour and carrying capacity following a bamboo die-off. *Biol Conserv* 49:85–104.

Rode, K. D., C. T. Robbins, and L. A. Shipley. 2001. Constraints on herbivory by grizzly bears. *Oecologia* 128:62–71.

Roff, D. A. 1992. *The evolution of life histories. Theory and analysis*. New York: Chapman and Hall.

Rogers, L. L. 1987. Effects of food supply and kinship on social behavior, movements, and population growth of black bears in northeastern Minnesota. *Wildl Monogr* 97.

Rosenthal, M. 1986–1999. *International studbook for the spectacled bear (Tremarctos ornatus)*. Chicago: Lincoln Park Zoo.

Rosenthal, M. 1989. Spectacled bears—An overview of management practices. In *Proceedings of the first international symposium on the spectacled bear*, edited by M. Rosenthal, pp. 287–95. Chicago: Lincoln Park Zoo.

Rosing, N. 2000. Bear beginnings. New life on the ice. *Nat Geogr* 198(6):30–39.

Rowan, W. 1947. A case of six cubs in the common black bear. *J Mammal* 28:404–5.

Samson, C., and J. Huot. 2001. Spatial and temporal interactions between female American black bears in mixed forests of eastern Canada. *Can J Zool* 79:633–41.

Sæther, B.-E., S. Engen, J. E. Swenson, Ø. Bakke, and F. Sandegren. 1998. Assessing the viability of Scandinavian brown bear, *Ursus arctos*, populations: The effects of uncertain parameter estimates. *Oikos* 83:403–16.

Santiapillai, A., and C. Santiapillai. 1996. The status, distribution and conservation of the Malayan sun bear *(Helarctos malayanus)* in Indonesia. *Tiger Paper* 23(1):11–16.

Schaller, G. B. 1977. *Mountain monarchs. Wild sheep and goats of the Himalaya*. Chicago: University of Chicago Press.

Schaller, G. B. 1979. *Stones of silence. Journeys in the Himalaya*. Chicago: University of Chicago Press.

Schaller, G. B. 1993. *The last panda*. Chicago: University of Chicago Press.

Schaller, G. B., J. Hu, W. Pan, and J. Zhu. 1985. *The giant pandas of Wolong*. Chicago: University of Chicago Press.

Schaller, G. B., Q. Teng, K. G. Johnson, X. Wang, H. Shen, and J. Hu. 1989. The feeding ecology of giant pandas and Asiatic black bears in the Tangjiahe Reserve, China. In *Carnivore behavior, ecology, and evolution*, edited by J. L. Gittleman, pp. 212–41. Vol. 1. Ithaca: Cornell University Press.

Schoen, J. W., J. W. Lentfer, and L. Beier. 1986. Differential distribution of brown bears on Admiralty Island, southeast Alaska: A preliminary assessment. *Int Conf Bear Res Manag* 6:1–5.

Schwartz, C. C., and A. W. Franzmann. 1991. Interrelationship of black bears to moose and forest succession in the northern coniferous forest. *Wildl Monogr* 113.

Schwartz, C. C., and A. W. Franzmann. 1992. Dispersal and survival of subadult black bears from the Kenai Peninsula, Alaska. *J Wildl Manag* 56: 426–31.

Schwartz C. C., K. A. Keating, H. V. Reynolds III, V. G. Barnes Jr., R. A. Sellers, J. E. Swenson, S. D. Miller, B. N. McLellan, J. Keay, R. McCann, M. Gibeau, W. F. Wakkinen, R. D. Mace, W. Kasworm, R. Smith, and S. Herrero. 2003. Reproductive maturation and senescence in the female brown bear. *Ursus* 14:109–19.

Schweinsburg, R. E. 1979. Summer snow dens used by polar bears in the Canadian high Arctic. *Arctic* 32:165–69.

Seidensticker, J. 1993. The sloth bear. In *Bears. Majestic creatures of the wild*, edited by I. Stirling, pp. 128–31. Emmaus, Penn.: Rodale.

Sellers R. A., S. D. Miller, T. S. Smith, and R. Potts. 1999. Population dynamics of a naturally regulated brown bear population on the coast of Katmai National Park and Preserve. Final Report. NPS/AR/NRTR-99/36. Anchorage: U.S. National Park Service.

Servheen, C. 1989. The status and conservation of the bears of the world. *Int Conf Bear Res Manag Monogr Ser* 2.

Servheen, C. 1999. Status and management of the grizzly bear in the lower 48 United States. In *Bears. Status survey and conservation action plan*, edited by C. Servheen, S. Herrero, and B. Peyton, pp. 50–54. Gland: IUCN.

Servheen, C. S., Herrero, and B. Peyton. 1999. *Bears. Status survey and conservation action plan*. Gland: IUCN.

Shaffer, M. L. 1983. Determining minimum viable population sizes for the grizzly bear. *Int Conf Bear Res Manag* 5:133–39.

Smith, R. B., and L. J. Van Daele. 1990. Impacts of hydroelectric development on brown bears, Kodiak Island, Alaska. *Int Conf Bear Res Manag* 8:93–103.

Sorensen, V. A., and R. A. Powell. 1998. Estimating survival rates of black bears. *Can J Zool* 76:1335–43.

Southwood, T. R. E. 1988. Tactics, strategies and templets. *Oikos* 52:3–18.

Stearns, S. C. 1976. Life-history tactics: A review of the ideas. *Q Rev Biol* 51:3–47.

Stearns, S. C. 1992. *The evolution of life histories*. Oxford: Oxford University Press.

Stirling, I., N. J. Lunn, and J. Iacozza. 1999. Long-term trends in the population ecology of polar bears in Western Hudson Bay in relation to climatic change. *Arctic* 52:294–306.

Suchy, W. J., L. L. McDonald, M. D. Strickland, and S. H. Anderson. 1985. New minimum viable population size for grizzly bears of the Yellowstone ecosystem. *Wildl Soc Bull* 13:223–28.

Swenson, J. E., P. Wabakken, F. Sandegren, A. Bjärvall, R. Franzén, and A. Söderberg. 1995. The near extinction and recovery of brown bears in Scandinavia in relation to the bear management policies of Norway and Sweden. *Wildl Biol* 1:11–25.

Swenson, J. E., F. Sandegren, and A. Söderberg. 1998. Geographic expansion of an increasing brown bear population: Evidence for presaturation dispersal. *J Anim Ecol* 67:819–26.

Swenson, J. E., A. Jansson, R. Riig, and F. Sandegren. 1999. Bears and ants: Myrmecophagy by brown bears in central Scandinavia. *Can J Zool* 77:551–61.

Swenson, J. E., B. Dahle, and F. Sandegren. 2001a. Intraspecific predation in Scandinavian brown bears older than cubs-of-the-year. *Ursus* 12:81–92.

Swenson, J. E., F. Sandegren, P. Segerström, and S. Brunberg. 2001b. Factors associated with loss of brown bear cubs in Sweden. *Ursus* 12:69–80.

Swihart, R. K., N. A. Slade, and B. J. Bergstrom. 1988. Relating body size to the rate of home range use in mammals. *Ecology* 69:393–99.

Tait, D. E. N. 1980. Abandonment as a reproductive tactic—The example of grizzly bears. *Am Nat* 115: 800–808.

Talbot, S. L., and G. F. Shields. 1996. A phylogeny of the bears (*Ursidae*) inferred from complete sequences of three mitochondrial genes. *Mol Phylog Evol* 5:567–75.

Taylor, A. H., D. G. Reid, Z. Qin, and J. Hu. 1991. Bamboo dieback: An opportunity to restore panda habitat. *Environ Conserv* 18:166–68.

Taylor, M., ed. 1994. Density-dependent population regulation of black, brown, and polar bears. *Int Conf Bear Res Manag Monogr Ser 3*.

Trivers, R. L., and D. E. Willard. 1973. Natural selection of parental ability to vary the sex ratio of offspring. *Science* 179:90–92.

Tschanz, V. B., M. Meyer-Holzapfel, and S. Bachmann. 1970. Das Informationsystem bei Braunbären. *Z Tierpsychol* 27:47–72.

Vaisfeld, M. A., and I. E. Chestin, eds. 1993. Bears. Brown bear, polar bear, Asian black bear. *Game animals of Russia and adjacent countries and their environment* (series). Moscow: Nauka.

Veitch, A. M., and F. H. Harrington. 1996. Brown bears, black bears, and humans in northern Labrador: An historical perspective and outlook to the future. *J Wildl Res* 1:245–50.

Wang, Y., S. Chu, and U. S. Seal, eds. 1994. *Asiatic black bear population and habitat viability assessment*. Apple Valley, Minn.: Conservation Breeding Specialist Group and Taipei Zoo.

Watts, P. D., and S. E. Hansen. 1987. Cyclic starvation as a reproductive strategy in the polar bear. *Symp Zool Soc Lond* 57:305–18.

Wei, F., Z. Feng, and J. Hu. 1997. Population viability analysis computer model of giant panda population in Wuyipeng, Wolong Natural Reserve, China. *Int Conf Bear Res Manag* 9(2):19–23.

Weigand, T., J. Naves, T. Stephan, and A. Fernandez. 1998. Assessing the risk of extinction for the brown bear (*Ursus arctos*) in the Cordillera Cantabrica, Spain. *Ecol Monogr* 68:539–70.

Welch, C. A., J. Keay, K. C. Kendall, and C. T. Robbins. 1997. Constraints on frugivory by bears. *Ecology* 78:1105–19.

Wielgus, R. B., and F. L. Bunnell. 1994. Sexual segregation and female grizzly avoidance of males. *J Wildl Manag* 58:405–13.

Wiig, Ø. 1998. Survival and reproductive rates for polar bears at Svalbard. *Ursus* 10:25–32.

Wilk, R. J., J. W. Solberg, V. D. Berns, and R. D. Sellers. 1988. Brown bear, *Ursos arctos*, with six young. *Can Field-Nat* 102:541–43.

Wong S. T. 2002. The ecology of Malayan sun bears (*Helarctos malayanus*) in the lowland tropical rainforest of Sabah, Malaysian Borneo. M.Sc. thesis. University of Montana, Missoula.

Wong S. T., C. Servheen, and L. Ambu. 2002. Food habits of Malayan sun bears in lowland tropical forests of Borneo. *Ursus* 13:127–36.

Xu, L., Y. Ma, Z. Gao, and F. Liu. 1994. Characteristics of dens and selection of denning habitat for bears in the south Xiaoxinganling mountains, China. *Int Conf Bear Res Manag* 9(1):357–62.

Yudin, V. G. 1993. The Asian black bear. In *Bears. Brown bear, polar bear, Asian black bear. Game animals of Russia and adjacent countries and their environment* (series), edited by M. A. Vaisfeld and I. E. Chestin, pp. 479–91. Moscow: Nauka.

Zhang, G. Q., R. R. Swaisgood, R. P. Wei, H. M. Zhang, H. Y. Han, D. S. Li, L. F. Wu, A. M. White, and D. G. Lindburg. 2000. A method for encouraging maternal care in the giant panda. *Zoo Biol* 19:53–63.

Zhu, X., D. G. Lindburg, W. Pan, K. A. Forney, and D. Wang. 2001. The reproductive strategy of giant pandas *(Ailuropoda melanoleuca):* Infant growth and development and mother–infant relationships. *J Zool Lond* 253:141–55.

BRIEF REPORT 4.1

Life History Traits and Reproduction of Giant Pandas in the Qinling Mountains of China

Dajun Wang, Xiaojian Zhu, and Wenshi Pan

LONG-TERM RESEARCH on wild living giant pandas in the Qinling Mountains of Shaanxi Province, China, has been conducted since 1985. In its early stages, the research was focused on panda habitat; Pan et al. (1988) have published the results from these initial efforts. Subsequently, the results from studies of population structure, movement patterns, and social structure (Lü 1991) and the mother-infant relationship (Lü et al. 1994) were published, mainly in the Chinese language. In 2001, the first detailed analysis of the reproductive strategy of giant pandas was published (Zhu et al. 2001). Here we summarize our data on the mating season, gestation periods, birth dates, and sex ratios of cubs at birth, and present an initial comparison with data from captive studies.

Ten male and six female giant pandas in our study area were tranquilized and fitted with radiocollars (Pan et al. 1988; Lü 1991). Between 1989 and 1996, four of the radiocollared females produced a total of nine litters (ten cubs).

Giant pandas are known to be solitary (Schaller et al. 1985). However, males and females congregate for purposes of mating during the females' spring estrus. For purposes of this study, if an adult panda of either sex appeared in such an assemblage and displayed an interest in mating, we regarded it as a participant in an estrous event. From the spring of 1989 through that of 1996, we observed a total of twenty mating aggregations. All of these occurred between the dates of March 7 and April 11, but with a notable peak during the last ten days of March (figure BR.4.1). Thus, in the Qinling Mountains, mating activity appears to occur during a very limited season. It is well known that a female giant panda has but a single estrus in any season (Zhu et al. 2001). Accordingly, none of the females seen in an aggregation in a given spring made more than a single appearance in the estrous events of that spring.

Because there was no way to determine the precise timing of conception in the field, we defined the gestation period as extending from the last observed mating event to the day of parturition. Gestation in the Qinling pandas ranged from 129 to 157 days. Using midpoints where necessary, an average gestation of 146.2 days was calculated (table BR.4.1).

By extrapolation from our data on annual births, an interbirth interval (sensu Craighead et al. 1976) was calculated: intervals of 3, 2, 2, and 2 years for female Jiaojiao, 1 and 2 years for female Momo, and 2 years for female Ruixue. Although failure of cubs to survive is usually not

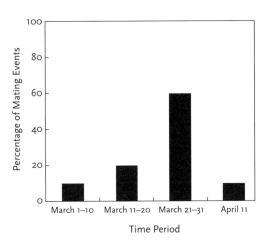

FIGURE BR.4.1. Temporal distribution of observed mating events in the Qinling Mountains from 1989 through 1996.

recorded in studies of ursids (Stringham 1990), we noted that Momo's cub of 1993 did not survive. She therefore had an effective interbirth interval (the interval between two successfully bred litters) at this time of 2.33 years. Allowing for this adjustment for the seven paired years in our sample, the effective interbirth interval for all females we observed was 2.17 years.

For the seven litters whose birth dates are known, the range of birth dates was only 10 days (table BR.4.2). All nine births known to us during the study period were in the second half of August. Despite the greater variance in mating times for these females, it would seem that births in the Qinling range are tightly clustered.

Ten litters comprising eleven cubs were born to four females during the study period. Among the seven cubs that we could sex, the female:male ratio was 3:4 (table BR.4.2). Excluding a male twin that died shortly after birth, the sex ratio of the surviving cubs was 1:1.

Zhang et al. (1993) summarized the mating times for five female pandas, Dongdong, Jiajia, Tangtang, Taotang, and Jiasi, at the Wolong Giant Panda Breeding and Research Center. Although these females were of different provenance, they were all residing at the same breeding center, and we use their data for comparative purposes. These five females mated naturally on thirteen occasions from 1991 to 1993. Five of the thirteen matings occurred in 1993, the earliest one on February 27 and the last two on June 28 and 29, covering a period of 121 days. However, over half of all captive matings reported by Zhang et al. (1993) occurred in May and June. In comparison with our results from Qinling, captive pandas appear to have a substantially longer mating season than those in the wild, as well as being somewhat out of phase with those in the Qinling

TABLE BR.4.1
Time of Last Mating Events, Parturition, and Gestation Period of Four Female Pandas

DAM ID	LAST MATING EVENTS OBSERVED	GESTATION PERIOD (DAYS)
Jiaojiao	March 11, 1989	157
	March 9–13, 1992	155–159
	April 10, 1994	129
	After April 8, 1996	135–136 (minimum)
	March 1998	—
Momo	March 31–April 1, 1993	146–147
	March 25, 1994	145
	March 7, 1996	—
Ruixue	March 25–27, 1993	147–149
	March 19–20, 1995	—
Nuxia	March 23–24, 1996	148–155

NOTE: —, No data available.

TABLE BR.4.2
Birth Date and Sex of Cubs

DAM ID	PARTURITION DATE	SEX OF CUB
Jiaojiao	August 15, 1989	Male
	August 8, 1992	Female
	August 17, 1994	Male
	August 20–21, 1996	Female
	August ?, 1998	Unknown
Momo	August 25, 1993	Female, male
	August 17, 1994	Male
Ruixue	August 21, 1993	Unknown
	August ?, 1995	Unknown

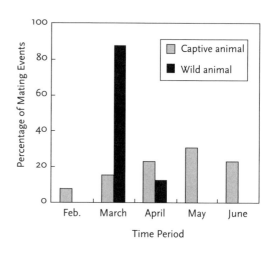

FIGURE BR.4.2. Comparison of the temporal distribution of mating in wild (Qinling Mountains) and captive (Wolong) giant pandas. Data for captive pandas are from Zhang et al. (1993).

population (figure BR.4.2). Further comparative study is required to determine whether the duration and timing of the mating season is site-specific.

As we have earlier noted, the time of parturition observed in the wild over an eight-year period (1989–1996) fell between August 15 and August 25. The reason for this apparent synchrony in parturition across the years of study is unknown.

Because both mating and birth seasons occurred in a relatively short period compared with the captive population, it is to be expected that the length of gestation for wild living pandas will show less variation. The average gestation period of 145–46 days (range, 129 to ~157 days; $N = 8$) was about 10 days longer than that reported for captive females ($\bar{X} = 135$; range, 97–161 days; see Schaller et al. [1985]). The reasons for this difference might be primarily nutritional, but further study is required to establish causal factors.

REFERENCES

Craighead, J. J., F. C. Craighead Jr., and J. Sumner. 1976. Reproductive cycles and rates in the grizzly bear, *Ursus arctos horribilis*, of the Yellowstone ecosystem. *Int Conf Bear Res Manag* 3:337–56.

Lü, Z. 1991. The population dynamics, movement pattern and social structure of the Qinling's giant panda. Ph.D. dissertation. (In Chinese.) Peking University, Beijing.

Lü, Z., Pan Wenshi, and J. Harkness. 1994. Mother-cub relationships in giant pandas in the Qinling Mountains, China, with comment on rescuing abandoned cubs. *Zoo Biol* 13:567–68.

Pan, W., Z. Gao, and Z. Lü. 1988. *The Qinling giant panda's natural refuge*. (In Chinese.) Beijing: Peking University Press.

Schaller, G. B., J. Hu, W. Pan, and J. Zhu. 1985. *The giant pandas of Wolong*. Chicago: University of Chicago Press.

Stringham, S. F. 1990. Black bear reproductive rate relative to body size in hunted populations. *Int Conf Bear Res Manag* 8:425–32.

Zhang H., K. Zhang, R. Wei, and M. Chen. 1993. Studies on the reproduction of captive giant pandas and artificial den at the Wolong Nature Reserve. In *Chengdu Zoo, Chengdu Giant Panda Research and Breeding Center, minutes of the International Symposium on the Protection of the Giant Panda*, edited by A. Zhang and G. He, pp. 221–25. Chengdu, China: Sichuan Science and Technology Press.

Zhu, X., D. Lindburg, W. Pan, K. Forney, and D. Wang. 2001. The reproductive strategy of Giant pandas: Infant growth and development and mother-infant relationships. *J Zool Lond* 253:141–55.

PART TWO

Studies of Giant Panda Biology

In part 1, we noted that a preponderance of the evidence supports an ursine origin for the giant panda. Biologically speaking, we are thus led to wonder how the giant panda—an "ancient" bear, according to one of the more recent evolutionary trees (O'Brien 1993)—compares with its ursine cousins. In chapter 4 (part 1), Dave Garshelis provides a superb beginning in drawing together comparative information on pandas vis-à-vis the other ursids, but the study of panda biology is too recent and too limited to lead to definitive characterizations. From the Schaller team's original work in the Qionglai Mountains of China in the early 1980s (Schaller et al. 1985) to the more recent decade of observations by Pan Wenshi and students in the Qinling region (Pan et al. 2001), it is abundantly clear that even where radiotelemetry has been utilized, the panda's elusiveness and relative scarcity have greatly limited direct observation. Furthermore, those reports having habitat features as their primary research objective (see part 3) reveal aspects of the panda's biology only indirectly. As a result, the information flow from wild-living pandas has been scanty, and the two reports (chapters 5 and 6) included here are a welcome indication that the situation is gradually being remedied. In addition, numerous aspects of panda biology are being discovered through the study of captive individuals; many reports of such studies were presented at *Panda 2000*, the international conference on pandas held in San Diego (see the Preface). Here, we present two reports (chapters 7 and 8) that tie in nicely with field studies and meet our criteria of offering in broad outline a sampling of the new information now becoming available from captive research.

The first chapter in this part comes from one of China's most distinguished observers of the giant panda, Professor Pan Wenshi, and students Long, Wang, Wang, Lü, and Zhu. Although their chapter stresses conservation in addressing the prospects for long-term survival of giant pandas, those prospects rest on data that are among the first to be obtained from a wild population since the report of Schaller et al. (1985) nearly two decades ago. Over thirteen years, Pan's all-Chinese team repeatedly tracked twenty-one radiocollared individuals living in a community of pandas in the Changqing Forestry District of Shaanxi Province. Demographic findings revealed a population of more than thirty individuals at the primary study site, expanding at a rate of 4.1% annually. In this mountain range (Qinling), the pandas ascend and descend annually, according to the seasons, and following on the loss of tree dens from logging of most old growth, females today utilize limestone caves as birthing sites. Details on the mating system and parturition over several seasons owe their acquisition to locating adults of both sexes via radiotelemetry. From blood samples collected during capture for collaring, Pan et al. found a level of genetic diversity that, along with demographic and habitat parameters, is used to predict long-term survival for the Qinling pandas.

Complementing the genetic data reported by Pan et al. in chapter 5 is a review by Zhang and Ryder (workshop report 5.1) of the current status of genetic work on both captive and wild populations. Noting the substantial benefits to promoting conservation (and to understanding the giant pandas' evolution and migration patterns) that can be realized from genetic data, these authors call for the establishment of a genome resource base as being essential to current and future studies. Other requirements are to improve molecular genetic tools for genotype assessment and the establishment of long-term genetic monitoring programs.

In chapter 6, Long, Lü, Wang, Zhu, Wang, Zhang, and Pan build on dietary and nutritional information initially reported by Schaller et al. (1985) for giant pandas from Wolong. Taking into account the pandas' heavy reliance on bamboo for food, they look to food selection processes, efficiency in feeding behavior, and limited utilization of special nutrients as techniques for maximizing nutrient extraction. Feeding bouts were divided into carefully timed phases of preparation, ingestion, and "intermission" (i.e., time between selection of individual items for consumption) to measure feeding efficiency. Unlike most other carnivores, giant pandas can obtain food value from hemicellulose and small quantities of cellulose as supplemental sources of nutrients. They also scavenge opportunistically, although the value of this latter propensity to nutrient health is not yet clear.

In a preliminary but beautifully crafted set of experiments on captive pandas, Tarou, Snyder, and Maple (brief report 6.1) describe the cognitive processes used in locating food. Finding food is one of the most important problems animals face each day, requiring the employment of appropriate sensory modalities and, in many cases, a mental map of where and when suitable foods may be found. Success in locating hidden food in an eight-arm radial maze revealed that pandas used a "least distance" strategy. Accuracy in choosing baited feeders improved above chance levels with time, declined to random levels on reversal of baits, then gradually rose above chance levels again, clearly demonstrating the use of spatial memory in locating food. Their findings indicate that giant pandas would rank fairly high on any animal intelligence scale, and these authors cite research showing that such perceptions strongly influence support for their conservation.

In chapter 7, Swaisgood, Lindburg, White, Zhang, and Zhou summarize recent studies revealing the role of chemical communication in regulating many giant panda interactions. Based on experimental work with captive subjects at the Wolong Center in China, these authors were able to shed light on the functional significance of a number of chemical signals. The identity of marking individuals, their sex and reproductive condition, and their social rank in the community are examples of the message that a panda may transmit via scent. Pandas also appear to be able to judge the age of a scent for up to three months, and presumably use this information in deciding a course of action, particularly during the spring mating season. The results from this study underscore the importance of chemical communication to giant pandas and suggest useful applications of scent in the conservation and management of the species in both the wild and in captivity.

What does a panda scent look like? In brief report 7.1, Hagey and MacDonald take us further into the realm of this highly olfactory carnivore's signaling system through identification of the volatile components of scent. Analysis with a gas chromatography–mass spectrometer generated graphics of the type, concentration, and pattern of the volatile components of scent—in effect, a picture of a particular scent. Not surprisingly, given each sex's differential use of scent, this study revealed a major difference in the chemical complexity of the type of scent (urine or glandular exudate) emphasized by each. In a further, imaginative analysis of the distribution of scent over the body, preliminary data suggest ways in which pandas may advertise their scent more effectively.

In chapter 8, Snyder, Lawson, Zhang, Zhang, Luo, Zhong, Huang, Czekala, Bloomsmith, Forthman, and Maple explore behavioral factors associated with reproductive failure in a captive population housed at the Chengdu Research Base of Giant Panda Breeding and the Chengdu Zoo. In so doing, their study adds significantly to a growing body of data on the association of estrous behaviors with underlying hormonal conditions. In addition, comparisons of reproductively successful with less reproductively successful females point to behavioral deficiencies in the latter, including the failure to exhibit sexual receptivity. Initial evidence of an increase in androgen concentrations that correlated with increases in scent marking, bleat vocalizing, and locomotion adds to earlier hints of a "rut-like" event in males during the mating season. Increased familiarization through scent exposure ameliorated aggressiveness of males toward estrous females, suggesting that an emulation of sexual contacts in wild pandas may facilitate breeding in the captive sector.

REFERENCES

O'Brien, S. J. 1993. The molecular evolution of the bears. In *Bears. Majestic creatures of the wild*, edited by I. Stirling, pp. 26–29. Emmaus, Penn.: Rodale.

Pan, W., Z. Lü, X. Zhu, D. Wang, H. Wang, Y. Long, D. Fu, and X. Zhou. 2001. *A chance for lasting survival*. (In Chinese.) Beijing: Peking University Press.

Schaller, G. B., J. Hu, W. Pan, and J. Zhu. 1985. *The giant pandas of Wolong*. Chicago: University of Chicago Press.

5

Future Survival of Giant Pandas in the Qinling Mountains of China

Wenshi Pan, Yu Long, Dajun Wang, Hao Wang, Zhi Lü, and Xiaojian Zhu

IN THE SPRING of 1985, we began developing a comprehensive research program to address giant panda *(Ailuropoda melanoleuca)* conservation issues. Initially, it was extremely difficult to ascertain population numbers and distribution patterns at our site in the Qinling Mountains (Shaanxi Province, Changqing Forestry District); yet obtaining these data was of utmost importance to a conservation effort. The best method for proceeding was to fit a radio-transmitting device on pandas for long-term monitoring. Between June 1986 and March 1999, we obtained direct and reliable data from twenty-one radiocollared individuals. These data included activity locations and frequencies, seasonal migrations, dispersal routes, home ranges, core areas, and provided opportunities to observe foraging behavior, mating behavior, copulations, and mother-infant relationships. In February 1994, we used a specially designed closed-circuit infrared miniature video recorder and audio recording system to document the behavioral relationships of mothers and newborn cubs. To analyze the population's genetic diversity, we gathered hair samples from tree trunks or bamboo branches and also obtained a small number of blood samples from radiocollared individuals.

We collected fecal samples from giant pandas and compared the biochemical composition of their feces with that of the bamboo species consumed by the animals. In addition to biochemical analysis, we examined the bamboo cell contents and cell walls, using light and electron microscopy, to determine the extent of bamboo utilization as food. We combined extensive ground truthing with satellite remote sensing technology in analyzing the spatial structure of panda habitat. We regularly carried out surveys at our research area in Qinling to calculate population density and numbers.

Based on our fifteen years of research on the Qinling giant pandas, we used the parameters of population structure to construct a population dynamics model to estimate the viability of the population. We found that the giant pandas of Qinling lived in an environment rich in biodiversity, with many flora and fauna that not only exist today, but also once occupied much broader spatial ranges. The giant panda stands as a flagship species in this biological community. Thus, protecting the giant panda means protecting the millions of life forms that existed in the distant past and still persist today. We realize that many of the interrelated factors in any ecological system are constantly changing. Therefore, we have recommended that the conservation of giant pandas in Qinling should concentrate on monitoring rather than forecasting.

During our study, we systematically compiled information on every major ecological event and the concurrent behavioral changes in the pandas. Even though we still cannot declare that we clearly understand every detail of their evolution, we have gradually developed a comprehensive outline of how this regional population can continue to survive. Therefore, we believe that presenting a summary of our fifteen years of results is very timely. These findings will augment our knowledge of the lives of giant pandas and the crises they face. At the same time, the results will help us choose the best conservation strategies for protecting local populations.

Our findings have been published in the Chinese language in an 800,000-word monograph entitled *A Chance for Lasting Survival* (Pan et al. 2001). Due to the importance of this information to scholars outside China, we summarize here the specifics of several topics covered in this work. The summaries presented here are based on our own research findings.

HABITAT OF GIANT PANDAS AT THE QINLING STUDY SITE

The giant pandas of Qinling are distributed mainly in the upstream region of three rivers (the Xushuihe, Youshuihe, and Jinshuihe), located on the southern slopes of the mountains at 1300–3000 m above sea level. At this elevation, the habitat is classified as mid-mountain to subalpine. This zone constitutes a natural refuge for giant pandas, with many unique features. The mountain range, running east to west, forms a natural barrier against cold northern winds. The montane-temperate and frigid-temperate climate at Qinling is favorable to a thriving bamboo forest ecosystem. Because the 1350-m elevation is the upper limit for farming, humans are precluded from year-round habitation, thus preserving for giant pandas the last remaining habitat in this region.

Our research revealed that the pandas of Qinling are situated on four "mountain islands" (figure 5.1). Within three of these (Xinglongling, Niuweihe, and Taibaishan), the potential for intermountain migration exists, and the individuals in these regions may be considered a metapopulation. However, roads in the area separate the fourth mountain island (Tianhuashan) and its panda population from the others.

An analysis of panda habitat, using satellite remote sensing, indicated that the dominant vegetation cover in their winter range is conifer/broadleaf and oak forests, with an undergrowth of *Bashania fargesii* bamboo. The dominant cover in the summer range is Chinese larch, fir, and birch forests, but with *Fargesia spathacea* as the prevailing bamboo. With respect to logging activities in this region, we indicated to the Changqing Forestry District that planned and reasonably scaled selective logging probably would not affect the survival of the pandas. However, unrestricted clearcutting and the subsequent replanting of commercial, fast-growing, high-yield forest would lead to massive curtailment of their habitat.

FARMLAND-FOREST/WILDLIFE ECOSYSTEM BOUNDARY

The dividing line between farmlands and the forest/wildlife ecosystem in the mid-mountain to alpine zones is situated at about 1350 m elevation. This line ensures the existence of a relatively stable habitat for the pandas. Since historical times, the Qinling region has served as a shortcut that connects Guanzhong (in the central part of Shaanxi Province) to Sichuan Province. Many ancient trails run through this area. Cultivation of the land along the two sides of these trails and the development of trade activities can be traced back 1600 years. As a result, panda habitat in this region has been subjected to human disturbance for a very long time.

Beginning in the 1930s and to the present time, the development of roads in the Qinling range has created a crisscrossing network of access to the area. This is a key factor in causing

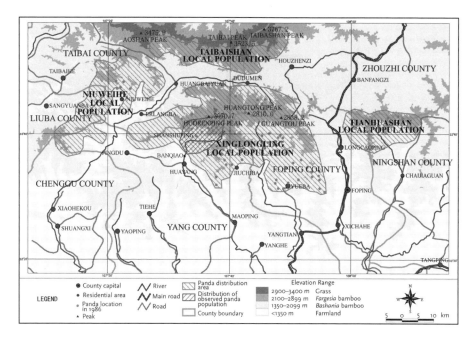

FIGURE 5.1. Distribution of giant pandas in the Qinling Mountains of China.

habitat fragmentation. The growth and development of Banqiao Village, located at 1200–1300 m, is the epitome of human exploitation and transformation of the southern slope of the Xinglongling "island." Its history demonstrates how farmland and wildlife have become increasingly separated through the process of village development. Farmers at Xinglongling were formerly dispersed throughout the area occupied by wildlife, but after 1950, they converged on the edge of this area, forming the village of Banqiao and farming virtually all of the land below about 1350 m.

DEMOGRAPHIC STUDIES OF THE QINLING POPULATION

Previous researchers have been unable to develop an effective method for surveying wild-living giant pandas. In 1986 and 1995, we conducted panda surveys throughout the 15 km^2 of our study site at Xinglongling, using direct sightings along transect lines. The results indicated a population density of two individuals/km^2, and a total population numbering about twenty-eight individuals. In the springs of 1993, 1994, and 1995, using the mark-recapture method, we estimated that seventeen of these were adults. Based on the population's apparent age structure, we recalculated the total number of pandas to be thirty-four individuals. From 1989 to 1997, use of the absolute population number estimate placed the size at twenty-seven. Assuming stability over the years of our investigations, we used a best estimate of approximately thirty pandas as constituting our study population. From the twenty-one radiocollared individuals, location data acquired between 1989 and 1997 led us to place their total range at 92 km^2, using the convex polygon method. Accordingly, the density would be 0.228 individuals/km^2. By extrapolation from this figure, we concluded that the giant pandas in the three connected "islands" of Xinglongling, Taibai, and Niuweihe numbered between 167 and 331 individuals.

By combining our research techniques with those of previous workers, we developed an integrated methodology for surveying giant pandas. We used the distance between locations of activity and length of bamboo stem fragments found

TABLE 5.1
Partial Results of Population Analyses of the Qinling Giant Pandas

CLASS	AGE (YEARS)	AGE STRUCTURE (%)	SURVIVAL RATE (PER YEAR)	REPRODUCTIVE RATE (PER YEAR)
Infant	0–1.5	23.7	0.894	
Subadult	1.5–4.5	23.7	0.909	
Reproductive age				
Female	4.5–20	50.0	0.930	0.5
Male	4.5–17	50.0	0.930	0.5
Postreproductive age				
Female	20–22	2.6	0.587	—
Male	17–19	2.6	0.587	—

NOTE: —, No data.

in panda fecal matter to determine population size. Implementation of this method proceeds in three stages: (1) laying out the survey routes, (2) conducting the field survey, and (3) analyzing fecal material in the laboratory. This methodology, through actual field testing and perfecting, was subsequently used in China's third national survey of giant pandas (see Yu and Liu, panel report 15.1), and it has now become the standard for surveying giant pandas by Chinese researchers.

POPULATION DYNAMICS

Our findings for age classes, age structure, and annual survival rate are presented in table 5.1. The sex ratio at Xinglongling was found to be 1:1. The average reproductive cycle of females was two years, and their annual reproductive rate was 0.5 offspring/year. Based on these findings, we calculate that females could contribute, on average, 1.53 female progeny over their reproductive life span.

Annual population growth, based on Leslie's matrix model, was 4.1%. Using a stochastic simulation model for population viability analysis, we mimicked the population dynamics of giant pandas in our study area. By holding current environmental conditions constant and assuming an extinction probability of less than 0.05, we determined that the minimal number of individuals needed for the population to sustain itself (i.e., minimal population viability) for two hundred years would be twenty-eight to thirty individuals. If we place the extinction probability at less than 0.01, the minimal number rises to forty individuals. Both density-dependent and density-independent factors increase the probability of extinction. Under the influence of density-dependent factors, the minimal viable population would number fifty to sixty individuals. We found that pressure from density-independent limitations should not exceed 0.01/year, which meant that the poaching or legal capture of giant pandas in the entire Qinling region should not exceed two individuals annually.

From our analysis of population parameters at Xinglongling, we conclude that if natural and social environmental conditions in the larger area of giant panda distribution do not vary greatly, we believe the population in the Qinling Mountains may be able to sustain itself continuously. Even though the numbers undergo temporal and spatial fluctuation, the population is basically stable. To maintain sustainability, existing resources must be preserved and long-term monitoring of the habitat and its pandas must be an ongoing activity.

MATING SYSTEM AND MOTHER-CUB FAMILY UNIT

As shown in table 5.1, the age at which sexual maturity is attained was determined to be 4.5 years for both sexes. Between 1985 and 1996, we observed a total of twenty-one mating events. The mating season began in early March and ended in mid-April (March 7–April 11). In 1993, the mating season lasted at least 33 days. In the spring of 1993 and 1994, the average daily temperature during this season exceeded 0°C.

The mating scene can be divided into two types, based on the number of males present: one female–one male groupings ($N = 3$) and one female–multiple male groupings ($N = 15$). For three mating events, we were unable to determine the number of males involved. Other females were never present, although an estrous female's 2-year-old cub was occasionally nearby. We regarded the most dominant male to be the one that maintained a position closest to the estrous female. Males established their dominance status through ways not yet determined. Female giant pandas showed certain preferences for males. In the fifteen multiple male clusters, four males sustained injuries from fights with one another. In the Qinling Mountains, both sexes had multiple mates.

A survey of scent-trees in 1995 revealed that of the twenty-two tree species found on the three ridges of the study area, sixteen were used as marking sites. The main species in this region are environmentally hardy *(Pinus armandii, Quercus spinosa, Populus davidiana,* and *Q. aliena* var. *acuteserrata)*. The pandas showed a preference for trees having a clumped distribution, a large diameter at breast height, and rough bark. However, they tended to avoid the aromatic *P. armandii*.

From 1989 to 1998, we pinpointed the location of eleven births for five different radio-collared females. Pregnant females (five individuals, eleven records) began early migration to the low to mid-mountain zones in the autumn. They also had smaller activity ranges than non-pregnant females. Births occurred between August 15 and August 25 ($N = 10$). The average length of gestation was 145–146 days (range, 129–157 days, $N = 10$). The average interbirth interval was 2.17 years ($N = 6$). The sex ratio of newborns was 1:1 ($N = 6$). Between 24 and 102 days of age, the average daily weight gain of three cubs that were weighed during dams' absence from the den were 64 g, 60 g, and 46 g. Cubs began opening their eyes between 40 and 49 days. At about 88–90 days, they began to gain sight. At 75–88 days, the first dental eruptions occurred. In addition, we opportunistically recorded changes in the cubs' body length and hair color.

We constructed an ethogram for recording the activities of dams and their cubs during the den phase. During the first five days after parturition, dams did not leave the den. Although they began leaving the den to defecate when cubs were at 6–14 days of age, they continued to fast during this time. The earliest departure for foraging was when the cubs attained 15 days of age.

We analyzed the behavioral characteristics of the dams from birth of the litter to the end of den life at 94–125 days. In analyzing the characteristics of seven dens, we found that they provide essential warmth and protection from the weather. Following the abandonment of dens, cubs were placed by their mothers in lairs (dense thickets) until they reached 5–6 months of age. After this age, they began to climb trees and stayed above ground while their mothers foraged. At this time, foraging trips began to increase in duration.

The behavioral development of cubs from birth to 8 months of age was divided into five stages:

1. Birth to 9–10 days (highest mortality risk);
2. 9–10 to 40–60 days (the period of rapid growth);
3. 40–60 days to 90–130 days (eyes open and increased movement);
4. 90–130 days to 5 months (living in lairs); and
5. 5–8 months (living in trees).

TABLE 5.2
Measure of Molecular Genetic Diversity of mtDNA in Three Panda Populations

DNA MARKER	QINLING	MINSHAN	QIONGLAI	ALL PANDAS
mtDNA RFLP				
Sample size	10	1	7	19
Haplotype	2	1	4	5
π	0.12	—	0.24	0.22
mtDNA sequences				
Sample size	14	7	15	36
Haplotypes	10	4	7	17
π	0.04	0.01	0.03	0.06
DNA fingerprints				
Sample size	12	—	6	18
HaeIII/Fcz8				
He (%)	39.0	—	41.4	39.8
ADP	36.9	—	41.3	38.3
HinfI/Fcz8				
He (%)	46.5	—	35.5	42.8
ADP	39.3	—	39.0	39.2
HaeIII/Fcz9				
He (%)	27.7	—	30.1	28.5
ADP	24.7	—	34.7	28.0
HinfI/Fcz9				
He (%)	26.3	—	36.3	29.6
ADP	25.1	—	38.2	29.4
MAPD (%)	31.5	—	38.3	33.7
Microsatellite DNA				
Sample size	14	7	15	36
Number of alleles	33	35	38	106
Observed heterozygosity (%)	57	58	49	44
Average number of alleles per locus	3.3	3.5	4.3	3.7
Number of signature alleles	15	7	14	36

SOURCE: Data are from Lü et al. (2001), with permission.

NOTE: —, No data.

If we use 1.4 as the mean litter size, based on what is known from captive breeding records, the birth rate in our study area would be 0.654 cubs per year, and the cub survival rate at attaining one year of age would be 59.5%. These findings indicate that the population of giant pandas at this location is self-sustaining. On the basis of its reproductive potential, the giant panda therefore remains an evolutionarily successful species. The key to protecting the giant panda is to protect its wild populations.

GENETIC DIVERSITY OF GIANT PANDA POPULATIONS

We used four molecular genetic markers (mtDNA restriction fragment length polymorphisms [RFLP], mtDNA D-Loop sequencing, nuclear

DNA fingerprinting, and microsatellite DNA variation) to estimate the level of genetic diversity of panda populations from the Qinling, Qionglai, and Minshan regions. We also collected historical evidence for the evolution of giant panda populations. The specific data are detailed in table 5.2.

The four genetic markers indicate that the three giant panda populations we sampled have not shown a noticeable loss in genetic diversity, and at present, maintain a fixed level of diversity compared with other carnivores (Lü et al. 2001). Of the three populations, genetic diversity was the highest in the Qionglai range. A low, although detectable, divergence appeared between each population, and the genetic distance corresponded to the chronological order of their geographical separation. The Qinling population was the most unique.

On the basis of genetics alone, either as a species or three separate populations, the giant panda has retained the main components of its historical population genetic diversity (Lü et al. 2001). Therefore, its future is promising. Nevertheless, the effects of habitat loss and fragmentation are becoming evident. Through habitat conservation and restoration (see Zhu and Ouyang, panel report 11.1) we may be able to facilitate the expansion of existing populations. In addition, promoting and maintaining gene flow may be a necessary and effective strategy for preventing a further decline of this endangered species.

CONCLUSIONS

Based on more than a decade of tracking twenty-one radiocollared pandas, we have been able to assess long-term demographic trends, analyze genetic structure, assess habitat conditions, and measure the reproductive potential of giant pandas in the Changqing Forestry District of Shaanxi Province. Our work leads to the following conclusions:

- The giant pandas of the Qinling Mountains have the potential for continued survival. Separation between existing local populations has occurred only in the past several hundred years, a brief period in this species' long evolutionary history.
- The reproductive potential, population structure, and genetic diversity of the species are characteristics favoring its continued survival.
- Human activities, resulting in habitat destruction, constitute the greatest threat to wild populations at this time. The key to preventing the extinction of giant pandas is to protect their natural habitat.
- Population management efforts should concentrate on the boundary between the low- to mid-mountain zones of farmlands and the mid-mountain to alpine zones of its forest/wildlife ecosystem.
- If humans approach the conservation of wild giant pandas with loving concern and scientific rigor, these animals will most likely be able to enjoy a continued existence.

ACKNOWLEDGMENTS

We thank Chia L. Tan, Zoological Society of San Diego, for translation of our work into English. We also thank an anonymous reviewer and the editors for suggestions for improving the manuscript.

REFERENCES

Lü, Z., W. Johnson, M. Menotti-Raymond, N. Yuhke, J. Martenson, S. Mainka, S.-Q. Huang, Z. Zheng, G. Li, W. Pan, X. Mao, and S. O'Brien. 2001. Patterns of genetic diversity in remaining giant panda populations. *Cons Biol* 15:1596–1607.

Pan, W., Z. Lü., X. Zhu, D. Wang, H. Wang, Y. Long, D. Fu, and X. Zhou. 2001. *A chance for lasting survival*. (In Chinese.) Beijing: Peking University Press.

WORKSHOP REPORT 5.1

Genetic Studies of Giant Pandas in Captivity and in the Wild

Yaping Zhang and Oliver A. Ryder

MODERN TECHNIQUES of genetic analysis make feasible the determination of the genotypes of individual giant pandas in captivity and, if a sufficient sample is available, in the wild. Such information would play an important supportive role in the conservation of the species in both milieux. For captive populations, the founders derived from the wild population provide a sample of genetic diversity within the species. Analysis of these samples provides an insight to the distribution of genetic variation within free-ranging populations without requiring additional sampling.

Assessments of genetic diversity derived from captive individuals, in addition to providing information about phylogeography, may also be utilized to estimate historic levels of migration between adjacent giant panda *(Ailuropoda melanoleuca)* ranges. Such knowledge contributes usefully to reserve design. Currently, genetic data from captive giant pandas has been successfully utilized to establish paternity of captive-born individuals under circumstances where several males could potentially be the biological father. This information not only helps to maintain an accurate studbook but also contributes to assessment of the success of artificial insemination techniques.

Genetic studies of giant pandas in wild populations are at their earliest stages. (Pan et al., chapter 5). However, the opportunity to use genetic data, along with geographic reference for use in conservation, allows us to incorporate information on the spatial aspects of genetic structure of populations in genealogical data. These integrated data have clear implications for design of reserves and establishment of corridors. Genetic data from wild populations may also be used to assess the extent of inbreeding and effective population size and can shed light on the social structure and behavioral components of fitness. Increased knowledge in these areas will contribute to a more informed process of management and lead to greater assurance of population persistence.

There are some data on the genetics of the giant panda, including its position in the phylogeny of the arctoid carnivores and the distribution of nuclear and mtDNA variation. Opportunities for increasing the contribution of genetic studies to the overall conservation effort for giant pandas and exploration of the outlook for developing routine field-based genetic assessment tools must be identified.

RESOLUTIONS AND RECOMMENDATIONS

The following six items are regarded as the highest priorities for future activity in global genetic analysis of giant pandas:

- Develop a genome resource base (GRB) that would be available for future and current genetic, reproductive, and biomedical evaluation. This requires the planning and implementation of a bioresource collection of biological materials (blood, serum, skin fibroblasts, sperm, hair, feces) from both captive and free-living specimens.
- Develop a curated genotype database of individual molecular genetic and biomedical parameters for all animals studied at present and in the future, with identification data linked to the giant panda studbook.
- Develop a long-term genetic monitoring program for the captive population and, as feasible, for the wild population.
- Develop a workshop around the GRB proposal, possibly under facilitation of the Conservation Breeding Specialist Group of the IUCN.
- Develop better molecular genetic tools for genotype assessment, with emphasis on expanding the number of loci analyzed, multiplex amplification and analysis, validating methodology, and ease of implementation.
- Recommend and implement sufficient radiocollar tracking of free-living giant pandas for behavioral and ecological monitoring. At the time of collar attachment, collect biomedical materials for the GRB.

ACKNOWLEDGMENTS

Stephen J. O'Brien and Kurt Benirschke made significant contributions to this workshop report.

6

Nutritional Strategy of Giant Pandas in the Qinling Mountains of China

Yu Long, Zhi Lü, Dajun Wang, Xiaojian Zhu, Hao Wang, Yingyi Zhang, and Wenshi Pan

ACCORDING TO Schaller et al. (1985), despite the wide distribution of bamboo throughout the tropics and subtropics, the animals that utilize this grass for food are quite rare. These authors noted that, in addition to the giant panda *(Ailuropoda melanoleuca)*, bamboo feeders in China include the fossorial bamboo rat *(Rhizomys sinense)* and the lesser or red panda *(Ailurus fulgens)*. In addition, Pan et al. (1988) have documented a small number of birds, insects, and small mammals that utilize bamboo shoots, leaves, or stems as food (see also Wei et al., chapter 13).

Sheldon (1937) may have been the first to suggest that giant pandas in the wild feed almost exclusively on bamboo. In adapting to bamboo as their main food source, giant pandas have undergone significant evolutionary changes. From a morphological standpoint, although pandas have retained the sharp claws, canines, and digestive system of a carnivore, they do not often prey on other species. Despite their large size, pandas' unwieldy deportment and loss of incentive in the midst of plentiful bamboo may have led to the dissipation of earlier predatory capabilities. Also, although a rich fauna exists in the area of their current distribution, potential prey species may not occur in sufficient numbers to provide a reliable basis for survival either as predators or scavengers, or for competing effectively with other predators, such as the Himalayan black bear *(Ursus thibetanus)*. Bamboo, however, provides the giant panda with an abundant and adequately nutritious food supply year-round. In addition to being widely distributed, bamboo has a relatively rapid annual growth rate and a long life cycle. These traits make bamboo suitable as the food of choice for giant pandas.

EARLY REPORTS ON THE DIET OF GIANT PANDAS

Schaller et al. (1985) summarized the knowledge available at the time on giant panda diets. They noted that several studies have revealed regional differences in the species of bamboo used as food. Pandas on the Wolong Reserve, for example, fed mainly on *Sinarundinaria fangiana* and *Fargesia spathacea*, and to a lesser extent on *S. nitida*. In the Liang Mountains, they reportedly consumed five different species, but only one in parts of the Min Mountains (see also Qin 1990). It was noted that stems, leaves, shoots and culms, but not underground rhizomes, were consumed.

Besides bamboo, Schaller et al. (1985) noted that giant pandas eat other plant species, fungi, and carrion, but in small quantities (see also Zhu 1983). In the Min Mountains, they reportedly fed on the stems, leaves, and bark of a variety of plants. At Wolong, they were known to consume the bark of wild parsnip (*Angelica* sp.), tree fungus *(Polyporus frondosus)*, fir *(Abies fabri)*, hemlock *(Tsuga chinensis)*, and pine *(Pinus armandii)*. These authors cite McClure (1943) as having reported nine plant species that were consumed by pandas in captivity. Additionally, pandas at the Chengdu Zoo were observed eating three kinds of grass, and at Wolong's Yingxionggou Captive Center, they ate the leaves of two Gramineae (grasses) *(Deyenxia scabrescens* and *Buddoeia davidii)*.

From these sources, it was concluded that giant pandas eat over twenty-five species of wild plants but usually in quantities of less than 1% of their total food intake. However, it has not been determined that food selection in captive pandas is a sufficient basis for explaining dietary flexibility in the wild.

The dietary summary provided by Schaller et al. (1985) mentions villagers living on the Wolong Reserve who reported finding the remains of small rodents in the stomachs of giant pandas and of observing one panda in the act of catching a bamboo rat. Schaller's team once found the hair of a golden monkey *(Rhinopithecus roxellana)* in a panda dropping at Wolong, and at the Wanglang Nature Reserve they found hair, flesh, bones, and hooves of musk deer *(Moschus chrysogaster)* in droppings. When mutton was offered to captive individuals at Yingxionggou, four out of seven immediately ate the meat. Domestic pig bones and goat heads were used by the Schaller team to entice pandas into live traps for radio-collaring (Schaller 1993). It seems, therefore, that giant pandas living under natural conditions will opportunistically feed on nonherbivorous foods.

The Schaller team once found gray, claylike soil in the droppings of a panda female at Wolong, and at the Yingxionggou Captive Center, another female was observed eating soil in her outdoor enclosure.

DIETS OF THE GIANT PANDAS IN THE QINLING MOUNTAINS

On the southern slopes of the Qinling Mountains, where our study took place, there are nine species of bamboo belonging to five genera. Among these, *Bashania* and *Fargesia* are the dominant native bamboos and the staple food of giant pandas (Pan et al. 1988). We also observed the consumption of introduced *Phyllostachys nigra* on two occasions. Yet, in more than a decade of field research at this location, we did not find the broad spectrum of plant foods in the diet that is suggested by other reports. Although grasses were occasionally ingested during feeding on bamboo, none were selected by themselves for consumption. As at Wolong, we observed giant pandas ingesting small quantities of soil and a limited amount of animal foods.

The differences in the food habits of wild-living giant pandas are undoubtedly attributable to local availability for the most part, but it is also possible that in any study, the consumption of some of the more infrequent foods will be missed by observers. Captive feeding, however, perhaps demonstrates that pandas may be capable of more flexibility in food consumption than one might see at a given field site. In the remaining part of this chapter, we provide an in-depth analysis of animal and bamboo feeding from data collected in the Qinling Mountains during more than ten years of effort.

CONSUMPTION OF ANIMAL FOODS AND SOIL

We noted eleven instances of feeding on carrion or the bones of five different mammals and two unidentified sources, either directly or from scats (table 6.1). Most of the observed incidents of carrion or bone eating occurred when females were either gestating or lactating. For example, in 1990, we found "Jiaojiao" eating the bones with meat of unknown origin, and in 1994, she fed on a hide of the wild boar *(Sus scrofa)*. She was with her cubs on both occasions. Also, in July 1997, around the time of her parturition, female

TABLE 6.1
Nonbamboo Food Items Consumed by Wild Giant Pandas

DATE	INDIVIDUAL	SEX	LOCATION	FOOD	REMARKS
March 1985	Dandan	Female	Foping	Domestic chicken and pig's skin	Found in feces
March 1987	Jiaojiao	Female	Tudigou	Hair and hoof nail of musk deer (*Moschus berezovski*)	Found in feces
February 1989	Shuilan	Female	Shuidonggou	Takin carcass (*Budorcas taxicolor*)	
December 1990	Huzi	Male	Shuidonggou	Bone with meat of unknown origin	
July 1991	Nüxia	Female	Liaojiagou	Domestic chicken	
October 1991	074	Male	Shuidonggou	Meat of unknown origin	
1992–1993	Jiaojiao	Female	Shuidonggou	Domestic chicken	
February 1993	Hope	Female	Shuidonggou	Bone of unknown origin	
March 1994			Shuidonggou	Bone of tufted deer (*Elaphodus caphalophus*)	
October 1994			Shanshuping	Skin, rib and femur of wild pig (*Sus scrofa moupinensis*)	
July 1997				Bones of domestic pig	

"Hope" fed on pig bones discarded by humans many months before. We hypothesized that these incidents may have been in response to the need for certain nutrients (possibly essential vitamins and amino acids). Because we have not determined whether carnivory is directly related to gestation and lactation, we will need further research to test this hypothesis.

Our findings may also be biased by the concentration of our research efforts during the phases of birth and cub rearing. Nor can we exclude the possibility that chance encounters with carrion influenced panda feeding rather than the condition of the food itself. There may be an evolutionary relationship between the carrion-eating behavior of extant bears and that of giant pandas. Having descended from carnivorous ancestors, the interest of giant pandas in carrion would then be vestigial. However, we may explain pandas' utilization of bamboo as a novel adaptation that evolved as a result of food scarcity. Therefore, when they encounter meat or bones in the wild, they will consume them without hesitation, even in the presence of fresh and tender bamboo.

In May 1995, we observed a male panda eating sand in the course of crossing a river. It is possible that geophagy in giant pandas derives from the need to acquire certain minerals or to facilitate digestion.

BAMBOO SELECTION BY SEASON

Based on the annual diet of giant pandas, Schaller et al. (1985) were able to distinguish three ecological seasons at their study site in Wolong: spring (April–June), when pandas fed mainly on the new shoots of *Fargesia*; summer (July–October), when the primary food was the leaves of *Sinarundinaria*; and winter (November–March), when the old shoots, stems, and leaves of *Sinarundinaria* were the predominant food. Beginning in the autumn of 1984, we obtained feeding data from radiotracking and by direct observations. As was found at Wolong, we also detected a distinct seasonal pattern in food selection by pandas.

In the Qinling area, giant pandas fed mainly on two dominant bamboo species: *Bashania*, occurring at 900–1900 m, and *Fargesia*, occurring at 1800–3000 m. The altitudinal separation of these bamboos enabled us to infer which species an animal was eating from radiotransmitted signals of its location. Furthermore, long-term radiotracking revealed seasonal differences in the movement of pandas between the two elevations. The majority moved from the *Bashania* occurring at lower elevations up to *Fargesia* between late May and early June, but a few delayed moving up until late June or early July (cf. Yong et al., chapter 10).

Most of the pandas remained in the *Fargesia* region for about 2–3 months (until late August–early September). At this time, they began the annual descent to resume feeding on *Bashania*. Again, a few individuals lingered on, delaying their descent until late September or early October. Occasionally, even as late as December, we found feeding signs and droppings at this higher elevation, suggesting that some individuals might have remained there throughout the year. Generally, pregnant females moved to the lower area earlier than did males and nonpregnant females (in late July or early August) to search for dens (Zhu et al. 2001). These variable patterns of migration could be an indication that individuals were adopting different resource utilization strategies.

To more effectively compare feeding at Qinling with that reported for Wolong, we conducted a fine-grained analysis of feeding in the different seasons (table 6.2). We found that from January to February, the Qinling pandas consumed *Bashania* leaves almost exclusively, and from March to April they ate both the leaves and stems of *Bashania*. New shoots of *Bashania* became available to them in April, and reached a peak in consumption in May. Throughout July and August the new shoots of *Fargesia* were the main food of pandas, but they also fed on shoots that had emerged in the previous year. From September to the following February, they fed mainly on *Bashania* leaves, only occasionally feeding on stems and old shoots of *Fargesia*.

TABLE 6.2
Monthly Feeding Observations and Percentage of Various Bamboo Species and Parts Consumed by Giant Pandas

MONTH	BASHANIA SHOOTS NUMBER OF OBSERVATIONS	BASHANIA SHOOTS PERCENTAGE	BASHANIA LEAVES NUMBER OF OBSERVATIONS	BASHANIA LEAVES PERCENTAGE	BASHANIA STEMS NUMBER OF OBSERVATIONS	BASHANIA STEMS PERCENTAGE	FARGESIA SHOOTS NUMBER OF OBSERVATIONS	FARGESIA SHOOTS PERCENTAGE	SUM OF ALL OBSERVATIONS
1			19	100					19
2			7	100					7
3			26	70.7	10	27	1	2.7	37
4	5	11.9	28	66.7	8	19	1	2.4	42
5	25	64.1	7	17.9			7	17.9	39
6	1	33.3					2	66.7	3
7							4	100	4
8							1	100	1
9			10	100					10
10			15	100					15
11			18	100					18
12			20	90.9	1	4.55	1	4.55	22

NOTES: Monthly percentage = the number of times we observed giant pandas feeding on the particular food item / the total number of feeding observations x 100. Data are for the years 1985–1996.

The data in table 6.2 reflect seasonal changes in food quality. *Bashania* leaves thrive from September to the following February, and throughout much of this time, giant pandas generally selected them in large quantities. These leaves are fresh and succulent, relatively nutritious, high in water content, and easily accessible. By March–April, after a period of winter growth, the cell wall in the leaves gradually thickens. Lignin and cellulose content also increase, and there is a decrease in the ratio of cell content in the leaves. Because of the dry climate at this time, most leaves lose water and become dehydrated. At this time giant pandas also feed on *Bashania* stems, although stems may not be the best food choice in terms of taste and nutrition. Nevertheless, they may be utilized to compensate for any nutritional deficiency.

When new *Bashania* shoots emerge in April, giant pandas immediately switch to this food source, probably to make up for any nutritional loss incurred in the winter and spring months. In May, these new shoots reach their growth peak and become the primary food supply. Between May and June, *Fargesia* shoots begin to sprout at the higher elevations and gradually replace *Bashania* shoots as the main food source. During this time, pandas sometimes consume old *Fargesia* shoots from the previous year and continue to do so until late August, when shoots mature and become lignified. Meanwhile, the lower elevation *Bashania* is flourishing again and the new leaves of this bamboo become the food of choice.

SELECTION FOR PARTS, AGE, AND THICKNESS OF STEMS

From the foregoing descriptions, we were able to discern those attributes of bamboo stems and leaves that determined the selection process of the pandas. Through examinations of the feeding tracks and feces, we found that giant pandas utilized *Bashania* leaves throughout the year. They consumed stems only in the spring (March–April). During this period, droppings contained 85.3% leaves and 14.7% stems. When feeding on *Bashania* stems, giant pandas concentrated on the middle section. Subsequently, they consumed the leaves before discarding the remaining tips. When exploiting new stems, they consumed the entire stem directly, without eating any leaves that had fallen to the ground. Similar feeding behavior was observed when they fed on the old shoots (or new stems) of *Fargesia*. Because the leaves of *Fargesia* are slender and sparse, they may be too time-consuming for giant pandas to utilize routinely.

Giant pandas selected bamboo according to its age and thickness. They mainly selected the 1- or 2-year-old *Bashania* shoots that were relatively thick, but 1-year-old shoots were especially preferred. Although we know relatively less about their preferences in feeding on *Fargesia,* our preliminary results indicate that they also favored the 1- or 2-year-old shoots and stems. It was unclear whether thickness affected their choices when feeding on *Fargesia*.

The selectiveness seen in their feeding behavior reflects in part the adaptation of giant pandas to a bamboo niche. Given the unusual amount of time required to harvest bamboo, they tend to exploit foods that are the most nutritious and the easiest to digest and obtain (e.g., carrion, bones, and the fresh and tender parts of the bamboo plant). Giant pandas select thick shoots because the proportion of the core to the outer sheath is much higher than is found in thin shoots. By preferring thick shoots, they are able to obtain more energy per unit of time expended in feeding.

BEHAVIORAL ADAPTATIONS TO FEEDING ON BAMBOO

A brief description of feeding activity by giant pandas is required to understand our analysis of behavioral adaptations to the consumption of bamboo. When feeding on stems, after a stem is selected, it is either bitten off in bite-sized segments while being grasped with the paw (between the distal end of the plantar surface of the paw and the pseudothumb), or its outer layer is peeled away with the teeth as the stem is

TABLE 6.3
Background Information on Videotaped Analyses of Feeding Behavior

INDIVIDUAL	SEX	DATE	AGE (YEARS)	DEVELOPMENTAL PHASE	FOOD TYPE	DURATION OF RECORDING (S)
Hope	Female	March 24, 1994	1.58	Infant	S	1014
		May 1994	1.75	Infant	L	917
		December 18, 1994	2.33	Subadult	L	468
		March 1995	2.58	Subadult	L	82
		May 1995	2.75	Subadult	Sh	23
		March 21, 1997	4.58	Adult	S, L	1738
Huzi	Male	August 1991	2	Subadult	L	3883
		October 30, 1994	5.17	Adult	L	78
		May 1995	5.75	Adult	L	11
		May 1995	5.75	Adult	Sh	121
Jiaojiao	Female	December 1994	~9.33	Adult	L	491
		May 1995	~9.75	Adult	Sh	59
Sun	Male	September 20, 1996	2.08	Subadult	S, T	581
Dashun	Male	January 1993	~19	Postreproduction	L	22

NOTES: Cub: 0–2 years; subadult: 2–4.5 years; adult male: 4.5–16.5 years; adult female: 4.5–18 years; postreproductive male: >16.5 years; postreproductive female: >18 years. Video segments that were taped in the same year and month were combined.
S, stem of *Bashania*; L, leaf of *Bashania*; Sh, shoot of *Bashania*; T, twig.

grasped. Peeled portions are then bitten off and chewed in the same manner as those that are not peeled. Any unwanted portion is discarded.

In feeding on leaves, the panda begins collecting the leaves by stripping them off a branch with the incisors until a leaf wad 7–10 cm in length is formed in the corner of the mouth. This wad is then grasped in the same manner as stems and held in the corner of the mouth while a segment is bitten off. Wads are then held in the paw during chewing. Additional bites followed by chewing reduce the wad to nothing or perhaps to a small tip that is discarded. This activity is then repeated with the same or another branch of bamboo.

Over the seven years between 1991 and 1997, we videotaped the feeding behavior of five wild pandas at different developmental stages and while eating different foods (table 6.3). From the clearest segments of tape, we conducted a careful and detailed analysis of feeding behavior. We input the video images into a computer, and for their analysis, we used a microVIDEO DC30 video card, 30 p/s (NTSC format) or 25 p/s (PAL format), to capture the desired visual elements and obtain an AVI image file. Using Adobe Premiere LE, we could then open the file and view images at a speed of 30 p/s. Images could be viewed continuously or one by one, and either sequentially or in a reverse mode.

After preliminary observations of feeding sequences, we developed a catalog of feeding acts (events, table 6.4), whose frequency and duration were measured. A further analysis was of the types of activities occurring between bouts of feeding on a particular food item (i.e., activity between reaching for one item of bamboo to reaching for another) and amount of time lapsed for each type (intermission time). As a final step, we classified acts as either food preparation or ingestion, and summed the time spent in these two components and in intermission.

As an example, our analysis of feeding by a 4-year old female (named "Hope") revealed that she spent 38.1% in such food preparation activities as searching for bamboo, collecting leaves,

TABLE 6.4
Conventions Used in the Analysis of Feeding Bouts

EVENT	DEFINITION
Reaching	A paw is raised to grasp bamboo
Grasping	Bamboo is taken into paw
Biting off stem	Severing stem from its base with the teeth
Stem consumption	From stem severance to end of consumption
Peeling of stems	Removal of tough outer layer of stem with teeth
Leaf collection/wad formation	Gathers leaves in corner of mouth to form a wad
Bites leaf wad	Jaws close in severing slice from leaf wad
Masticating	Chewing of food item
Wad consumption	Time from beginning to end of wad consumption
Activities between bouts of feeding (intermission)	
Resting	Short pauses between acts of feeding
Shifting posture	Change in sitting, lying, standing, etc.
Short-distance moving	Relocates while feeding at a site
Discarding	Abandons bamboo not selected for consumption
Classification of feeding acts according to type	
Reaching for a piece of bamboo	Preparation
Biting off a bamboo stem	Preparation
Peeling a stem	Preparation
Collecting a wad of leaves	Preparation
Consuming a bite from the wad	Ingestion
Consuming an entire wad	Ingestion
Biting off portions of a wad	Ingestion
Chewing an entire wad	Ingestion
Consuming a piece of stem	Ingestion

and peeling off tough layers of stems. She spent 56.8% ingesting, and the remaining 5.1% in brief rests, short-distance moving, posture shifting, and discarding movements.

FEEDING EFFICIENCY AS AN ADAPTIVE BEHAVIOR

Schaller et al. (1985) suggested that giant pandas meet their energy requirements by short retention time in the gut of large quantities of relatively nonnutritious bamboo. Modes of feeding that reduce the amount of time spent in this activity might therefore be an important part of their adaptation to a bamboo niche. Based on the time allocations seen in the female "Hope," we suggest that one way giant pandas could increase feeding efficiency is through a reduction of time spent peeling stems; that is, by selecting the most tender stems. However, given their availability, overspecialization on tender stems risks increasing search time and the time between bouts.

Efficiency could also be achieved by reducing the interval between bouts of feeding, perhaps through selecting sites where edible items are most plentiful. Pandas seldom change locations during feeding, and it seems to us that they try to eat as much as possible at one sitting.

A third possibility is that feeding economy is achieved by the manner in which giant pandas select, bite off, and masticate their food. Data collected on a 9-year-old female indicated that she required an average of 15.5 ± 2.5 s to consume a wad of *Fargesia* leaves.

Feeding actions were rapid and continuous, seemingly without any unnecessary steps. When selecting a stem with few leaves, pandas collect the leaves by bending the stem to the mouth rather than by biting it off. A complete wad of leaves was invariably collected before the shearing off of bite-sized morsels for mastication. If the leaves remaining on a stem were not sufficient to make a wad, it would be discarded as a food source. The stems of 1-year-old bamboo were preferred, presumably because the outer layer was less lignified and therefore easier to peel than those of older stems. Our observations also suggest that giant pandas adopted energy-conserving postures during feeding (i.e., sitting or reclining on a hillside) as a way of reducing energy expenditure.

UTILIZATION OF SPECIAL NUTRIENTS
HEMICELLULOSE

Besides using protein, soluble sugar, and fatty acids like other carnivores, giant pandas can also utilize some constituents of the plant cell wall, which is an adaptation for processing an herbivorous diet. Our calculations showed that in bamboo digested by giant pandas, the amount of hemicellulose decreased dramatically after digestion. Because the lignin in the cell wall is indigestible for many animals (Schaller et al. 1985), we can use it as a constant marker in calculating dry matter digestibility; and on this basis, calculate the digestibility of each dietary constituent. This obviates the need to measure fecal output, but requires measurement of the daily intake of bamboo by weight. We used the formula $E = t(v/p)$, where E = bamboo intake (kg/day), t = time spent feeding per day, v = feeding rate, and p = percentage of leaves in panda food (88.3%).

Feeding time (t) was derived from subtraction of travel and maintenance time from measures of total daily activity. By multiplying the percent leafy material times the number of leaves per handful, times leaf weight, times consumption time per handful, we were able to determine the feeding rate (v).

Table 6.5 shows that the utilization of hemicellulose was relatively high. During winter and spring, giant pandas consumed 11.7% of *Bashania* stems and 88.3% of the leaves (Pan et al. 2001). The digestibility of hemicellulose was therefore 18.1% ([% stems × % digestible] + [% leaves × % digestible] = digestibility for leaves and stems combined). In the summer, when pandas mainly fed on *Fargesia* stems, the hemicellulose digestibility was only 15.2%. The digestibility of hemicellulose in pandas housed at the National Zoo was 27%, and at Wolong, the percentage was calculated separately according to seasons: 21.5% for the spring, 26.0% for the summer, and 18.2% for the winter (Schaller et al. 1985).

Our calculations indicated that pandas in the Qinling Mountains utilized 479 g of hemicellulose from the cell wall per day when eating *Bashania*. This quantity can provide 1755 kcal of energy or about 27.4% of the total daily energy requirement. The ability of giant pandas to utilize hemicellulose in bamboo indicates an adaptation to a food source that has relatively low nutrient values. Among the various species of bamboos, the soluble sugar content is relatively low, averaging 2–3% (calculated from the percentage of dry weight: Schaller et al. [1985]). Therefore, giant pandas may be required to seek other alternatives to obtain sufficient energy to maintain daily activities. In this light, hemicellulose utilization may be regarded as one of the panda's important nutrient sources.

CELLULOSE

Using the hemicellulose methodology, we determined the digestibility of cellulose by giant pandas (table 6.6). Digestibility of cellulose in 1-year old *Bashania* stems (11%) was surprisingly high. Since the cellulose content in the surface layer of stems was much higher than in the internal layers, our high values could be from failure to peel off the tough surface layer in calculating the total cellulose content.

TABLE 6.5
Digestibility of Hemicellulose and Dry Matter by Giant Pandas

SEASON	SAMPLES ANALYZED		HEMICELLULOSE CONTENT (g/100 g)	HEMICELLULOSE DIGESTIBILITY (%)	DRY MATTER DIGESTIBILITY (%)
Winter and spring	*Bashania* leaf	Young leaf Feces	29.972 29.856	17.5	17.2
	Bashania stem	Young stem Feces	36.262 33.460	22.4	15.9
Summer	*Fargesia* stem	Stem Feces	45.561 44.232	15.2	12.7

TABLE 6.6
Digestibility of Cellulose and Dry Matter by Giant Pandas

SEASON	SAMPLES ANALYZED		CELLULOSE CONTENT (g/100 g)	CELLULOSE DIGESTIBILITY (%)	DRY MATTER DIGESTIBILITY (%)
Winter and spring	*Bashania* leaf	Young leaf Feces	20.217 24.290	0.48	17.2
	Bashania stem	Stem Feces	30.779 32.583	10.95	15.9
Summer	*Fargesia* stem	Young stem Feces	32.595 34.650	7.18	12.7

The digestibility of cellulose in *Bashania* leaves is substantially different from that found in *Fargesia* stems. A possible explanation may be found in the nutritional content of the two food items. The leaves of 1-year old *Bashania* have low cellulose (20.2%) and high protein (9.6%) content. In contrast, *Fargesia* stems contain high cellulose (32.6%) and low protein (2.1%) content. It seems that the high protein content in the *Bashania* leaves provides most of the energy that giant pandas need. In addition, leaves have a higher cell content than stems, among which soluble carbohydrates and lipids are important sources of energy. A certain amount of hemicellulose (271 g/day) also supplements as an energy source. Hence, the animal may be able to obtain sufficient energy without having to depend on cellulose, which is difficult to digest.

When eating *Fargesia* bamboo, due to the plant's low protein and cellulose content, giant pandas must seek other nutritional sources. They utilize much hemicellulose (534 g/day) and some cellulose (1791 g/day) in the cell wall. This may explain why cellulose digestibility is relatively high. Further research is needed to verify the above explanations regarding the ability of pandas to digest cellulose and determine whether the digestibility of cellulose in pandas varies with food type.

CONCLUSIONS: SUCCESSFUL FEEDING STRATEGIES

Through morphological, physiological, and behavioral adaptations, giant pandas in the Qinling Mountains are able to obtain sufficient nutrients from bamboo, a plant with relatively low nutritional values, to meet their metabolic requirements. They do so by:

1. Choosing *Bashania* and *Fargesia* as their main food sources. Abundant and fast growing, these bamboo species ensure ample food supplies for the giant pandas throughout the year.
2. Consuming carrion opportunistically to acquire additional protein and minerals.
3. Choosing the most nutritious foods available. Giant pandas feed on different bamboo species and different parts of the bamboo in different seasons.
4. Maximizing feeding rates to avoid unnecessary movement/activity and increasing feeding (biting and chewing) time.
5. Utilizing hemicellulose and possibly small amounts of cellulose that typically cannot be used by carnivores.

REFERENCES

McClure, F. 1943. Bamboo as panda food. *J Mammal* 24:267–68.

Pan, W., Z. Gao, and Z. Lü. 1988. *The giant panda's natural refuge in the Qinling Mountains.* (In Chinese.) Beijing: Peking University Press.

Pan, W., Z. Lü, X. Zhu, D. Wang, H. Wang, Y. Long, D. Fu, and X. Zhou. 2001. *A chance for lasting survival.* (In Chinese.) Beijing: Peking University Press.

Qin, Z. 1990. Bamboo food resources of giant pandas and the regeneration of the bamboo groves in Sichuan. In *Research and progress in biology of the giant panda,* edited by J. Hu, pp. 103–10. Chengdu: Sichuan Publishing House of Science and Technology.

Schaller, G. B. 1993. *The last panda.* Chicago: University of Chicago Press.

Schaller, G. B., J. Hu, W. Pan, and J. Zhu. 1985. *The giant pandas of Wolong.* Chicago: University of Chicago Press.

Sheldon, W. 1937. Notes on the giant panda. *J Mammal* 18:13–19.

Zhu, J. 1983. Rise and decline of the giant panda. (In Chinese.) *Acta Zool Sin* 29:93–104.

Zhu, X., D. Lindburg, W. Pan, K. Forney, and D. Wang. 2001. The reproductive strategy of giant pandas: Infant growth and development and mother-infant relationships. *J Zool Lond* 253:141–55.

BRIEF REPORT 6.1

Spatial Memory in the Giant Panda

Loraine R. Tarou, Rebecca J. Snyder, and Terry L. Maple

OPTIMAL FORAGING theory postulates that animals use efficient foraging strategies when searching for and processing food. Efficient strategies are those that maximize the energetic gains and minimize the energetic costs of foraging (Pyke et al. 1977). Some of the issues facing a foraging animal that affect efficiency include prey choice, handling time, encounter rate, and staying time within a given patch. Scientists interested in optimal foraging theory have attempted to incorporate these factors into both patch and prey models designed to predict resource maximization (Kamil and Sargent 1981; Stephens and Krebs 1986). However, most models assume that the animal is familiar with its environment and has already found a patch of food. Very little attention has been given to the initial search process for the patch itself or the ways in which foragers can decrease search time for nonrandomly distributed food (Krakauer and Rodriguez-Girones 1995). Furthermore, few foraging models consider how the spatial distribution of food influences foraging decisions (Sherry 1998). Because vegetation often occurs in patches that are predictable in both space and time (Milton 1988), herbivores can achieve nonrandom foraging and decrease search time by using sensory cues (visual or olfactory) associated with the vegetation itself, spatial memory, or a combination of these types of information to recognize and locate viable patches of food.

The giant panda *(Ailuropoda melanoleuca)* is somewhat of an anomaly in the animal kingdom, in that it is a carnivore that subsists almost exclusively on a herbivorous diet. Bamboo comprises approximately 99% of its diet (Schaller et al. 1985). Because bamboo is such a low-quality food source, optimal foraging may be particularly important for them (see Long et al., chapter 6). Pandas have several morphological, physiological, and behavioral adaptations that presumably help them increase their foraging efficiency for bamboo. For example, they have a unique pseudothumb that helps decrease handling time. They also are large in size, which gives them a lower metabolic rate and allows them to subsist on a poor diet. Schaller et al. (1985) state that pandas have adapted behaviorally to their food source by remaining inactive for many of their waking hours to conserve energy. However, little is known about the perceptual/cognitive abilities of the giant panda and what role these abilities may play in foraging.

A growing body of research concerning foraging and choice behavior shows that spatial memory plays an important role in foraging in a variety of animals, particularly rodents and birds

(Sherry 1998). It is unknown whether pandas are capable of using spatial information to locate food sources. Given the abundance and close proximity of food patches in their environment, giant pandas may have little need for spatial memory. Schaller et al. (1985) observed that pandas in the Wolong Nature Reserve traveled an average of only about 420 m each day. Although feeding sites were sometimes well spaced and the pandas seemed to selectively choose only certain patches even when surrounded by potential food, they were observed to travel an average of only 2.3–2.8 m to the next nearest patch of bamboo. This suggests that after depletion of one patch, pandas are almost always in sight of another one. Wei et al. (2000) reported that of the five species of bamboo that occurred in the Yele Natural Reserve (southwestern Sichuan Province), *Bashania spanostachya*, the preferred bamboo species of giant and red pandas *(Ailurus fulgens)*, was the most prolific, covering entire hillsides. Therefore, pandas might not have much need for spatial memory in locating bamboo.

We designed a series of studies to examine the foraging strategies used by giant pandas, with the goal of determining the type of information they use to make foraging decisions. Three studies have been completed with the pandas housed at Zoo Atlanta and findings are summarized here. The first foraging task examined the pattern of foraging used when depleting food sites and the ability of pandas to avoid revisiting sites during a given session. The second and third foraging tasks explored the ability of pandas to use spatial cues alone to locate food.

Our subjects were a female, "Lun Lun," and a male, "Yang Yang," approximately 3 years of age. We constructed sixteen feeders of PVC pipe, each having an opaque cover attached with a spring hinge. The pandas were able to obtain a food reinforcer by lifting this lid with either the muzzle or a paw. Eight feeders were arranged in a semicircular array approximately 1 m apart along the walls of each of two indoor dayrooms.

Efficient foragers will often visit adjacent food sites to decrease travel time between sites (Pyke et al. 1977). In our first experiment, we baited all eight feeders with a high-fiber leaf-eater biscuit and paid special attention to the frequency with which the pandas visited previously depleted sites. The animals could also increase foraging efficiency by avoiding feeders that had been depleted during a test session. The ability to avoid such sites has been referred to as "working memory" (Honig 1978).

In all tasks, the pandas were tested individually in separate rooms. We recorded each visit to a feeder, as well as the duration and frequency of several general behaviors (inactive, locomote, feed, scent mark, self-directed behavior, and object investigation). A visit to a feeder was defined as the placement of a subject's muzzle or paw under the lid of the feeder. In this test, efficient foragers should use a least-distance strategy when traveling between food sites, so we determined the percentage of visits in which the pandas bypassed zero, one, two, or three feeders to visit successive feeders. If no feeders were bypassed (i.e., the pandas visited adjacent feeders), the foraging pattern was considered optimally efficient. The greater the number of feeders bypassed, the less efficient the foraging pattern.

During the first 5 days of testing, both pandas showed zero separation in successive choices of feeders far more than they showed separation by visiting one, two, or three feeders (figure BR.6.1A). This was evident even in the early stages of trials. With eight possible choices, a randomly foraging animal would be expected to visit five nonadjacent feeders for every two adjacent ones. Over five sessions, an animal would be expected to make thirty-five total choices, twenty-five to nonadjacent feeders and ten to adjacent ones. The observed thirty-two visits to adjacent feeders and three visits to nonadjacent ones differed significantly from expectations for a randomly foraging animal ($\chi^2 = 67.76, p < .001$). In the final 5 days of testing (figure BR.6.1B), the pandas were exclusively visiting adjacent feeders, performing significantly better than would be expected by chance visits ($\chi^2 = 122.5, p < .001$).

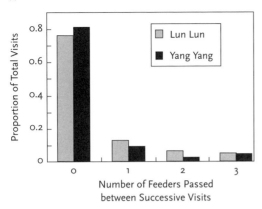

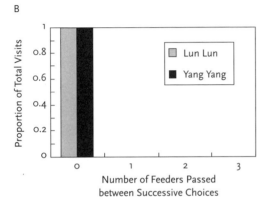

FIGURE BR.6.1. Mean proportion of total visits for which the pandas passed zero, one, two, or three feeders to reach the next site during the first 5 days (A) and the last 5 days (B) of testing. If zero sites were passed, the subject visited the site adjacent to the previous one. If three sites were passed, the subjects traveled across the test area to reach the next site.

Would pandas avoid feeders they had already depleted during a given trial? To answer this question, we calculated the frequency of visits to depleted sites during each trial. Both pandas retrieved the food items in all eight feeders. On the first day of testing, they did not stop foraging after visiting the eight feeders; instead, they returned to many of the feeders already depleted. Lun Lun revisited eleven sites, and Yang Yang revisited six sites on the first day. However, the number of revisits decreased significantly over time for both pandas ($R = -0.85, p < .001$). By the ninth day of testing, Yang Yang was visiting each of the eight feeders only once, never going back to feeders that he had already visited.

By designing additional foraging tasks, we wanted to determine if our pandas could use spatial cues alone to locate food sites. Four of the eight feeders were baited with food, and the pandas were allowed to locate and deplete them. Of interest in this task was their ability to learn to visit the baited sites and ignore the unbaited ones. This task is often described as a "win-stay" foraging task, because efficient foraging requires the animal to learn to visit only those sites that were baited in the preceding trial (Olton et al. 1981). This type of strategy is appropriate when prey or food is clumped in nature, such that one item is associated with other items in that particular location (Olton et al. 1981). The four feeders that were unbaited in the first task were baited in the second one. If the first four visits by the pandas were to the sites that had been baited in the previous task rather than the newly baited ones, then it can be concluded that they were not choosing based on visual or olfactory cues associated with the feeders. However, if the first four choices were to the newly baited sites, then it cannot be concluded that they were using spatial memory.

In the first task, we randomly selected four of the eight feeders for baiting. Placement of the baited feeders remained constant in each of the remaining trials. Of interest was the number of correct choices made during a session. Visits to nonbaited food sites and revisits to previously depleted sites were considered to be incorrect choices. One trial was conducted each day for 15 consecutive days or until the subject reached a criterion level of performance (i.e., a performance level of at least 75% correct in the last five sessions); also, there was no greater than a 5% difference in accuracy between these sessions and there were no increasing or decreasing trends in the data. The maximum number of sessions in each experiment was set at thirty trials.

In the reversal foraging task, the four feeders that were unbaited in the first task were baited for the next one. One session was conducted each day for 15 days or until the performance criteria were met. Correct visits were defined as those to baited feeders and incorrect visits as those to empty ones. We compared the total number of correct visits out of the first four visits across the first five days of testing with the number expected by chance if the pandas were visiting adjacent feeders consecutively. The results revealed that on the first five days, both pandas were performing at chance level ($\chi^2 = 0.0$, $p > .05$). By the sixth day for the male and the ninth day for the female, they were beginning to bypass feeders that were unbaited. In the last five days of the task, 90% of their first four visits were to baited feeders. This was significantly better than what would be expected for chance visits ($\chi^2 = 12.8$, $p < .01$).

On the first day of the reversal task, the performance of both pandas dropped significantly and was no different than that expected from a random forager ($\chi^2 = 2.25$, $p > .05$). In fact, the female traveled to three of the feeders that had been baited the day before in her first four choices, and the male traveled directly to the four feeders that had been baited the day before, even though those feeders no longer contained food. These results strongly suggest that the pandas were not using visual or olfactory cues to locate food. By the second day of the reversal task, both pandas were again performing at chance level, visiting approximately two baited feeders of the first four choices.

In both the spatial memory and reversal tasks, the pandas only needed to travel to four feeders to obtain all of the food provided. Any more than four visits represented unnecessary visits. The pandas visited many feeders unnecessarily in the first several days of testing. But this pattern decreased significantly over the course of the task ($R = -0.96$, $p < .001$). After the ninth day of testing, the male was visiting only the four correct feeders and none of the others. This behavior continued on the first day of the reversal task. Despite obtaining no food in the first four visits, Yang Yang visited only one of the other feeders that contained food. These results suggest that his behavior had become quite inflexible. Yang Yang's behavior changed dramatically on the second day of the reversal. He visited all of the feeders in a clockwise pattern, after which he revisited more feeders than he had in any other session.

The results of our research indicate that the two pandas housed at Zoo Atlanta were able to forage efficiently in a modified version of the radial arm maze, traveling to adjacent feeders more often than would be expected if they were foraging randomly. The ability to avoid revisiting depleted feeders during a given trial is sometimes interpreted as evidence of memory for places visited (Olton et al. 1981). However, these results must be interpreted with caution. The pandas traveled in a very systematic manner, often visiting the feeder located closest to the door of the test area first and then traveling in a circle until they reached the door again, suggesting that they may have been using a simple rule of thumb, rather than spatial memory, to end their search and avoid revisiting sites they had already depleted. Both pandas were able to rely on spatial memory alone to locate food, as evidenced by the disruption of their performances caused by the reversal.

Why might pandas exhibit spatial memory when locating food? Given the abundance of bamboo in their environment (Schaller et al. 1985; Wei et al. 2000) and the close proximity of bamboo patches (Schaller et al. 1985), it might seem surprising that pandas were able to return daily to baited food sites using spatial memory. Bamboo does not replenish itself quickly when depleted, and a patch would not be expected to regenerate over the course of a night, as simulated in this study.

There are several possible answers. First, giant panda habitats today may not reflect those in which pandas evolved. Their range has become severely restricted over the past hundred years (Schaller et al. 1985). The distribution of bamboo in earlier environments may have favored the evolution of spatial memory. Second, spatial memory may not be necessary for locating food, but may have developed for the efficient localization of other resources such as water, den sites, or mates.

Finally, pandas may use spatial memory to selectively forage for preferred species, parts (stems, leaves, shoots), or ages of bamboo (new, 2-year-old growth, or old). These preferences are known to change seasonally throughout the year (Schaller et al. 1985; Reid and Hu 1991; Wei et al. 2000). Therefore, it is possible that pandas use spatial memory to remember the location of preferred patches of bamboo, thereby increasing foraging efficiency throughout the year. This has been shown to be true for such domesticated grazers as sheep and cattle. According to Bailey et al. (1989) and Edwards et al. (1997), pastures are not homogenous entities, and grazing animals spend the majority of their time foraging where high-quality and large-quantity concentrations of preferred grasses occur. They exploit their environment in a way that suggests that they might be able to remember the location of these resources, allowing them to return to preferred sites later in foraging. In both open-field and laboratory experimental tests, spatial memory has been found to be used by such large herbivores as sheep, cattle, and deer in locating preferred food patches when grazing (Bailey et al. 1989; Dumont and Petit 1998). Like bamboo, pasture grasses do not replenish quickly after grazing. However, a forager may not fully deplete a patch in one visit, making return visits worthwhile. Moreover, spatial memory has been found to be long-lived in both sheep (3 days) and cattle (15 days) (Edwards et al. 1996), suggesting that they may be capable of returning to previously depleted patches following an interval long enough for regrowth of the vegetation to occur.

The findings summarized here are part of a larger project designed to examine the ability of pandas to use visual, olfactory, and spatial information in their search for food. Little is known about the visual characteristics of giant pandas, but, according to Schaller et al. (1985), their vision may not be very acute. Because they inhabit densely forested areas that limit visual range, finely tuned vision may not be necessary. However, such visual cues as leaf shape, size, or color may help in distinguishing bamboo from other types of vegetation, thus improving foraging efficiency.

REFERENCES

Bailey, D. W., L. R. Rittenhouse, R. H. Hart, and R. W. Richards. 1989. Characteristics of spatial memory in cattle. *Appl Anim Behav Sci* 23:331–40.

Dumont, B., and M. Petit. 1998. Spatial memory of sheep at pasture. *Appl Anim Behav Sci* 60:43–53.

Edwards, G. R., J. A. Newman, A. J. Parsons, and J. R. Krebs. 1996. The use of spatial memory by grazing animals to locate food patches in spatially heterogeneous environments: An example with sheep. *Appl Anim Behav Sci* 50:147–60.

Edwards, G. R., J. A. Newman, A. J. Parsons, and J. R. Krebs. 1997. Use of cues by grazing animals to locate food patches: An example with sheep. *Appl Anim Behav Sci* 51:59–68.

Honig, W. K. 1978. Studies of working memory in the pigeon. In *Cognitive processes in animal behavior*, edited by S. H. Hulse, H. Fowler, and W. K. Honig, pp. 211–48. Hillsdale: Erlbaum.

Kamil, A. C., and T. D. Sargent, eds. 1981. *Foraging behavior: Ecological, ethological, and psychological approaches.* New York: Garland.

Krakauer, D. C., and M. A. Rodriguez-Girones. 1995. Searching and learning in a random environment. *J Theoret Biol* 177:417–19.

Milton, K. 1988. Foraging behaviour and the evolution of primate intelligence. In *Machiavellian intelligence: Social expertise and the evolution of intellect in monkeys, apes, and humans*, edited by E. W. Byrne and A. Whiten, pp. 285–305. Oxford: Clarendon.

Olton, D. S., G. E. Handelmann, and J. A. Walker. 1981. Spatial memory and food searching strategies. In *Foraging behavior: Ecological, ethological and psychological approaches,* edited by A. C. Kamil and T. D. Sargent, pp. 333–77. New York: Garland.

Pyke, G. H., H. R. Pulliam, and E. L. Charnov. 1977. Optimal foraging: A selective review of theory and tests. *Q Rev Biol* 52:137–54.

Reid, D. G., and J. Hu. 1991. Giant panda selection between *Bashania fangiana* bamboo habitats in Wolong Reserve, Sichuan, China. *J App Ecol* 28:228–43.

Schaller, G. B., J. Hu, W. Pan, and J. Zhu. 1985. *The giant pandas of Wolong.* Chicago: University of Chicago Press.

Sherry, D. F. 1998. The ecology and neurobiology of spatial memory. In *Cognitive ecology: The evolutionary ecology of information processing and decision making,* edited by R. Dukas, pp. 261–96. Chicago: University of Chicago Press.

Stephens, D. W., and J. R. Krebs. 1986. *Foraging theory.* Princeton: Princeton University Press.

Wei, F., Z. Feng, Z. Wang, and H. Jinchu. 2000. Habitat use and separation between the giant panda and the red panda. *J Mammal* 81:448–55.

7

Chemical Communication in Giant Pandas

EXPERIMENTATION AND APPLICATION

*Ronald R. Swaisgood, Donald Lindburg, Angela M. White,
Hemin Zhang, and Xiaoping Zhou*

> How would I ever understand pandas? They moved from odor to odor, the air filled
> with important messages where I detected nothing.
>
> GEORGE B. SCHALLER (1993: 99)

IN THEIR SEMINAL WORK on giant pandas *(Ailuropoda melanoleuca)* in the wild, Schaller et al. (1985) brought to the fore chemical communication as a fundamental aspect of the panda's behavioral ecology. Without knowledge of this chemical communication system, we cannot understand how pandas locate and choose mates, regulate their use of space, and assess potential competitors; moreover, responses to conspecific odors might determine whether pandas use certain habitats that appear otherwise suitable. Despite the long-recognized importance of scent for pandas (see also Morris and Morris 1966; Kleiman 1983), until recently, little effort has been made to address this poorly understood aspect of giant panda biology. Giant pandas are solitary by nature, rarely meeting face-to-face as they traverse their home ranges through dense bamboo forests. Indeed, they appear to make great effort to avoid encountering one another throughout most of the year. Yet pandas will seek out conspecific odors, and have established a system that maximizes the communicatory potential of scent deposition. Several neighboring pandas with overlapping home ranges will utilize specific sites, called scent stations, where they deposit scent and investigate odors left behind by previous visitors. These stations can be likened to "community bulletin boards," wherein messages are left at prominent locations where others are likely to find them. Pandas possess a specialized anogenital gland that secretes a waxy substance comprised largely of short-chain fatty acids and aromatic compounds (see Hagey and MacDonald, brief report 7.1). They also utilize urine as a chemical signal. Pandas adopt four distinct postures to deposit scent at varying heights above the ground (Kleiman 1983). As with bulletin boards used by people, messages accumulate over time, making for a potentially confusing array of signals, new and old. Pandas must surely possess sophisticated assessment abilities to extract information from these scent stations. But what exactly does a visitor to one of these stations learn about other pandas? In recent years, we have undertaken a series of systematic investigations aimed at unraveling the mysteries of chemical communication in this species (e.g., Swaisgood et al. 1999, 2000, 2002).

An obvious starting point for such studies is to develop hypotheses regarding what sorts of information pandas would need to signal and extract from conspecific scent. First and foremost, perhaps, is the need to determine the message afforded by the scent. Important components of the message might include classification of the signaler in terms of species, sex, and age (reviews in Brown 1979; Macdonald 1985). Clearly, pandas need to discern whether the signal emanated from a same-sexed individual (representing a potential competitor) or member of the opposite sex (representing a potential mating partner). Of course, male pandas must not only identify females, but also determine the female's reproductive condition; that is, is she or will she soon be fertile? If males experience temporal changes in sexual motivation and reproductive ability, females may need to make similar distinctions between reproductively active and inactive males (as is seen in elephants: Schulte and Rasmussen [1999]), a possibility supported by observations of a seasonal change in testicular size, ejaculate volume, sperm quality, and behavior (Kleiman 1983; Platz et al. 1983). These functions would be greatly enhanced if pandas could also distinguish the age of the signaler, providing information regarding whether the signaler has reached adulthood. Having attained reproductive maturity, adults are capable of breeding, but might also pose a greater risk by virtue of their larger body size and motivation to defend resources. Following the same line of reasoning, pandas might need to make further assessments regarding the competitive ability or the territorial or dominance status of the signaler. Such information might be important not only for avoiding risky encounters with high-status individuals, but also provide valuable information about mate quality (see also Hurst 1993; Rich and Hurst 1999).

Even after assessing these and other components of the chemical message, the panda must extract additional information to make this signaling system truly useful and efficient. As with community bulletin boards, it is also important to determine which individual left the signal ("identity"), as well as the amount of time that has transpired since the mark was deposited ("date"). Consider the following scenario that underscores the need for multifunctional chemical signals. An adult male is searching for potential mates and comes across a scent station. He discovers an anogenital mark left by an adult female 3 days in advance of her fertile period. He recognizes the scent as belonging to a female whose core home range lies on the next ridge immediately south of the station. The male also ascertains that the female left the mark 2 days ago, thus making further pursuit worthwhile.

Scent plays a prominent role in the regulation of reproduction in numerous mammalian species, for example, in the attainment of puberty, mate choice, mate location, priming sexual motivation, and reducing aggression (Vandenbergh 1983; Doty 1986; Bronson 1989; Johnston 1990; Hurst and Rich 1999). Knowledge of this important aspect of giant panda behavioral biology could substantially enhance our ability to manage the wild population for conservation purposes and facilitate efforts to breed pandas in captivity, which has proven to be an elusive goal. Take the case of the cheetah (*Acinonyx jubatus*) for comparative purposes. Captive reproduction of this species at the San Diego Wild Animal Park floundered until efforts were made to provide potential breeding pairs with opportunities for olfactory communication prior to staging the breeding introduction (Lindburg 1999). Similarly, efforts to reintroduce captive mammals to the wild may be seriously compromised by ignorance of the animal's use of scent for territory settlement, establishing social relationships, and courting and mating. It is therefore incumbent upon us to learn as much about chemical communication in this species as possible, and to seek avenues for its application to conservation measures.

METHODS

For detailed descriptions of housing, husbandry, and experimental methods, see Swaisgood et al. (1999, 2000, 2002) and White et al. (2002, 2003). These studies are relatively unique in giant panda research, in that they utilized relatively large sample sizes (range, 11–28), and often included individuals of all age-sex categories. Earlier studies do not include subadult subjects because of inadequate sample size, but thanks to recent breeding success at Wolong, later studies have been able to incorporate younger animals. To test responsiveness to odor stimuli, we used two distinct experimental designs. In the first, we presented pandas with two odors simultaneously and recorded their behavioral responses to each. This method, often called a "preference test," has the advantage of controlling for temporal variation in motivation, in that the panda's "mood" is the same when it interacts with each of the two scent stimuli. Thus, this test is most sensitive to detecting a preference between two scent types. In the second design, we presented two different scent stimuli at two different times, separated by at least 24 hours to minimize carryover effects from exposure to the previous scent. This design, although less sensitive, has the advantage of allowing the experimenter to attribute subsequent behavior to the exposure to one or the other of the scent stimuli. For example, in the simultaneous exposure design, if the panda moves off and vocalizes and scent marks, one cannot ascertain which scent may have caused these behaviors. We used three different protocols to expose pandas to scent:

1. Placement of scent on wood blocks and presentation of the scent in the subject's home pen;
2. Direct placement of the scent on the walls or ground in the enclosure of the subject's home pen; and
3. Relocation of the subject into the pen of another individual (pen swapping), which exposes the subject to all the odors left by the inhabitant of the pen.

In all cases except for pen swapping, anogenital gland secretions or urine were collected fresh from naturally deposited scents from known individuals and were either presented to subjects immediately or frozen at −20°C for later use.

WHAT HAVE WE LEARNED ABOUT GIANT PANDA OLFACTORY COMMUNICATION?

RESPONSES TO CONSPECIFIC ODORS

These studies have provided clear evidence of the overriding importance of conspecific odors to giant pandas, as evidenced by the dramatic and often prolonged investigation of experimentally placed scent. For example, during the pen-swapping experiment, pandas increased chemosensory investigation (sniffing), flehmen, and licking of scent stimuli approximately tenfold above baseline levels (Swaisgood et al. 2000). Prior to these studies, neither the flehmen response nor the presence of a vomeronasal organ (VNO) had been described for the species. Flehmen in pandas is somewhat more subtle than other mammalian species. It is, however, easily recognized by deep inhalation (as evidenced by rapid expansion of the thoracic cavity), an abrupt upward movement of the head, and a slight upward curl of the upper lip, exposing the teeth. This position is only held for a second or two, and lacks the stereotyped appearance seen in ungulates and felids. It is often accompanied by prolific drooling. We have observed this behavior only when pandas are exposed to relatively novel conspecific odors, a rare occurrence at most captive facilities and difficult to observe in the wild. We were also unable to find any reference to the presence of a VNO in the giant panda, but we suspected its existence because flehmen functions to deliver socially significant airborne chemicals to the VNO (Hart 1983).

Subsequent investigation of the palate of the pandas residing at the San Diego Zoo confirmed the presence of incisive papillae, strongly suggestive of a VNO (D. Janssen, pers. comm.). Although flehmen is observed most frequently during males' investigation of female odors, it is also seen in response to other odors and by other individuals, suggesting that it serves important functions beyond the traditional assessment of female reproductive condition (Estes 1972). Licking of scent stimuli appears to be primarily a male behavior that may aid in the delivery of less volatile chemical constituents to the VNO or be important for gustatory analysis.

Other responses to conspecific scent include scent marking, countermarking over the experimental scent, scent rubbing (rubbing the pelage—especially the back, shoulders, and nape—over the scent), foot scraping (dragging the hindclaws backward, creating claw marks on the substrate), vocalization, and avoidance. These responses occurred less frequently and only occasionally were sufficiently common to merit analysis.

IDENTIFICATION OF SEX AND REPRODUCTIVE CONDITION

Outside the mating season, many solitary species avoid or respond aggressively or fearfully to conspecifics. Individuals of such species must overcome these tendencies during the female's fertile period if successful mating is to occur. Scent often plays a critical role in this process, most prominently in discerning the sex of a conspecific. Male mice, for example, will attack females doused with male urine and mount males scented with female urine (Dixson and Mackintosh 1971; Connor 1972). Female odors can also enhance male copulatory performance (Goodwin et al. 1979; Johnston 1990). Similarly, male odors are often important for activation of female sexual arousal (Floody et al. 1977; Johnston 1990). In addition, the ability of male mammals to assess female reproductive condition via chemical cues is undoubtedly important for reproductive success. Yet only a few species have been tested for this ability, and in several instances, tests failed to provide evidence of discrimination. These shortcomings led Taylor and Dewsbury (1990) to conclude that the common assumption that males can discriminate female reproductive condition on the basis of chemical cues is unwarranted. This ability should be of utmost importance, as it is nearly a prerequisite for effective intersexual communication during the mating season. A male should be able to discern reproductive condition via chemical cues, so that once he has detected a female's odor, he can determine whether it is profitable to pursue her for reproductive purposes at that time. Exposure to estrous female odors may, therefore, instigate mate searching and prime males for sexual behavior (see Rasmussen et al. 1997).

Sex recognition was readily evident among male pandas in one of our experiments using the pen-swapping protocol, in which subjects are exposed to all odors produced by the enclosure occupants (Swaisgood et al. 2000). Males strongly preferred female odors over male odors, investigating, licking, scent marking, and bleating more in response to female than to male odors. Subsequent experiments have confirmed these results in different experimental settings (White 2001), suggesting that this preference is robust. Males also bleated more after investigating estrous odors compared with their response to nonestrous odors. These findings demonstrate that males can indeed tell the difference between male versus female and estrous versus nonestrous odors. Because bleats are related to sexual motivation (Kleiman and Peters 1990), this study suggests that female odors, especially estrous female odors, promote sexual arousal in males. We suggest that in nature, bleating in response to female odors alerts the female to the male's presence, signals affiliative intent, and promotes association for mating purposes. Estrous females apparently adopt a similar strategy, as suggested by the higher rate of bleating and chirping in response to male compared with female odors. Male odors had a dramatically different effect on males, most notably in the complete absence of any bleat vocalizations, and high frequency of foot scraping, a behavior most often observed during intramale aggressive interactions. Thus, in contrast to female odors, male odors appear to promote aggressive motivation in other male pandas.

Although we measured several behavioral variables, we found, surprisingly, that male pandas only discriminated estrous from nonestrous female scent on the basis of vocalizations (Swaisgood et al. 2000). We therefore conducted a second study using simultaneous presentations in the hope of obtaining more robust evidence for this ability (Swaisgood et al. 2002). We presented eight male and ten female pandas with urine from estrous and nonestrous females. Males but not females spent significantly more time investigating estrous than they did nonestrous female urine. Males also investigated, displayed flehmen, and licked female urine in general (estrous and nonestrous samples combined) more than did females. Positive findings for estrus status discrimination by males but not females provides more compelling evidence that male preference for estrous female urine is part of a reproductive strategy, and not just the result of an arbitrary preference. For example, estrous females may produce stronger odors that were investigated more, because of their greater olfactory valence, not their salience for reproduction. If true, however, pandas of both sexes should prefer estrous female odors.

The prevalence of flehmen by males during chemosensory investigation of female odors in both studies, coupled with the elevated rate of flehmen to estrous female urine in the second study, indicate an important role for the VNO in the assessment of female reproductive condition. Because flehmen facilitates transport of nonvolatile chemicals to the VNO, flehmen is believed to be a good behavioral index of VNO utilization (Hart 1983). VNO analysis has been implicated for estrus discrimination and activation of sexual arousal in other mammalian species (Estes 1972; Beauchamp et al. 1983; Hart 1983; Johnston 1990). In addition, in both studies male pandas licked female odors more often than did females, suggesting that direct contact with female odors may be important for estrus discrimination, as has been shown for other species (Johnston 1990; Rivard and Klemm, 1990). Although licking may be important for gustatory analysis, it is perhaps more plausible that it functions to deliver less volatile chemical constituents to the VNO (Ladewig and Hart 1982).

Taken together, these two studies indicate that males can readily discriminate female reproductive condition on the basis of chemical cues, and that urine contains chemical constituents that change in the course of the estrus cycle. These findings are consistent with the hypothesis that female urine and perhaps other scents (e.g., anogenital gland secretions) function as reproductive advertisement. Male odors probably also affect female reproductive strategies, as suggested by:

1. Higher rates of sexually motivated vocalizations in response to male than female odors (Swaisgood et al. 2000);

2. Female chemosensory preference for adult male over adult female urine (White 2001);

3. Greater responsiveness to male urine by females just prior to their fertile period (White et al. 2003);

4. Female preference for odors from reproductively mature males over odors from immature males (White et al. 2003); and

5. Female preference for male odors plausibly indicating high competitive status (White et al. 2002).

These results suggest a clear role in the application of chemical communication for captive breeding programs. Indeed, "communication breakdown" is considered to be a primary determinant of reproductive failure in many captive species, especially those that are generally solitary (Lindburg 1999). Managers of captive pandas have struggled for years to overcome apparent behavioral deficits that hinder successful mating (He et al. 1994; Lindburg et al. 1997; Zhang et al. 2004). The most frequently cited reasons for reproductive failure are excessive and sometimes injurious aggression, lack of sexual interest, or fear and avoidance of the opposite sex. Our experiments have shown that conspecific scent is important for recognition of sex and female reproductive condition, and that exposure to these scents may enhance sexual motivation and mitigate aggressive tendencies. In nature, pandas would undoubtedly encounter such scents prior to direct contact with potential mates, and scent may play an important role in reducing the tendency to avoid or aggressively confront opposite-sexed conspecifics, while priming sexual motivation. A captive breeding program that incorporated such considerations should, therefore, achieve greater success.

Because of the highly endangered status of the panda, researchers are justly reluctant to deprive animals of opportunities that may promote mating success; it is therefore impractical to compare experimental groups given access to conspecific odors with control groups given no such exposure, a necessary procedure to test adequately the hypothesis that scent exposure and reproductive success are related. However, the breeding program at the Wolong breeding center provides compelling evidence consistent with this hypothesis. In recent years, scent exposure has figured prominently in the behavioral management program (Swaisgood et al. 2000, 2003; Zhang et al. 2004). At the onset of behavioral estrus, the female is brought to an enclosure adjacent to a potential mating partner, and over the next several days, the male and female are given frequent opportunities to investigate each other's odors. This is achieved by swapping the male and female into one another's enclosures. Concomitant with the development of this strategy, Wolong has increased the rate of successful breeding, and now about 90% of estrous females mate naturally each year (Swaisgood et al. 2003; Zhang et al. 2004), a rate unequaled by any other facility. Although other aspects of the management program also have changed during these years, the staff believe that this policy made a major contribution, in part because of the obvious and immediate effect that scent has on behavioral indicators of sexual motivation.

Our experiments do not provide a precise "recipe" for optimal scent exposure methods. However, it is worth noting that the experiment in which we presented a small amount of a single type of scent (i.e., female urine) did not promote behavioral indicators of sexual motivation to the same degree achieved by the pen-swapping method. Pen swapping exposes the panda to a whole suite of odors that saturate the environment, and it seems probable that such exposure to a complex gestalt of odors will greatly enhance its effects on sexual motivation, as has been found for some rodent species (Johnston 1990). In addition, if our hypothesis of VNO involvement in assessment of sexual odors is correct, simply allowing pandas to smell one another through cage bars between adjoining pens may not provide sufficient stimulation (Swaisgood et al. 2002). Thus, it may behoove managers at captive breeding facilities to allow pandas direct access to one another's scent prior to mating introductions.

ASSESSMENT OF SIGNALER AGE AND COMPETITIVE ABILITY

In another series of investigations, we examined the role that chemical cues play in assessment of reproductive maturity and competitive ability (e.g., dominance, territorial status) (White et al. 2002, 2003). Although it is debatable exactly where pandas lie on the territorial continuum, they appear to be somewhat intolerant of intruders into core areas, and some aggression appears to be governed by spatial considerations. Escalated fights in pandas can have severe consequences, and pandas appear to make an effort to avoid direct encounters. Undoubtedly, confrontations with adults or high-status individuals carry higher risks than those with subadults or low-status individuals. Therefore, selection should favor individuals who adopt strategies for assessing maturity and fighting ability, and use this information to avoid risky encounters with individuals of high competitive status. Perhaps most useful in this regard is the extraction of information from chemical signals, as this allows the animal to make assessments without risking direct confrontation. Previous research with mice has shown that social odors may contain chemical cues related to signaler age (Brown and Macdonald 1985; Ma et al. 1999), and that discrimination of such odors may structure competitive interactions (Hurst 1989). Similarly, mice possess surprisingly sophisticated chemosensory assessment strategies for identifying individuals of high competitive status. Males use this information in decisions regarding avoidance, submission, or challenging the chemical signaler (Gosling et al. 1996a,b; Drickamer 1997; Hurst 1990, 1993), whereas females base their mate preferences on this same information (Johnston et al. 1997a; Rich and Hurst 1998, 1999; Hurst and Rich 1999). It also seems plausible that chemical signals play an important role in governing reproductive behavior in the giant panda, especially given the brief 1- to 3-day female fertile period. With several female ranges overlapping a given male's range, it is important that the male identify which females should be monitored prior to the onset of the mating season, allowing him to concentrate his efforts on reproductively mature females. Similarly, females might use chemical cues to identify potential mates and assess male quality as it relates to age, dominance, and territory ownership.

In our first study, twenty-five pandas of all age and sex categories were presented with scent stimuli (male urine, male scent mark, female urine) collected from adults and subadults (White et al. 2003). The results clearly show that pandas from all age and sex categories prefer to investigate adult odors more than subadult conspecific odors of all three types, suggesting that this assessment task is fundamental to many giant panda social functions. The preference for odors from mature same-sexed individuals documented in this study may serve to gather information about the identity and status of competitors in the area (sensu Hurst 1989). Preferences for odors from adult opposite-sexed individuals might relate to identification of potential mates, but the possibility of competitive interactions between males and females should not be discounted. Given the high rate of intersexual aggression seen in captivity (Lindburg et al. 1997; Zhang et al., 2004)—perhaps a consequence of misidentification of estrous females—it seems possible that aggressive competition between males and females occurs in wild pandas outside of the mating season. However, adult females showed no overt investigative preferences between adult and subadult female urine, suggesting that adult females have no need to assess the age of a same-sexed potential competitor. As these tests were conducted during the mating season, it is possible that such competition is important during other times of the year; for example, when pregnant females are competing for birth dens.

A second factor that may be correlated with risk and/or mate quality is competitive ability or status. Competitive ability could be conveyed through inherent chemical properties of the scent and/or through the pattern or frequency of scent marking (reviewed in Hurst and Rich 1999), as well as deposition posture. For example, because larger animals can deposit scent higher than can smaller ones, the height of a scent mark may afford reliable information regarding signaler body size, a known determinant of competitive ability (Huntingford and Turner 1987). If assessors begin to use the height of scent deposition as a cue of competitive ability, fitness benefits should accrue to signalers that exaggerate this cue by using postures to place scents as high as possible. This selective advantage offers one explanation for the curious use of the handstand posture in several mammalian species (Rasa 1973; Macdonald 1979, 1985; Kleiman 1983). Subsequent selection on the assessor may favor skeptical assessment of these exaggerated cues, as such bluffing will not remain an evolutionary stable strategy (see Krebs and Dawkins 1984). However, the maximum height of scent deposition in the handstand posture may still correlate with body size. Moreover, the use of such postures may be co-opted to advertise competitive status, such as territory ownership or dominance. Such a shift from advertisement of competitive ability to aggressive motivation has been proposed for vocal communication (Morton 1977).

It is well known that pandas use four different postures to deposit scents at varying heights, but the functions of these postures remain unknown. Reverse and leg-cock postures, in which the panda backs into a vertical surface to deposit scent, have been seen in both males and females. Squats, where scent is deposited on the ground, are seen in all individuals, but mostly in subadults and females. Handstands have only been observed in adult males and usually only as a urine mark. Advertisement of competitive status (via selection to exaggerate apparent body size) may offer the most plausible explanation for the efforts that males go to when depositing urine in this handstand position. Indeed, males are less likely to use the handstand position when trespassing on another panda's home area, perhaps to avoid assertion of dominance in an area that they do not normally occupy (Swaisgood et al. 2000).

In our second study, we examined the effect of the height of the chemical signal on behavioral responsiveness (White et al. 2002). Adult male scent mark was placed at 0 m and 0.5 m to mimic squat and reverse/leg-cock postures frequently employed by adult males. Adult male urine was placed at 0 m and 1 m, reflecting the common placement of urine by males in either the squat or handstand position. Because females usually scent mark in either the squat or reverse positions, we placed female scent marks at 0 m and 0.5 m. Females infrequently urinate in any posture other than the squat posture, and therefore female urine was not used in this experiment. The results demonstrated that the height of scent stimuli had a pronounced and highly significant effect on several measures of behavioral responsiveness. For example, males and females of all ages spent more time investigating scent stimuli placed high than those placed low, regardless of whether the scent was male scent mark, male urine, or female scent mark. Thus, pandas may perceive conspecific odors placed high on vertical surfaces as indicating the presence of a more "important" animal, and therefore invest more time in memorizing the scent, in order to recognize it again in the future. As hypothesized earlier, one reason for this perceived importance might be the relationship between height of scent mark deposition and competitive ability. Also consistent with this hypothesis is the finding that pandas show a significant tendency to subsequently avoid the area where male urine was deposited at a height mimicking the handstand, presumably governed by the perceiver's attempts to avoid confrontation with a high-status male (see also Hurst 1990, 1993; Gosling et al. 1996a,b). This aversive effect of high male urine was most pronounced in subadult males, who are arguably most at risk of injurious aggression from adult males of high competitive status. Females also showed a tendency to avoid male urine placed to mimic the handstand posture, which on the surface appears inconsistent with the hypothesis that females might mate preferentially with these males. However, because these females were not in estrus at the time of testing, it is possible that they avoid high status males until the fertile period draws near. The conclusion most readily drawn from these results is that scent mark postures function in part to convey some aspects of competitive ability and/or status. Other hypotheses for the use of elevated postures, such as detection distance, and signaling sex or age status, are not supported by the evidence (see White et al. 2002).

MOTHER-INFANT RECOGNITION

Scent plays an important role in the mother-infant relationship in a number of species (Leon 1983), and it would be surprising if this were not true for the panda. We therefore have begun several investigations into the use of scent in mother-infant interactions in this species. Caretakers routinely wean the cub from the mother at about 6 months of age at the Wolong facility. One or two days after weaning, females were presented with the odor of their cub and the odor of a strange cub. In general, pandas prefer to investigate novel rather than familiar conspecific odors (Swaisgood et al. 1999). However, we hypothesized that a female that had recently been separated from her cub might actively prefer the odor of her cub to the odor of a strange cub. Our results were consistent with this hypothesis, in that females spent significantly more time sniffing their own cub's odor in comparison with the odor from a strange cub of the same age and sex. Interestingly, females also engaged in significantly more investigation of the environment after detecting their own cub's odor, perhaps because the odor stimulated further searching for the cub's whereabouts. That mothers are capable of recognizing their cub via odors may have implications for management. For example, it may be important to provision a new mother with her cub's odors if the cub is removed for a period of hand-rearing, a strategy we have successfully employed (Zhang et al. 2000). Also, it might be possible to encourage a female to adopt a surrogate cub rejected by its biological mother if the cub is first covered with the scent of her own cub. Although females rarely rear two offspring from birth, females will provide care for two older cubs simultaneously; in addition, two cubs may be swapped back and forth between the nursery and the mother every few days (Swaisgood et al. 2003). Thus, it is probably possible to encourage a panda mother to rear a second unrelated offspring along with her own.

In a second experiment, cubs were presented with odors either from their mothers or from a strange adult female. Odors were placed on a burlap stack stuffed with straw, to provide the opportunity for the cub to express comfort-seeking behavior. Odor presentations were sequential rather than simultaneous. Cubs spent significantly more time sniffing the odors of their mothers in comparison with the odors of strange females. Other responses suggested that the mother's odor produced a calming effect on a generally agitated, recently weaned cub. Although the cub left the sack more frequently to engage in environmental investigation (presumably searching for the mother), the cub also returned to the sack more often, and spent more time resting quietly in contact with the sack. In contrast, when given sacks scented with a strange female's odor, cubs appeared much more agitated, as demonstrated by their greater number of escape attempts, distress vocalizations, and bipedal postures. Because these signs of agitation did not differ substantively from the cubs' behavior with no scented sack present, it does not appear that the sack with the strange odor increased their distress. These results suggest a role for maternal odors in the mitigation of stress for a few days during the potentially traumatic transitional period following the rather abrupt weaning experienced by most captive panda cubs. In nature, olfaction probably plays a significant role in mother-cub recognition; because the mother and cub may be separated for hours or days (Lü et al. 1994; Zhu et al. 2001), such recognition should prevent misidentification upon reunion.

DISCRIMINATION OF INDIVIDUAL CHEMICAL SIGNATURES

In addition to these and other messages afforded by giant panda scents, it is important for pandas to be able to recognize the identity of the signaler through individually distinctive chemical signatures. Indeed, many of the functions already discussed require as a prerequisite the ability to discriminate individual differences in conspecific odors (Halpin 1986; Swaisgood et al. 1999). For example, mothers cannot reliably recognize their own cub's odor unless it contains a unique chemical profile differentiating it from other cubs' odors. Competitor assessment strategies might also be greatly enhanced by discrimination of individual odors: an individual that recognizes the scent of a high-status individual may benefit if it can subsequently match that odor to the odor of an individual encountered in the environment (cf. Gosling 1982). This ability allows the individual to avoid potentially injurious escalation with a superior competitor without engaging in more direct probing of the individual's fighting ability and/or motivation. Adaptive female mate choice may also depend on the ability to match scent found in the environment with potential mates. For example, they may prefer to mate with males whose scent is encountered most frequently in the environment because the predominance of a particular male's scent is a reliable cue of his ability to exclude rival males (Hurst and Rich 1999). Individual discrimination is so important in some species that there is evidence that odors have become evolutionarily specialized to signal individual identity (e.g., Johnston et al. 1993).

As early as 1966, Morris and Morris suggested that scent marking in pandas served as a method of stamping the individual's identity on its territory, yet this basic ability has never been tested. To determine if individual discrimination is indeed important in giant panda chemical communication, we conducted experiments using twenty adult pandas at the Wolong breeding center (Swaisgood et al. 1999). For pragmatic reasons, we used female urine and male anogenital gland secretions as odor stimuli. Using a habituation-discrimination paradigm (see Halpin 1986), we exposed each subject to the scent of "panda A" repeatedly for 5 days; the panda's response to this scent gradually diminished. On the sixth day, we presented the subject with both the scent of panda A and that of another panda ("panda B"). The pandas showed a clear preference for the novel scent from panda B over the habituated scent from panda A, as evidenced by the amount of time spent investigating the two scents. Thus, panda A must smell different from panda B, implying that pandas produce individually distinct chemical profiles. Pandas readily discriminated between individuals on the basis of male anogenital gland secretions, but the preference for novel female urine did not quite attain statistical significance, suggesting that male odors or anogenital gland secretions from either sex (most probably the latter) contain more individually distinctive chemical cues. We suggest that anogenital gland secretions are perhaps specialized for this function.

PERSISTENCE OF ODORS THROUGH TIME AND ASSESSMENT OF THE RELATIVE AGE OF ODORS

Chemical signals are unique in that they persist in the environment long after the signaler has left, but eventually even these long-lasting signals will fade (Wilson and Bossert 1963). This temporal persistence allows relatively solitary species to communicate in absentia, greatly enhancing the efficiency of the chemical communication system. To understand chemical communication in the giant panda, we must determine how long these chemical signals remain biologically active. Are scent marks ephemeral chemical signals, or can pandas detect them days or months later? In 1998, we conducted an experiment designed to test how long giant panda scent marks remain biologically active. We collected male anogenital gland secretions on wood blocks and aged them for approximately 120 days before presenting them to sixteen adult pandas. Although aged indoors, the window was open, and therefore temperature and humidity were comparable to that experienced in the panda's natural environment. Simultaneous presentations of these scent stimuli and identical unscented control blocks were used to assess whether pandas could discriminate between scented and unscented blocks. The results were dramatic: pandas spent only a few seconds investigating control blocks and more than ten times as much time sniffing scented blocks. It is clear from these results that pandas are able to detect scents that are at least 120 days old under these conditions. In a recent follow-up experiment, we aged scent stimuli for varying lengths of time outdoors, exposed to sun and rain. Preliminary findings reveal that pandas detect and respond to anogenital gland secretions aged for three months, but remain responsive to urine marks for only about two weeks. Clearly, the persistence of these scents in the environment has ramifications for the functional consequences of chemical communication for the species, as their biologically active period determines how soon another panda must locate the scent for communication to take place. Similarly, knowledge of persistence times will facilitate the use of scent for conservation efforts by providing temporal criteria for their effectiveness.

The ability to assess the relative age of a scent mark also may have important consequences for regulation of giant panda social behavior. As scent marks in the environment age, exposed to sun, air, and moisture, they can degrade in fairly predictable ways (Regnier and Goodwin 1977). Volatile constituents are slowly lost, weakening the signal, and the chemical structure of some compounds may be altered. With some mammalian scents lasting more than 100 days (Johnston and Schmidt 1979), an area frequently used for scent marking by many animals may become a confusing place unless perceivers possess the ability to determine the age of scent marks. With the "background noise" of dozens of marks of varying ages, the task of signal detection and interpretation may be rendered difficult. It is just such a challenge that giant pandas confront when visiting communal scent mark stations. The ability to accurately determine the age of scents might confer other advantages as well. Patrol intervals of the signaler might be deduced from the amount of time elapsing between bouts of marking in a particular area. How recently a scent has been deposited might also be important because fresh scents might indicate that the signaler is still in the area, enabling the assessor to avoid or challenge the signaler as appropriate. The ability to determine the age of scent might also come into play in the male's efforts to locate fertile females. For example, 1-week-old scent from an estrous female informs the male that the female has most likely ovulated, rendering further pursuit unprofitable.

In the spring of 1999, we conducted several experiments that show that pandas can indeed tell the difference between scents of different ages. Our experiment with fifteen adults and nine subadults demonstrated that adult pandas can discriminate between fresh and 1-day-old scent, fresh and 3-day-old scent, and 1-day-old versus 5-day-old scent. In all cases, adult pandas spent significantly more time investigating fresher scent marks than older ones. In contrast, subadults, although equally interested in investigating conspecific odors, did not show any tendency to prefer fresh odors over older odors. One interpretation of these findings is that young pandas must learn to discriminate between these odors through repeated exposure to conspecific odor. Alternatively, the ability and/or motivation to discriminate among these odors may be the result of a simple maturational process, perhaps mediated by reproductive hormones. Regardless, this research highlights the importance of exploring the ontogeny of the chemical communication system in giant pandas.

CONCLUSIONS

From these studies, we are beginning to gain meaningful insights into the pervasive role that social odors play in the social lives of giant pandas. The governing influence of conspecific odors appears to run the gamut across all aspects of social behavior, ranging from the mother-cub relationship to reproductive behavior and competition. Pandas possess a suite of sophisticated discriminatory abilities for distinguishing between odors emanating from various categories of conspecifics, and the pattern of response often varies with the sex, age, and reproductive condition of the perceiver in ways meaningful to the fulfillment of specific functional endpoints. Although our tests were conducted on captive animals, it seems reasonable that all of these abilities come into play in the regulation of social behavior among wild pandas. One might expect captive animals to lose some of these chemosensory abilities, but it seems highly implausible that these abilities are somehow the artifact of captivity, especially given their important functional implications. Moreover, it would be virtually impossible to conduct such tests on wild pandas; the researcher would have to monitor an experimentally placed scent for days or weeks before it was discovered, and by that time, the response would be confounded by the age of the scent. Nonetheless, it is important to obtain more data on giant panda social dynamics in the wild, to place these experiments in an appropriate context in nature.

Throughout this chapter, we have also suggested potential applications of these experimental results to management and conservation of the species in the wild and captivity. Ramifications of the results obtained are most readily evident in the use of scent to encourage natural mating in captivity. In addition to its potential role in activating sexual motivation discussed earlier, it might also be possible to use scent to recruit new breeders into the captive population that may well suffer from lack of genetic diversity. For example, we may be able to encourage a female to mate with a particular male that is genetically unrepresented by manipulating the female's mating preference. It has been shown in some rodent species that females will use olfactory cues to mate preferentially with males of high competitive ability (Johnston et al. 1997b; Hurst and Rich 1999). Females may use encounter rate, relative predominance of a male's scent over other males, or prevalence of countermarking on top of other males' scents as an indicator of a male's ability to exclude other males or effectively patrol and maintain his chemical signature as the predominant one in the area. We have recently used this idea to manipulate female mate preference in another threatened species, the pygmy loris *(Nycticebus pygmaeus)* (Fisher et al. 2003a,b). We presented six females with a particular male's odor repeatedly for 3 months and then provided the female with a choice between the familiar-smelling male and a novel-smelling male. In all six cases, the female displayed more sexual interest in the familiar-smelling male. In the case of pandas, it might also be important to consider the height of presentation of the male's scent, as this may also be used as a cue of male competitive ability. However, note that conspecific odors can have both suppressive or facilitative effects on reproduction (Doty 1986; Bronson 1989; Bartos and Rödl 1990; Lindburg 1999), and thus the effects of odors should be studied carefully and used judiciously.

We envision a day when it might be possible to apply what we have learned in captivity directly to the conservation of the panda in the wild. For example, one can make a strong case for a role for scent in habitat selection and settlement. There is growing evidence that many species use the presence of conspecifics as a cue to habitat quality, following the rule of thumb that if the area is safe and productive enough to support other members of the species, then it must be a good place to settle (Stamps 1988; Reed and Dobson 1993). This behavioral mechanism raises the possibility of planting conspecifics or cues from conspecifics to encourage settlement in unoccupied areas, such as new reserves or habitat corridors between islands of habitat. In pandas, for whom olfaction reigns supreme, it may be possible to place conspecific odors in these areas to encourage pandas to use them, with the caveat that some odors may be aversive rather than attractive. For example, there is the distinct possibility that odors from adult males placed on trees at the level of a handstand urine mark may actually deter pandas from entering the area, especially individuals of low competitive ability. A thorough understanding of how pandas of all age and sex classes respond to various conspecific odors is crucial for the successful application of scent in such conservation measures.

Scent may also provide a useful tool in proposed reintroduction efforts (see Swaisgood et al. 1999). Many reintroductions and translocations fail because the released animal fails to remain at the release site or because of excessive aggression between the reintroduced and resident conspecifics (Yalden 1993; Reading and Clark 1996). It may be possible for researchers to use the reintroduced animal's own scent to pre-establish a home range by depositing scent throughout the area for several months prior to release. This may have several beneficial consequences for the reintroduction. First, it might encourage the reintroduced animal to remain in the area where it encounters its own scent, as this may diminish the stress response to the novel environment. Second, it may discourage neighboring pandas from aggressively confronting the released panda. This second effect relies on the applicability to pandas of Gosling's (1982) scent-matching hypothesis for recognizing territory owners. Under this hypothesis, animals recognize territory owners by matching the scent that predominates in the environment with the scent of an individual that is encountered in the area. Because only an animal that is occupying and successfully defending an area will deposit its scent frequently in the area, this provides a reliable cue of territory occupation. Territory owners have proven their competitive ability in maintaining the territory and are also more motivated to fight in defense of this resource. Thus, it behooves other animals to recognize such individuals and avoid an escalated contest with them. If pandas use this strategy, it may be possible to "fake" territory ownership prior to the animal's release, thereby mitigating aggressive interactions with neighboring pandas.

We still have much to learn about chemical communication in giant pandas; in particular, the role it plays in regulating social dynamics in the wild. Continued research efforts in these areas hold promise for both a deeper understanding of the animal and more informed management of the species in the wild and captivity.

ACKNOWLEDGMENTS

We express our sincere gratitude to the State Forestry Administration of China and the Zoological Society of San Diego for their generous support, and to the Wolong animal care staff for their assistance. Ronald R. Swaisgood's position was supported by a donation from J. Dallas Clark.

REFERENCES

Bartos, L., and P. Rödl. 1990. Effect of a conspecific's urine on reproduction in arctic foxes. In *Chemical signals in vertebrates*, edited by D. W. Macdonald, D. Müller-Schwarze, and S. E. Natynczuk, pp. 352–59. Vol. 5. Oxford: Oxford University Press.

Beauchamp, G. K., G. I. Martin, J. L. Wellington, and C. J. Wysocki. 1983. The accessory olfactory system: Role in maintenance of chemoinvestigatory behaviour. In *Chemical signals in vertebrates*, edited by D. Müller-Schwarze and R. M. Silverstein, pp. 73–86. Vol. 3. New York: Plenum.

Bronson, F. H. 1989. *Mammalian reproductive biology.* Chicago: University of Chicago Press.

Brown, R. E. 1979. Mammalian social odors: A critical review. In *Advances in the study of behavior*, edited by J. S. Rosenblatt, R. A. Hinde, C. Beer, and M. C. Busnel, pp. 103–62. Vol. 10. New York: Academic.

Brown, R. E., and D. W. Macdonald, eds. 1985. *Social odours in mammals.* Oxford: Clarendon.

Connor, J. 1972. Olfactory control of aggressive and sexual behavior in the mouse *(Mus musculus)*. *Psychonomic Sci* 27:1–3.

Dixson, A. K., and J. H. Mackintosh. 1971. Effects of female urine upon the social behaviour of adult male mice. *Anim Behav* 19:138–40.

Doty, R. L. 1986. Odor-guided behavior in mammals. *Experientia* 42:257–71.

Drickamer, L. C. 1997. Responses to odors of dominant and subordinate house mice *(Mus domesticus)* in live traps and responses to odors in live traps by dominant and subordinate males. *J Chem Ecol* 23:2493–506.

Estes, R. D. 1972. The role of the vomeronasal organ in mammalian reproduction. *Mammalia* 36:315–41.

Fisher, H. S., R. R. Swaisgood, and H. Fitch-Snyder. 2003a. Countermarking by male pygmy lorises *(Nycticebus pygmaeus)*: Do females use odor cues to select mates with high competitive ability? *Behav Ecol Sociobiol* 53:123–30.

Fisher H. S., R. R. Swaisgood, and H. Fitch-Snyder. 2003b. Odour familiarity and female preferences for males in a threatened primate, the pygmy loris, *Nycticebus pygmaeus*: Applications for genetic management of small populations. *Naturwissenschaften* 90:509–12.

Floody, O. R., D. W. Pfaff, and C. D. Lewis. 1977. Communication among hamsters by high-frequency acoustic signals II. Determinants of calling by females and males. *J Comparative Physiol Psychol* 91:807–19.

Goodwin, M., Gooding, K. M., & Regnier, F. 1979. Sex pheromone in the dog. *Science* 203:559–61.

Gosling, L. M. 1982. A reassessment of the function of scent marking in territories. *Z Tierpsychol* 60:89–118.

Gosling, L. M., N. W. Atkinson, S. A. Collins, R. J. Roberts, and R. L. Walters. 1996a. Avoidance of scent-marked areas depends on the intruder's body size. *Behaviour* 133:491–502.

Gosling, L. M., N. W. Atkinson, S. Dunn, and S. A. Collins. 1996b. The response of subordinate male mice to scent mark varies in relation to their own competitive ability. *Anim Behav* 52:1185–91.

Halpin, Z. T. 1986. Individual odors among mammals: Origins and functions. *Adv Study Behav* 16: 39–70.

Hart, B. L. 1983. Flehmen behavior and vomeronasal organ function. In *Chemical signals in vertebrates*, edited by D. Müller-Schwarze and R. M. Silverstein, pp. 87–103. Vol. 3. New York: Plenum.

He, T., K. Zhang, H. Zhang, R. Wei, C. Tang, G. Zhang, and M. Cheng. 1994. Training male giant pandas for natural mating. In *Proceedings of the international symposium on the protection of the giant panda (Ailuropoda melanoleuca)*, edited by A. Zhang and G. He, pp. 188–92. Chengdu, China: Chengdu Foundation of Giant Panda Breeding.

Huntingford, F., and A. Turner. 1987. *Animal conflict*. London: Chapman and Hall.

Hurst, J. L. 1989. The complex network of olfactory communication in populations of wild house mice *Mus domesticus* rutty: Urine marking and investigation within family groups. *Anim Behav* 37:705–25.

Hurst, J. L. 1990. Urine marking in populations of wild house mice *Mus domesticus* rutty. I. Communication between males. *Anim Behav* 40:209–22.

Hurst, J. L. 1993. The priming effects of urine substrate marks on interactions between male house mice, *Mus musculus domesticus* Schwarz and Schwarz. *Anim Behav* 45:55–81.

Hurst, J. L., and T. J. Rich. 1999. Scent marks as competitive signals of mate quality. In *Advances in chemical communication in vertebrates*, edited by R. E. Johnston, D. Müller-Schwarze, and P. Sorensen, pp. 209–23. New York: Kluwer Academic.

Johnston, R. E. 1990. Chemical communication in golden hamsters: From behavior to molecules to neural mechanisms. In *Contemporary trends in comparative psychology*, edited by D. E. Dewsbury, pp. 381–409. Sunderland, Mass.: Sinauer.

Johnston, R. E., and T. Schmidt. 1979. Responses of hamsters to scent marks of different ages. *Behavioral Neural Biol* 26:64–75.

Johnston, R. E., A. Derzie, G. Chiang, P. Jernigan, and L. Ho-Chang. 1993. Individual scent signatures in golden hamsters: Evidence for specialization of function. *Anim Behav* 45:1061–70.

Johnston, R. E., E. S. Sorokin, and M. H. Ferkin. 1997a. Female voles discriminate males' overmarks and prefer top-scent males. *Anim Behav* 54:679–90.

Johnston, R. E., E. S. Sorokin, and M. H. Ferkin. 1997b. Scent counter-marking by male meadow voles: Females prefer the top-scent male. *Ethology* 103:443–53.

Kleiman, D. G. 1983. Ethology and reproduction of captive giant pandas *(Ailuropoda melanoleuca)*. *Z Tierpsychol* 62:1–46.

Kleiman, D. G., and G. Peters. 1990. Auditory communication in the giant panda: Motivation and function. In *Proceedings of the second international symposium on the giant panda*, edited by S. Asakura and S. Nakagawa, pp. 107–22. Tokyo: Tokyo Zoological Park Society.

Krebs, J. R., and R. Dawkins. 1984. Animal signals: Mind-reading and manipulation. In *Behavioural ecology: An evolutionary approach*, edited by J. R. Krebs and N. B. Davies, pp. 380–402. Sunderland, Mass.: Sinauer.

Ladewig, J., and B. L. Hart. 1982. Flehmen and vomeronasal organ function. In *Olfaction and endocrine regulation*, edited by W. Breipohl, pp. 237–47. London: IRL.

Leon, M. 1983. Chemical communication in mother-young interactions. In *Pheromones and reproduction in mammals*, edited by J. G. Vandenbergh, pp. 39–75. New York: Academic.

Lindburg, D G. 1999. Zoos as arks: Issues in ex situ propagation of endangered wildlife. In *The new physical anthropology: Science, humanism, and critical reflection*, edited by S. C. Strum, D. G. Lindburg, and D. Hamburg, pp. 201–13. Upper Saddle River, N.J.: Prentice Hall.

Lindburg, D. G., X. Huang, and S. Huang. 1997. Reproductive performance of giant panda males in Chinese zoos. In *Proceedings of the international symposium on the protection of the giant panda (Ailuropoda melanoleuca)*, edited by A. Zhang and G. He, pp. 67–71. Chengdu, China: Sichuan Publishing House of Science and Technology.

Lü, Z., W. Pan, and J. Harkness. 1994. Mother-cub relationships in giant pandas in the Quinling Mountains, China, with comment on rescuing abandoned cubs. *Zoo Biol* 13:567–68.

Ma, W., D. Wiesler, and M. V. Novotny. 1999. Urinary volatile profiles of the deermouse *(Peromyscus maniculatus)* pertaining to gender and age. *J Chem Ecol* 25:417–31.

Macdonald, D. W. 1979. Some observations and field experiments on the urine marking behavior of the red fox, *Vulpus vulpus* L. *Z Tierpsychol* 51:1–22.

Macdonald, D. W. 1985. The carnivores: Order Carnivora. In *Social odours in mammals*, edited by R. E. Brown and D. W. Macdonald, pp. 619–722. Vol. 2. Oxford: Clarendon.

Morris, R., and D. Morris. 1966. *Men and pandas*. New York: McGraw-Hill.

Morton, E. S. 1977. On the occurrence and significance of motivation-structural rules in some bird and mammal sounds. *Am Nat* 111:855–69.

Platz, C. C., D. E. Wildt, J. G. Howard, and M. Bush. 1983. Electroejaculation and semen analysis and freezing in the giant panda *(Ailuropoda melanoleuca)*. *J Reprod Fertil* 67:9–12.

Rasa, A. E. 1973. Marking behaviour and its social significance in the African dwarf mongoose, *Helogale undulata rufula*. *Z Tierpsychol* 32:293–318.

Rasmussen, L. E. L., T. D. Lee, W. L. Roelofs, A. Zhang, and G. D. Daves, Jr. 1997. Purification, identification, concentration and bioactivity of (Z)-7-dodecen-1-yl acetate: Sex pheromone of the female Asian elephant, *Elephas maximus*. *Chemical Senses* 22:417–37.

Reading, R. P., and T. W. Clark. 1996. Carnivore reintroductions: An interdisciplinary examination. In *Carnivore behavior, ecology, and evolution*, edited by J. L. Gittleman, pp. 296–336. Ithaca: Cornell University Press.

Reed, J. M., and A. P. Dobson. 1993. Behavioural constraints and conservation biology: Conspecific attraction and recruitment. *Trends Ecol Evol* 8:253–55.

Regnier, F. E., and M. Goodwin. 1977. On the chemical and environmental modulation of pheromone release from vertebrate scent marks. In *Chemical signals in vertebrates*, edited by D. Müller-Schwarze, and M. M. Mozell, pp. 115–33. Vol. 1. New York: Plenum.

Rich, T. J., and J. L. Hurst. 1998. Scent marks as reliable signals of the competitive ability of mates. *Anim Behav* 56:727–35.

Rich, T. J., and J. L. Hurst. 1999. The competing countermarks hypothesis: Reliable assessment of competitive ability by potential mates. *Anim Behav* 58:1027–37.

Rivard, G., and W. R. Klemm. 1990. Sample contact required for complete bull response to oestrous pheromone in cattle. In *Chemical signals in vertebrates*, edited by D. W. Macdonald, D. Müller-Schwarze, and S. E. Natynczuk, pp. 627–33. Vol. 5. Oxford: Oxford University Press.

Schaller, G. B. 1993. *The last panda*. Chicago: University of Chicago Press.

Schaller, G. B., J. Hu, W. Pan, and J. Zhu. 1985. *The giant pandas of Wolong*. Chicago: University of Chicago Press.

Schulte, B. A., and L. E. L. Rasmussen. 1999. Signal-receiver interplay in the communication of male condition by Asian elephants. *Anim Behav* 57:1265–74.

Stamps, J. A. 1988. Conspecific attraction and aggregation in territorial species. *Am Nat* 131:329–47.

Swaisgood, R. R., D. G. Lindburg, and X. Zhou. 1999. Giant pandas discriminate individual differences in conspecific scent. *Anim Behav* 57:1045–53.

Swaisgood, R. R., D. G. Lindburg, X. Zhou, and M. A. Owen. 2000. The effects of sex, reproductive condition and context on discrimination of conspecific odours by giant pandas. *Anim Behav* 60:227–37.

Swaisgood, R. R., D. G. Lindburg, and H. Zhang. 2002. Discrimination of oestrous status in giant pandas via chemical cues in urine. *J Zool Lond* 257:381–86.

Swaisgood, R. R., X. Zhou, G. Zhang, D. G. Lindburg, and H. Zhang. 2003. Application of behavioral knowledge to giant panda conservation. *Int J Compar Psychol* 16:65–84.

Taylor, S. A., and D. A. Dewsbury. 1990. Male preferences for females of different reproductive conditions: A critical review. In *Chemical signals in vertebrates*, edited by D. W. Macdonald, D. Müller-Schwarze, and S. E. Natynczuk, pp. 184–98. Vol. 5. Oxford: Oxford University Press.

Vandenbergh, J. G. 1983. *Pheromones and reproduction in mammals*. New York: Academic.

White, A. M. 2001. Chemical communication in giant pandas: The role of marking posture and age of signaler. M.Sc. thesis. San Diego State University, San Diego.

White, A. M., R. R. Swaisgood, and H. Zhang. 2002. The highs and lows of chemical communication in giant pandas *(Ailuropoda melanoleuca)*: Effect of scent deposition height on signal discrimination. *Behav Ecol Sociobiol* 51:519–29.

White, A. M., R. R. Swaisgood, and H. Zhang. 2003. Chemical communication in giant pandas: The role of signaler and assessor age. *J Zool Lond* 259:171–78.

Wilson, E. O., and W. H. Bossert. 1963. Chemical communication among animals. *Recent Prog Hormone Res* 19:673–716.

Yalden, D. W. 1993. The problems of reintroducing carnivores. *Symp Zoological Soc Lond* 65:289–306.

Zhang, G. Q., R. R. Swaisgood, R. P. Wei, H. M. Zhang, H. Y. Han, D. S. Li, L. F. Wu, A. M. White, and D. G. Lindburg. 2000. A method for encouraging maternal care in the giant panda. *Zoo Biol* 19:53–63.

Zhang, G., R. R. Swaisgood, and H. Zhang. 2004. An evaluation of behavioral factors influencing reproductive success and failure in captive giant pandas. *Zoo Biol* 23:15–31.

Zhu, X., D. G. Lindburg, W. Pan, K. A. Forney, and D. Wang. 2001. The reproductive strategy of giant pandas: Infant growth and development and mother-infant relationships. *J Zool Lond* 253:141–55.

BRIEF REPORT 7.1

Chemical Composition of Giant Panda Scent and Its Use in Communication

Lee R. Hagey and Edith A. MacDonald

THE GIANT PANDA *(Ailuropoda melanoleuca)* lives in dense bamboo forests and communicates with conspecifics via scent marks. Under the tail is a specialized gland that is used to deposit chemical messages in the environment. The first step in discerning the messages transmitted by these marks is to identify their chemical components. Using a swab, colleagues in the United States and China collected samples of scent directly from the gland and from marks left in the environment by adults of both sexes. Samples were transferred to a solid-phase matrix extractor, taken to the laboratory, and subjected to gas chromatography–mass spectrometry analysis.

CHEMICAL PROFILES OF GIANT PANDA SCENT

We found that chromatograms of the male scent, constructed from either the gland or a deposited mark, were similar, and were dominated by a high proportion of short-chain fatty acids (the four largest peaks in the profile, figure BR.7.1A). Many of the higher molecular weight compounds seen in the male profile are alkane hydrocarbons, which form a sticky, nonvolatile matrix used to stabilize and retain the volatile fatty acid signals.

The chemical profile from the gland of females is quite similar to that of males. However, female scent marks show a more complex chemical signature than males, consisting of over one hundred volatile and semivolatile compounds (figure BR.7.1B). A high proportion of the chemicals found in female marks originate in urine and are not detected in the scent gland.

The chemical signature of a 70-day-old female panda was determined from scent collected from various parts of the body, and showed a distinctive pattern that looked neither like an adult male nor adult female. Although many of the individual compounds in the infant mark are identical to those of adults, the combination and concentration of the chemicals formed a unique and uniform chemical message, regardless of collection site.

The differential pattern of scent marking is consistent with an interpretation of different communicative objectives by the two sexes, suggested by our initial observations of behavior in the pair of pandas maintained at the San Diego Zoo. Males use low-volatile nonphysiological scent marks to establish and control the areas

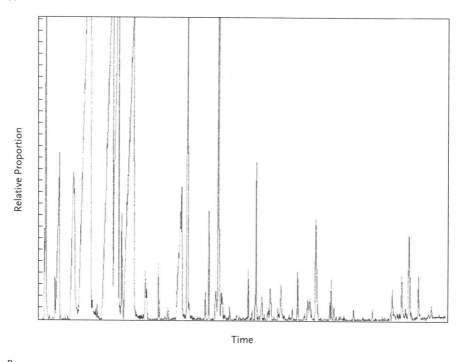

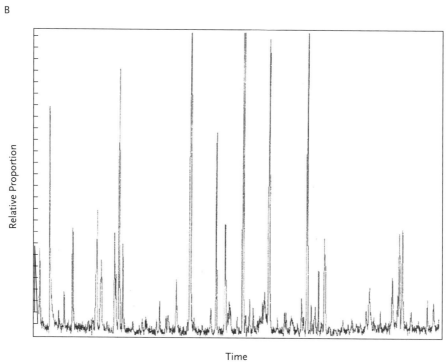

FIGURE BR.7.1. Representative chromatograms from male scent marks and glands (A) and female scent marks (B).

FIGURE BR.7.2. Schematic of scent distribution on the pelage of the giant panda.

in which they live, whereas females use their more volatile urine marks as short-term reports on their internal physiology and reproductive status. Young animals are covered with a unique full-body scent pattern that enhances recognition by their mothers.

MAPPING THE DISTRIBUTION OF SCENT ON THE PANDA'S PELAGE

In another set of investigations, we were interested in finding out if some part of the scent message is transferred to or rubbed off on other parts of the body surface, or if other kinds of scents are present in different body regions. To obtain samples, predetermined regions of the pelage from a male and a female giant panda were rubbed with swabs that were then stored in glass vials. These volatile organic compounds were subsequently transferred to a solid-phase matrix extractor and identified by gas chromatography–mass spectrometry.

Both sexes revealed low levels of volatile compounds on the forearms, legs, and back, where the dominant chemical signature was that of fatty lipids. The ears of the male were heavily scented with his urine. In contrast, the female's ears contained strong signals of plant aromatics. Giant pandas are often found sitting in an upright position and, as a result, the ears function as excellent transmission sites, as they are the high points on the body and allow scent to be picked up and carried by air currents. When presented with substances containing novel odors, females demonstrate anointing behaviors, featuring extensive head and ear rubbing.

The scent pattern of the male's eye patch is complex, unique, and not found anywhere else on his body. It is unusually free of urine signals. This finding is surprising, in as much as pandas are habitually observed rubbing their eyes. How is it possible for the male panda to transfer urine to his ears and not to his eyes during an eye-rub? The answer was found in an examination of the volatiles on the paws and forearms. The upper portion of the forepaw contained remnants of eye-patch scent, whereas the underside of the paw was contaminated with urine. Thus, the male avoids mixing the selective messages contained on its body through the actions of specific behaviors and movements.

The body of the giant panda is shown to be a kaleidoscope of scent patches and zones, each with a unique chemical makeup. In the schematic (figure BR.7.2), the different shades of gray represent a visualization of different odor profiles. There is the fresh odor of urine blowing off the ears, a dark splotch of scent deposited from the gland onto the tree, and the distinctive odor of the eye patches. Our data suggest that scent communication in the giant panda is more complex than the simple monitoring of scent marks deposited in the environment. Through the use of specific behaviors, movements, and applications, the panda uses its body as a kiosk, or message board, and can transmit selective short-range signals to nearby conspecifics.

8

Reproduction in Giant Pandas

HORMONES AND BEHAVIOR

*Rebecca J. Snyder, Dwight P. Lawson, Anju Zhang,
Zhihe Zhang, Lan Luo, Wei Zhong, Xianming Huang,
Nancy M. Czekala, Mollie A. Bloomsmith,
Debra L. Forthman, and Terry L. Maple*

ALTHOUGH ADULT GIANT PANDAS (*Ailuropoda melanoleuca*) are solitary and rarely come into contact with one another except to mate, they are part of a social community with behavioral complexities we are only now beginning to understand. To investigate giant panda reproductive cycles, it is necessary to first understand their social organization. Schaller et al. (1985) is still the predominant text on giant panda natural history, and much of our understanding of their behavior in the wild originates from this work. They discovered that giant pandas at the Wolong Nature Reserve occupy small (3.9–6.4 km²), relatively stable home ranges that may overlap among pandas of all ages and both sexes. Resident females and males have somewhat different land tenure systems. Females are generally more dispersed than males, and each female has a core area in which she concentrates most of her activity. This core area is not shared with other females or subadults. Males, however, roam widely, visiting the core areas of both females and subadults.

Social organization is greatly affected by the distribution of resources, particularly food, water, and shelter (Berger and Stevens 1996). Giant pandas feed primarily on bamboo, an abundant, evenly distributed food source with little seasonal variation in nutrition (Schaller et al. 1985, 1989; see also Long et al., chapter 6). Thus, it is not surprising that giant pandas have overlapping home ranges. However, that adult females maintain an exclusive core area suggests that some important resource may be present in these areas. Schaller et al. (1985) found that core areas tended to be well-wooded and of relatively level terrain, and presumed that females selected these areas for the following reasons: bamboo growing beneath a continuous canopy is more nutritious; large trees provide potential den sites; and travel over level terrain requires less energy. Conversely, the important resource influencing a male's range is the presence of females. Thus, males travel throughout their ranges, monitoring other pandas (Schaller et al. 1985).

Given that a male increases his chances of reproductive success by contacting as many females as possible, Schaller et al. (1985) noted that it is surprising that Wolong male ranges were not significantly larger than those of females. Pan and Lü (1993) found that males in the Qinling Mountains have larger home ranges (11.8 km^2) that encompass the smaller ranges (4.2 km^2) of three to five females. They also stated that fights between males at range boundaries occurred several times outside the mating season, suggesting that these ranges may actually be territories. Schaller et al. (1985), however, reported that male pandas in Wolong showed no exclusiveness, defense, or other indication of territoriality.

Giant pandas are reported to be seasonally monoestrous; however, a weak "autumn heat" has been reported for some individuals (Kleiman 1983; Schaller et al. 1985). According to Schaller et al. (1985), the entire estrous period lasts about 12–25 days, but peak receptivity is only about 2–7 days. Bonney et al. (1982), Hodges et al. (1984), Murata et al. (1986) Czekala et al. (1997), and Lindburg et al. (2001) have described the hormone profile of the giant panda estrous cycle. Over a 1–2-week period, females experience a gradual rise in estrogen concentrations to a peak, followed by a sudden drop in concentrations indicative of ovulation. A female panda's behavior also changes during estrus, and the expression of some behaviors correlate with changes in the hormone profile (Lindburg et al. 2001).

Throughout its adult life, the social milieu of the giant panda culminates in brief yearly mating seasons. Except for indirect chemical communication with neighbors, animals that have spent the vast majority of the year in virtual isolation are thrust together for 1–2 weeks by suites of hormone-mediated behaviors and physiological changes. The fate of this dwindling species rests on the outcomes of these brief interactions. To gain a better understanding of the relationship between behavior, hormones, and social environment, we examined these variables in a group of captive giant pandas over a 3-year period.

METHODS

For detailed descriptions of housing and husbandry conditions for this study, see Snyder (2000). Behavioral data were collected on nine adult females and three adult males over a 3-year period. All subjects were housed at either the Chengdu Zoo or the Chengdu Research Base of Giant Panda Breeding in Sichuan, China. Both sexes were observed from February–May in 1997, 1998, and 1999, but males were also observed outside the mating season (i.e., September–November in 1997 and June–September in 1998). Data were collected on a different subset of females each year, because some females had dependent cubs and did not cycle every year. Data were collected on the males in all three years. One male was exposed to females during the 1997–1999 mating seasons and given the opportunity to mate. This male was also housed with adult females at other times of the year. The other males were isolated from females in the spring of these same three years. They were usually housed adjacent to females at other times of year, and one of them was housed with an adult female during October–December in 1997.

Thirty-minute focal observation sessions were used to record behavioral information. Each subject was typically observed for two consecutive sessions every other day. For the females, all-occurrence sampling (Altmann 1974) was used to record scent marking, urination, dribbling urine, olfactory investigation, water play, rolling, body rubbing, tail up, lordosis, anogenital presenting, backwards walking, masturbating, somersaulting, observing males, affiliation with males, aggressiveness toward males, and aggressive vocalization toward males. For the males, all-occurrence sampling was used to record, scent marking, urine marking, urination, olfactory investigation, flehmen expression, substrate licking, body rubbing, affiliation with females, and aggressiveness toward females. Instantaneous sampling (Altmann 1974) at 1-min intervals was used to record feeding, locomotion, and resting for both sexes. One-zero sampling (Altmann

1974) during 1-min intervals was used to record bleating and chirping for both sexes and to record spontaneous tail-up occurrences in females. An average of 21.5 h of behavioral data was collected every year for each female included in that year's observations. An average of 22 h of data was collected in all 3 years during the mating season for each male and an average of 12 h of data was collected outside of the mating season for each male in 1997 and 1998.

Hormonal measures were obtained from urine samples collected during the period surrounding presumed ovulation for females. Urine samples were also collected from the males throughout most of 1997. Female samples were assayed for estrone conjugates and male samples for testosterone, using enzyme immunoassays (see Czekala et al. [1986] for procedures). Urine samples were difficult to collect frequently, because the animals were housed socially and were often on a grass substrate (outdoors). More frequent sampling would require separation and confinement indoors for long periods of time. The effects of long confinement on an individual's behavior made this procedure undesirable.

Because peak estrus lasts for only a few days, behavioral and hormonal data should be collected daily to establish a completely accurate estrous profile. Given the number of animals included in this study and that animals tended to be active for only a few hours a day, it was not possible to collect behavioral data daily. Some behaviors (e.g., body rubbing, water play, rolling, backwards walking, and masturbating) were rarely exhibited, and 1-h observation sessions were insufficient to provide a complete representation of all aspects of a female's estrous behavior.

RESULTS AND DISCUSSION

FEMALE HORMONES AND BEHAVIOR

Captive studies of giant panda reproductive behavior have been conducted on male-female pairs at the National Zoological Park, Washington, D.C. (Kleiman et al. 1979; Bonney et al. 1982; Kleiman 1983, 1985) and at the Zoological Society of San Diego (Czekala et al. 1997; Lindburg et al. 2001). Kleiman (1983) reported that behavioral changes began about 1–2 weeks before peak receptivity. For the female, these changes included decreased food consumption, increased scent marking, increased bleating, increased restlessness with frequent position changes, rolling, writhing, and water bathing. As peak receptivity neared, the female in her study exhibited backwards walking, anogenital rubbing or masturbation, increased chirping, and increased interactions at sites where the pair had visual and olfactory contact (Kleiman et al. 1979; Kleiman 1983). During peak receptivity, this female would present the anogenital region to the male and exhibit a tail-up lordosis. Other signs of estrus were swelling and reddening of both the vulva and nipples. Similar behavioral changes during estrus have been noted in other captive pandas (Hodges et al. 1984; Murata et al. 1986).

In our study, the behavior of seven females changed significantly over the course of the estrous cycle. Bleating, chirping, the tail-up posture, and looking at males occurred at the highest levels during the week surrounding presumed ovulation. Resting was lowest during this same period. Aggressive vocalizations toward a male (i.e., moaning and barking) were highest in the week prior to the estrogen peak, and feeding was highest during the week after the peak. Hormone concentrations also changed significantly during the estrous cycle, and were highest during the week of presumed ovulation. These findings fit the expected profile. That is, aggressive vocalizations were highest in the week prior to peak receptivity, when a male or males would be drawn closer to the female, but the female would not yet be receptive to males' advances. During the week surrounding the estrogen peak, behaviors that signal sexual interest and receptivity occurred at the highest levels. Feeding would be expected to return to pre-estrus levels at the end of the female's cycle.

All the females for whom hormone data were available displayed normal estrogen profiles consistent with ovulation. We found that changes in estrone conjugates correlated significantly with changes in behavioral values for nine of the behaviors measured. Affiliative behaviors (chirping and looking at males) and those associated with mate attraction (scent marking and urinating) were significantly positively correlated with estrone conjugates. Locomoting, resting, and aggressive vocalization toward a male were similarly correlated. Feeding and olfactory investigation were significantly negatively correlated with estrone conjugates.

Females varied substantially in the expression of behavior, and not all females exhibited significant correlations between hormones and behaviors. This indicates that hormones underlie many estrous behaviors, but are not sufficient to elicit a behavior, a phenomenon that has plagued captive breeding efforts. The social environment, including that experienced during the developmental period, as well as that experienced during a particular mating season or seasons, likely mediates an individual female's behavioral repertoire.

Five of nine females in our study displayed changes in behavior that significantly correlated with changes in hormone concentrations. No hormone data were available for one of the four remaining females. Another exhibited associations that approached significance. The two females failing to show any detectable associations were the only wild-caught females in the study—one captured at approximately 1 year of age and the other at about 2.5 years of age. Neither of these females had ever mated in captivity, despite being paired with experienced males. However, both had normal estrogen profiles, indicative of ovulation and had produced offspring through artificial insemination. Despite the absence of significant correlations, both did exhibit behavioral changes around the time of ovulation, but the changes were inappropriately expressed. For example, one female displayed tail-up and lordosis, but only toward other females, and was always aggressive toward the one male to whom she was introduced.

Very few matings have been observed in the wild, but it seems that a wild female is able to exercise a preference in mates. Schaller et al. (1985) observed aggregations of two to five males competing for access to an estrous female. The dominant male in these interactions is believed to be preferred as a mate by females (Schaller et al. 1985; Pan and Lü 1993). However, after the dominant male has mated, subordinate males also sometimes mate with the same female (Schaller et al. 1985), and this may be an important learning opportunity for young males. It is also possible that males, familiar with a female's reproductive condition from monitoring during the preceding days, can detect the time of follicle maturation; this would give dominant males an advantage (Schaller et al. 1985). The relative importance of females' preferences and dominant males' assertiveness remains to be determined.

Inability to exercise mate choice may explain why some females that displayed normal hormone profiles during estrus did not exhibit receptive behaviors. Another possibility is that exposure to agonistic behavior between males is important for a female's stimulation, and that receptive behavior is not elicited in some females without this stimulation. Females in this study, as in many captive setttings, were introduced to a single male for mating. The females that did not display receptive behaviors may have been incompatible with this male, and might have shown receptive behavior toward a different male or toward a male that had engaged in agonistic interactions.

MATE ATTRACTION AND LOCATION

In the wild, male giant pandas seem to locate estrous females by scent and vocalizations (see Swaisgood et al., chapter 7). That a male may remain near an estrous female for a few days prior to contacting her physically is probably important for synchronizing reproductive states and behaviors. "Such synchronization through scenting, calling, and other means, leading to compatibility, appears essential to successful mating; in any event the failure of many captive pairs to copulate can be traced to a seeming indifference or aggressiveness on the part of the male, even when the female is in heat and solicits his attention" (Schaller et al. 1985: 187).

Males are undoubtedly aware of the reproductive status of females in their area, based on the information left in scent marks by females. Females rarely mark except during estrus (Schaller et al. 1985; Pan and Lü 1993). Males, however, mark more frequently than do females and do so outside the mating season. This enables females in the area to become familiar with particular males, which seems to be important for mating. It also probably serves to prevent encounters between males who are intolerant of one another (Schaller et al. 1985), and supports Pan and Lü's (1993) observation of territoriality.

Schaller et al. (1985) reported that 87.5% of the total vocalizations recorded throughout the year occurred during the mating season (March–May). They described an instance in which a male called loudly (the type of vocalization is not indicated) from a tree near an estrous female with whom he mated the next day. They opined that such calls advertise a male's presence and readiness for mating.

Bleats and chirps are reported to be associated primarily with estrus (Kleiman et al. 1979; Peters 1982, 1985; Kleiman 1983). They are believed to promote close range, friendly contact (Peters 1985). Neither vocalization seems loud enough to serve as long-distance advertisement calls (Peters 1982). Therefore, they are not expected to occur as far in advance of peak estrus as longer-lasting scent marks. We found that bleats and chirps changed significantly during the course of the estrous cycle, occurring at the highest levels during the week surrounding presumed ovulation (see also Lindburg et al. 2001). Chemical communication is of primary importance throughout most of the year, allowing solitary pandas to monitor the activities and reproductive conditions of other individuals. Vocalizations seem to be used mainly when pandas are interacting at close range, providing information about changing motivational states (Schaller et al. 1985). Our findings support the conclusion that vocalizations are essential for mediating encounters between males and females.

MALE HORMONES AND BEHAVIOR

In the male giant panda, activity levels are higher and scent marking more frequent in spring than in fall, and behavioral changes during the mating season have been observed to precede those of the female, suggesting that males may experience an annual spring rut (Kleiman et al. 1979). Captive males have also been reported to undergo short-term behavioral changes associated with estrus, including decreased feeding, increased urination, increased bleating and chirping, increased scent marking, and decreased resting (Kleiman et al. 1979; Kleiman 1983). During peak receptivity, the male in Kleiman's study was observed to initiate contact more frequently and to mount repeatedly with thrusting (Kleiman et al. 1979; Kleiman 1983, 1985). Bonney et al. (1982) reported that the onset of sexual activity in this same male coincided with elevated androgen levels. Furthermore, a brief but dramatic rise in androgen excretion was reported as occurring the day before mounting and ejaculation occurred. The possibility that male pandas experience a spring rut is consistent with evidence of spermatogenic seasonality (Platz et al. 1983; Moore et al. 1984; Masui et al. 1985).

We found three behaviors—scent marking, bleating, and locomoting—to be significantly positively correlated with changes in androgen concentrations in one of three males during the mating season. This occurred in the only male that had contact with estrous females. We also found that affiliative behavior toward a female by this male was significantly negatively correlated with androgen concentrations. He exhibited, furthermore, the highest levels of most behaviors (i.e., scent marking, urine marking, olfactory investigation, licking, body rubbing, bleating, locomoting, and resting), but did not urinate or feed as frequently as other males in this study.

Because of overlapping ranges, direct and indirect chemical communication is ongoing and presumably enables males and females to synchronize receptivity during the mating season. Captive studies (Kleiman et al. 1979; Bonney et al. 1982; Kleiman 1983, 1985) suggest that exposure to estrous females may be important in influencing male hormone levels and the expression of courtship behaviors. Similarly, exposure to males may be important in influencing female hormone levels and the expression of courtship behavior.

In our study, the male that had contact with estrous females displayed a peak in androgen concentrations at the end of March, shortly before two females underwent estrus. A male isolated from females at this time had his highest androgen concentration in May, despite being housed with an adult female from October through December (no samples were collected from him before May). The other isolated male in this study peaked in androgen concentrations in October. Hormone data were available for him in April and May. A female was housed in an enclosure adjacent to this male from October through December, and her proximity may account for the peak androgen concentration in October. These observations from single males are not generalizable, but raise questions needing further study.

The males in our study were expected to display higher levels of several behaviors (i.e., scent marking, urine marking, urination, olfactory investigation, flehmen expression, licking, body rubbing, bleating, chirping, and locomoting) during the mating season than outside the mating season. Although we found no statistically significant differences between seasons for a given male, each differed in the temporal occurrence of behaviors and these differences were particularly evident in 1997. The two males isolated from females in the spring and exposed to females from October through December exhibited higher levels of most behaviors outside the mating season, and some occurred only at that time (e.g., substrate licking and aggressive interactions toward females). Conversely, the male that had access to estrous females displayed higher levels of all behaviors except flehmen and resting during the mating season. Differences in the amount of contact the males had with females most likely account for these differences in behavior.

These results suggest that contact with females encourages males to perform behaviors associated with sexual interest. Given that lack of sexual interest is one of the primary problems reported for captive males (Lindburg et al. 1997; Zheng et al. 1997), allowing males greater exposure to females even outside the mating season could help to alleviate this problem. Swaisgood et al. (2000) suggested that reproductive failure of captive giant pandas may be attributed to insufficient opportunities for chemical communication, and recommended that social odors be used to increase sexual interest and decrease aggressiveness prior to mating. When a pair was given access for mating during this study, the male usually spent the first few minutes smelling the female's enclosure before approaching her. He often walked past a bleating female in lordosis to first smell the enclosure, perhaps an indication that olfactory cues about a female's estrus are initially more important than visual or auditory ones. Frequent enclosure exchanges between males and females both in and outside the mating season might therefore increase familiarity and promote reproductive synchronization.

CONCLUSIONS

From ongoing studies of giant panda reproductive endocrinology and behavior, we are beginning to generate a model of those events that will refine our understanding of mating encounters in the wild. We have found that fluctuations in estrogen concentrations correlated significantly with some behavioral measures (i.e., chirping, observing the male, scent marking, urinating, aggressive vocalization toward the male, olfactory investigation, locomotion, resting, and feeding) in captive females. Androgen concentrations correlated significantly with scent marking, bleating, affiliative interactions with a female, and locomotion in at least one captive male. However, there was considerable individual variation in behavioral/hormonal correlations. Despite the importance of hormones in triggering specific estrous behaviors, it is becoming apparent that social setting and individual experience, including early development (Snyder et al., 2003), combine to culminate in species-appropriate behaviors that lead to successful mating.

Males and females both rely on behavioral cues and feedback from potential mates for reproduction to take place. For instance, contact with females encouraged males to perform actions associated with sexual interest, and exposure to females appeared to have a greater effect on male behavior than did time of year. Because male giant pandas in the wild would experience olfactory, auditory, and some visual contact with estrous females throughout the mating season, providing these experiences to captive males may increase sexual stimulation and counter the lack of sexual interest reported for many males. Allowing communication between males and females during estrus and throughout the year will increase familiarity, perhaps thereby decreasing male aggression toward potential mates. Similarly, observations in the wild indicate that estrous females are exposed to multiple, competing males (Schaller et al. 1985). This, and anecdotal evidence from captivity, suggests that some females may require exposure to competing males or the opportunity to select a mate before they can express receptivity. Combining these initial findings for males and females leads to a developing picture of the importance of seasonal sociality in the giant panda.

The captive panda population enables us to examine behavioral relationships and physiological processes that are extremely difficult to study in the wild. However, such assisted reproduction techniques as artificial insemination, although bolstering numbers, are enabling behavioral deficiencies to accrue in the captive population. As long as captive pandas serve as insurance for the endangered wild population and reintroduction remains a potential conservation tool, it is essential to maintain a complete repertoire of behaviors and to produce captive animals with the behavioral competence to reproduce. To accomplish this, wild and captive populations each need to serve as a laboratory for a better understanding of the other, and for an increased appreciation of the behavioral and social nuances of this supposedly solitary creature.

ACKNOWLEDGMENTS

We thank China's Ministry of Construction for its support and cooperation. We greatly appreciate the assistance that was provided by the directors, researchers, and animal care staff from the Chengdu Research Base of Giant Panda Breeding and the Chengdu Zoo. This research was funded by Zoo Atlanta, the Chengdu Research Base of Giant Panda Breeding, the Chengdu Zoo, and a grant (D00ZO-42) from the Morris Animal Foundation.

REFERENCES

Altmann, J. 1974. Observational study of behavior: Sampling methods. *Behaviour* 49:227–65.

Berger, J., and E. Stevens. 1996. Mammalian social organization and mating systems. In *Wild mammals in captivity*, edited by D. Kleiman, M. Allen, K. Thompson, and S. Lumpkin, pp. 344–51. Chicago: University of Chicago Press.

Bonney, R. C., D. J. Wood, and D. G. Kleiman. 1982. Endocrine correlates of behavioral oestrus in the female giant panda *(Ailuropoda melanoleuca)* and associated hormonal changes in the male. *J Reprod Fertil* 64:209–15.

Czekala, N., S. Gallusser, J. E. Meier, and B. L. Lasley. 1986. The development and application of an enzyme immunoassay for urinary estrone conjugates. *Zoo Biol* 5:1–6.

Czekala, N., D. Lindburg, B. Durrant, R. Swaisgood, T. He, and C. Tang. 1997. The estrogen profile, vaginal cytology, and behavior of a giant panda female during estrus. In *Proceedings of the international symposium on the protection of the giant panda (Ailuropoda melanoleuca)*, edited by A. Zhang and G. He, pp. 111–13. Chengdu: Sichuan Publishing House of Science and Technology.

Hodges, J. K., D. J. Bevan, M. Celma, J. P. Hearn, D. M. Jones, D. G. Kleiman, J. A. Knight, and H. Moore. 1984. Aspects of the reproductive endocrinology of the giant panda *(Ailuropoda melanoleuca)* in captivity with special reference to the detection of ovulation and pregnancy. *J Zool Lond* 203:253–67.

Kleiman, D. G. 1983. Ethology and reproduction of captive giant pandas *(Ailuropoda melanoleuca)*. *Z Tierpsychol* 61:1–46.

Kleiman, D. G. 1985. Social and reproductive behavior of the giant panda *(Ailuropoda melanoleuca)*. *Bongo (Berlin)* 10:45–58.

Kleiman, D. G., W. B. Karesh, and P. R. Chu. 1979. Behavioural changes associated with oestrus in the giant panda *(Ailuropoda melanoleuca)* with comments on female proceptive behavior. *Int Zoo Yrbk* 19:217–23.

Lindburg, D. G., X. Huang, and S. Huang. 1997. Reproductive performance of male giant panda in Chinese zoos. In *Proceedings of the international symposium on the protection of the giant panda (Ailuropoda melanoleuca)*, edited by A. Zhang and G. He, pp. 67–71. Chengdu: Sichuan Publishing House of Science and Technology.

Lindburg, D. G., N. M. Czekala, and R. R. Swaisgood. 2001. Hormonal and behavioral relationships during estrus in the giant panda. *Zoo Biol* 20: 537–43.

Masui, M., H. Hiramatsu, K. Saito, N. Nose, R. Nakazato, Y. Sagawa, N. Kasai, H. Tajima, K. Tanabe, and I. Kawasaki. 1985. Seasonal fluctuation of sperm count in the urine of giant panda. *Bongo* 10:43–44.

Moore, H. D. M., M. Bush, M. Celma, A. Garcia, T. Hartman, J. Hearn, J. Hodges, D. Jones, J. Knight, L. Monsalve, and D. Wildt. 1984. Artificial insemination in the giant panda *(Ailuropoda melanoleuca)*. *J Zool Lond* 203:269–78.

Murata, K., M. Tanioka, and N. Murakami. 1986. The relationship between the pattern of urinary oestrogen and behavioural changes in the giant panda. *Int Zoo Yrbk* 24/25:274–79.

Pan, W., and Z. Lü. 1993. The giant panda. In *Bears*, edited by I. Stirling, pp. 140–49. London: Harper Collins.

Peters, G. 1982. A note on the vocal behavior of the giant panda Ailuropoda melanoleuca (David, 1869). *Z Saugetierk* 47:236–46.

Peters, G. 1985. A comparative survey of vocalization in the giant panda (*Ailuropoda melanoleuca*, David, 1869). *Bongo* 10:197–208.

Platz, C. C., D. Wildt, J. Howard, and M. Bush. 1983. Electroejaculation and semen analysis and freezing in the giant panda *(Ailuropoda melanoleuca)*. *J Reprod Fertil* 67:9–12.

Schaller, G. B., J. Hu, W. Pan, and J. Zhu. 1985. *The giant pandas of Wolong*. Chicago: University of Chicago Press.

Schaller, G. B., Q. Teng, K. Johnson, X. Wang, H. Shen, and J. Hu. 1989. The feeding ecology of giant panda and Asiatic black bears in the Tangjiahe Reserve, China. In *Carnivore behavior, ecology and evolution*, edited by J. L. Gittleman, pp. 212–41. Ithaca: Cornell University Press.

Snyder, R. J. 2000. A behavioral and hormonal study of giant panda *(Ailuropoda melanoleuca)* reproduction. Ph.D. dissertation. Georgia Institute of Technology, Atlanta.

Snyder, R. J., A. J. Zhang, Z. H. Zhang, G. H. Li, Y. Z. Tan, X. M. Huang, L. Luo, M. A. Bloomsmith, D. L. Forthman, and T. L. Maple. 2003. Behavioral and developmental consequences of early rearing experience for captive giant pandas. *J Comp Psychol* 117:235–45.

Swaisgood, R. R., D. G. Lindburg, X. Zhou, and M. A. Owen. 2000. The effects of sex, reproductive condition and context on discrimination of conspecific odours by giant pandas. *Anim Behav* 60:227–37.

Zheng, S., Q. Zhao, Z. Xie, D. E. Wildt, and U. S. Seal. 1997. Report of giant panda captive management planning workshop, Chengdu, China, 10–13 December 1996. Apple Valley, Minn.: IUCN and SSC Conservation Breeding Specialist Group.

PART THREE

Pandas and Their Habitat

What does the giant panda *(Ailuropoda melanoleuca)* need, above all else, to survive? The answer, plain and simple, is habitat—habitat of sufficient extent, quality, and connectivity to support a genetically stable population. We cannot avoid asking—or being asked—"How many pandas is enough?" But the more relevant question is, "How much habitat is enough?" If subpopulations are too small and too segregated to remain viable, the species' survival odds are not improved by increasing its numbers. It is therefore more important to maximize the amount of suitable panda habitat by bringing as much of it under legal protection as possible, and by reconnecting isolated fragments through forest restoration.

So how are we doing? That depends on whether you are looking ruefully at the past or gazing optimistically into the future. What we have today is a precarious situation indeed: less than two thousand pandas scattered sparsely across five far-flung mountain ranges in southwest China, squeezed from all sides by human activities, their habitat pulverized by decades of intensive logging and agriculture, isolated from one another by roads, farms, railways, and cities. As several chapters in this part point out, habitat fragmentation, degradation, and outright destruction occurred at a blinding rate throughout much of the past four decades. Sichuan Province lost half of its panda habitat between 1975 and 1989, and habitat loss was comparably rampant in Shaanxi and Gansu Provinces. Although twenty-one new reserves were established in the past decade, bringing the total number to around thirty-four, about half of giant panda habitat remains unprotected.

To say that this all came to a screeching halt with the declaration in 1998 of a 10-year logging ban covering the giant panda's entire range is an exaggeration, but not by much. Implementation of the ban was swift and strict. To find proof of its totality, we need look no farther than Indonesia, Myanmar, and elsewhere in Southeast Asia, where forests are now being pillaged to satiate China's voracious demand for wood products, now that its domestic source has disappeared. Although this is wreaking ecological havoc in the affected countries, the silver lining is the desperately needed respite that the logging ban bestows on giant pandas and China's other forest-dwelling wildlife.

Thus we dare to hope for the future. The confluence of the logging ban, China's recently launched "Grain to Green" program for reforesting steeply sloping crop land, and a new plan for greatly expanding China's nature reserve system have ushered in an unprecedented period of conservation opportunities. A visionary strategy for setting aside new panda reserves, restoring panda habitat, and reconnecting forest fragments (and thereby isolated panda populations) must be woven into these new environmental policies as they roll out and take shape in the coming years. Meanwhile, China's push to modernize its lagging west through infrastructure development—along with industrial expansion and increased tourism under the Western China Development Plan—presents a potential threat of equally momentous weight.

How to ensure that its environmental policies maximize benefits to giant panda conservation—and its economic development policies minimize degradation of sensitive panda habitats, biodiversity-rich forests, and key watersheds—is one of the most complex challenges conservationists and the Chinese government face today. The results of China's Third National Survey of the Giant Panda and its Habitat, completed in 2002, will provide a blueprint to guide this process. The survey pinpoints more precisely than ever the most critical habitat areas for giant pandas, locations where this habitat is still not protected, and sites where restoring forest and reconnecting habitat fragments will have the greatest conservational impact. Equipped with this information, China can make prudent decisions about where to develop and where to protect. But the survey is surely not the only tool at China's disposal. The chapters included in this part describe many additional approaches and methods for understanding the giant panda's habitat use and needs and determining the most effective ways to protect and restore their forest home.

In chapter 9, Hu and Wei examine the ecological similarities and differences in giant pandas among the five mountain ranges they inhabit (Qinling, Minshan, Qionglai, Xiangling, and Liangshan) in terms of habitat requirements, diet, reproduction, population and genetic status, seasonal movements, and predation threats. They identify areas in each region that are known to have high concentrations of pandas and describe the highly fragmented nature of much of the habitat, noting the looming threat of inbreeding and warning that the Xiangling and Liangshan populations are most at risk. They assert that protecting existing habitat and relinking isolated panda populations through forest restoration are the paramount priorities for protecting not only giant pandas but also other endangered species, such as the takin *(Budorcas taxicolor)* and golden monkey *(Rhinopithecus roxellana)*.

To protect and restore habitat, it is essential to map existing and potential habitat as accurately as possible. In a report from one of the *Panda 2000* panels (panel report 9.1), Loucks and Wang present an overview of past efforts (including the Third National Survey) and recommend actions to identify habitat and ascertain panda distribution. These include expanding the role of remote sensing, GIS, and artificial intelligence techniques in identifying potential remaining habitat and potential corridors linking habitat blocks, assessing threats to habitat integrity, delineating bamboo distribution, and guiding government decisionmaking. They also advocate developing a central clearinghouse of panda-related information for public use and incorporating human demography and socioeconomic data into habitat analysis. Loucks and Wang suggest that, although the spatial analyses they describe are a highly cost-effective planning tool, decisionmakers need more thorough natural history data on giant pandas to fine-tune conservation approaches. To this end, they contend that radio-/GPS telemetry is the most accurate and efficient way to determine movement patterns, feeding and habitat preferences, and reproductive success of wild pandas.

Another powerful method yielding a wealth of natural history data that can have direct implications for evaluating habitat needs is DNA analysis. In brief report 9.1, Zhang, Zhang, Zhang, Ryder, and Zhang demonstrate the feasibility of routine mtDNA amplification by using DNA extracted from fecal samples. This technique is attractive, in part, because it obviates the need for blood and tissue samples and the analysis can be carried out in the host country.

As described in chapter 10 by Yong, Liu, Wang, Skidmore, and Prins, radiotelemetry was used in Foping Nature Reserve to confirm the existence of the pandas' seasonal ranges and the characteristics of the habitat in each, as well as the temporal patterns of movements between them. These authors address the relationships between movement patterns and certain habitat parameters, such as vegetation type, elevation, slope, and aspect. Their data reveal that pandas in Foping ascend from winter to summer ranges quickly, in as little as 2 or 3 days, as there is very little bamboo available in the transition zones. Higher-elevation summer ranges were smaller than winter ranges—probably due to the high density of bamboo in the summer range—and pandas in Foping appeared to prefer gentle slopes over steep ones.

In the continuing search for noninvasive research techniques that yield key information on habitat use, as well as on behavior and population dynamics, Durnin, Huang, and Zhang (brief report 10.1) tested whether remotely activated camera systems and barbed-wire hair snares, widely used in studying other bear species, hold promise for panda research. Cameras were deployed at communal scent marking stations, and hair snares were baited with meat to attract pandas. The authors conclude that remote cameras and hair snares are of limited value, due to the low rate of return per sampling effort, but conceded that their utility might increase with some refinements to the experimental design. Furthermore, they postulate that hair snares may be effective in censusing and monitoring pandas at larger spatial scales than used in their study.

In chapter 11, Liu, Bronsveld, Skidmore, Wang, Dang, and Yong echo Loucks and Wang's (panel report 9.1) positive appraisal of the value of mapping for habitat assessment and monitoring. They introduce a novel mapping approach that integrates expert system and neural learning techniques, remote sensing, ground survey, and radio tracking data together within a GIS environment. This integrated method produced habitat classifications for Foping Nature Reserve with 83% accuracy. It showed that 52% of the reserve serves as suitable winter habitat, 16% as suitable summer habitat, and 20% as transitional habitat for pandas to move between winter and summer habitats. The authors suggest that the transitional and summer habitats need to be given more attention for conservation, and that human impact in and around Foping should be reduced.

Habitat restoration is also a relatively new and experimental endeavor, still more art than science. Zhu and Ouyang summarize a study on the subject in panel report 11.1 and offer some promising approaches. Their project aimed to reveal the main causes of forest degradation, develop and test a set of criteria for assessing forest status in terms of both degradation and potential for restoration, and determine the habitat needed by pandas in relation to existing and potentially restorable areas. Based on the results of this study, as well as a discussion by a panel of experts, the authors recommend restricting human access to mountainous regions to allow natural forests to recover. They would also increase protection of the main bamboo feeding areas, establish corridors that facilitate giant panda migration and dispersal, conduct large-scale planting of appropriate tree and bamboo species, and enhance protection of forested areas lying outside the giant panda reserves.

Several chapters urge us not to contemplate giant pandas in a vacuum; human activities in panda habitat must also be taken into consideration. Aside from the giant panda's interaction with people, it is also illuminating to observe its interaction with other species, as Wei, Li, Feng, Wang, and Hu do in chapter 12, with their study of coexistence between giant pandas and red pandas *(Ailurus fulgens)*. Both species specialize on bamboo and consume the same bamboo species in regions where they are sympatric. However, as fundamental ecological principles dictate, food resource overlap between the two species is very limited, due to significantly different patterns of resource and habitat utilization. For instance, although both species eat shoots and leaves, leaves make up 89.9% of the annual diet of red pandas, but only 34.7% of that of giant pandas. As for bamboo shoots, red pandas prefer short ones, which giant pandas mostly avoid in favor of taller ones. In terms of habitat use, giant pandas occur at sites on gentle slopes with a lower density of fallen logs, shrubs, and bamboo culms, whereas red pandas occur at sites on steeper slopes with a higher density of fallen logs, shrubs, and bamboo culms.

In chapter 13, Wei, Yang, Hu, and Stringham address the availability of bamboo shoots, a major food of giant pandas, in light of competition from humans, bamboo rats, and insects. The authors adapted a predation model with logistic submodels to explore the population dynamics of predators (panda) and prey (bamboo). This chapter will be rough going for all but the mathematically sophisticated, but its message is clear and sobering. Although the models used are simplified and need a great deal more work to verify their accuracy, the study points out the difficulty in setting biologically appropriate limits to human harvesting in areas where such key resources are shared. Crop management of this sort is difficult without specific information on how much bamboo is actually needed to sustain the pandas while simultaneously assuring perpetuation of the bamboo crop.

Simply calling a place a nature reserve does not magically transform it into a haven for pandas. Quite a bit more goes into ensuring that pandas and their habitat are shielded from threats. Yu and Deng outline (panel report 13.1) several important aspects of effective reserve management: reserve design, training of personnel, long-range planning, and implementation of effective protection programs. Essential features of reserve management are embodied in Sichuan Province's management plan, which emphasizes strict protection of natural resources; specifically, protection from timbering, fires, and poaching. The authors, in summarizing a discussion on reserve management that took place at *Panda 2000*, state that management planning, community-based conservation (involving local people in conservation efforts), and development of ecotourism as a source of additional funding are three underlying issues critical to achieving more effective conservation activities in panda reserves. They cite the Foping and Wanglang Nature Reserves as places that are making progress toward implementing effective reserve management.

9

Comparative Ecology of Giant Pandas in the Five Mountain Ranges of Their Distribution in China

Jinchu Hu and Fuwen Wei

GIANT PANDAS (*Ailuropoda melanoleuca*) are found today in the Qinling area of Shaanxi Province and in the high mountain ranges of Gansu and Sichuan Provinces, including the Qingzang Plateau (Hu 2001). The entire distribution of the species falls at roughly 102°00′–108°11′E longitude and 27°53′–35°35′N latitude. Fifty years ago, their more-or-less continuous distribution in this area resembled a large C in shape, extending from the Qinling Mountains to the Sichuan Basin. The development of roads and transportation, intensification of human activities, and introduction of large-scale logging have led to fragmentation of the habitat into five mountain regions: Qinling, Minshan, Qionglai, Xiangling, and Liangshan (figure 9.1) (Hu et al. 1985, 1990a; Schaller et al. 1985; Hu 2001). Giant pandas have gradually retreated to the mid-elevations of these ranges, resulting in ecological separation of the different populations.

To clarify the ecological differences and similarities of giant pandas among the different mountain ranges, beginning in 1978, we established field observation stations in four of the ranges (Minshan, Qionglai, Xiangling, Liang- shan) and have carried out long-term field studies in all five. The first station was established on the Wolong Nature Reserve (Qionglai Mountains, at Wuyipeng, a station that is still in use) by the first author in 1978. From 1980 to 1986, Wuyipeng became the site of a collaborative study with Dr. George Schaller (World Wildlife Fund) and other colleagues (Hu 1985, 1986a; Hu et al. 1985, 1990c; Schaller et al. 1985; Johnson et al. 1988a,b; Taylor and Qin 1988a,b,c, 1989; Reid et al. 1989, 1991; Reid and Hu 1991; Wei and Hu 1994; Wei et al. 1997).

A second station was established at Baixiongping, Tangjiahe Nature Reserve (Minshan Mountains) in 1984, with the same collaborators as at Wolong (Hu 1986b, 2001; Hu et al. 1990b,c; Schaller et al. 1989; Wei et al. 1990). A third station was established by our group in 1990 at Dafengding, Mabian-Dafengding Natural Reserve in the Liangshan Mountains, and a 3-year project had been carried out at this location (Yang et al. 1994, 1998; Wei et al. 1995, 1996a,b; Zhou et al. 1997). Our fourth station was established in 1994 at Shihuiyao, Yele Natural Reserve in the Xiangling Mountains, and 4 years of field work had been conducted at this site on

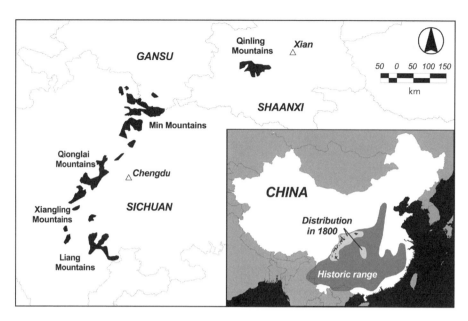

FIGURE 9.1. Approximate distribution of giant panda habitat remaining in China today. Map courtesy of the World Wildlife Fund—China.

giant pandas, and also on the sympatric red panda *(Ailurus fulgens)* (Wei et al. 1999a,b,c,d,e, 2000a,b; He et al. 2000; Wei et al., chapter 12). In addition to these long-term studies, colleagues in our research group (especially J. Hu) have visited nearly all giant panda reserves and counties outside reserves in the five ranges where giant pandas are found, and have collected supplemental ecological data.

In this chapter, we describe different aspects of giant panda ecology (geographic features, distribution, habitat, diet, demography, reproduction, genetic status, predation threats, and conservation issues) in the five areas, based primarily on studies by our group, but also including information obtained by other researchers, especially in the Qinling range (Pan et al 1988, 2001; Yong et al. 1993, 1994; Liu 2001). Except for the references specifically cited, results presented in this report are from our own observations.

MOUNTAIN RANGES OF PRESENT-DAY GIANT PANDA HABITATS

The main features of each of the five mountain ranges are presented in table 9.1. Most giant pandas occupy only the more favorable slopes in these ranges; that is, those with gentle gradients and warmer, moister conditions than are found on the opposite slopes.

The Qinling Mountain portion of the pandas' range constitutes the northernmost area of the present-day distribution of the species. These mountains run from east to west for a total distance of about 500 km. Most of the panda population is found on the southern slopes, in the upper reaches of the Hanshui River. The terrain in this area is relatively flat and covered with well-preserved evergreen broadleaf and upland conifer forests. Because the weather is mild and moist, the bamboo species *Bashania fargesii* and *Fargesia qinlingensis* thrive in this region. Giant pandas are distributed in eight counties (table 9.2), mainly confined to the southern slopes, such as those in Yangxian and Foping counties, but they are sometimes found on northern slopes such as those in Zhouzhi and Taibai counties. It is there, in the Qinling area, that one finds the density of giant pandas to be the highest of any of the five main distributions.

The Minshan range is located to the south of the Qinling Mountains. It extends from north to

TABLE 9.1
Features of the Five Mountainous Areas of Giant Panda Distribution

MOUNTAIN RANGE	MAXIMUM ELEVATION (m)	GRADIENT/ MOISTURE[1]	ASPECT	ELEVATIONS OCCUPIED BY PANDAS	ESTIMATED NUMBER OF PANDAS	HABITAT SIZE (km^2)
Qinling	3767	S-gentle/moist N-steep/dry	S	1200–3100	250	3000
Minshan	5588	E-gentle/moist W-steep/dry	E	2100–3600	600	9000
Qionglai	6250	SE, NW-gentle/moist SW-steep/dry	SE	2500–3600	400	6000
Xiangling Greater	3522	SE-gentle/moist NW-gentle/moist	SE, NW	2800–3800	25	800
Xiangling Lesser	4791	E-gentle/moist NW-steep/dry	E	2800–4100	25	800
Liangshan	>4000	E-gentle/moist NW-steep/dry	E	2000–3800	>100	2000

[1]Compass readings give direction of gradient.

south for a distance of about 500 km. Here, too, the opposite slopes of the range offer contrasting habitats, the west side being steep and dry, whereas the eastern slopes are more gentle and receive more moisture. In the northern part of the Minshan, giant pandas are found in the drainage of the Baishuijiang River, a part of the larger Jialingjiang watershed. Farther south, they are found in the upper Peijiang River drainage, and in the southernmost portion of the range, they occur in the relatively flat reaches of the Tuojiang River. The pandas are distributed over sixteen counties, but primarily in the counties of Songpan, Pingwu, Beichuan, Maoxian, Qingchuan, and Wenxian (table 9.2). In all of these areas, pandas are found mainly on the climatically more favorable eastern slopes. The population estimate of six hundred constitutes the largest of any of the five ranges.

The Qionglai range is located to the west of the Minshan, and extends for about 250 km from northeast to southwest. Toward the east, pandas are found on the flat slopes of the upper Minjiang River. Farther to the west and south, they select the flat reaches of the upper Dadu River and its tributaries. Upland conifer forest shades the bamboo in this region. In the Qionglai range, pandas occur in ten counties, but mainly in the counties of Wenchuan, Baoxing, and Tianquan (table 9.2).

Located to the south of the Qionglai Mountains, the Xiangling area is heavily fragmented and is divided into large and small sections. In the Lesser Xiangling, pandas utilize the upper reaches of the Yuexi and the Anning Rivers, and although found in four different counties, occur mainly in Mianning and Shimian (table 9.2). The conifer forests in this area shade the bamboo. In the Greater Xiangling, pandas are found in the upper Zhougong and Xingjing Rivers, which are tributaries of the Dadu River. They occupy two counties, but are mainly found in the county of Hongya (table 9.2). The habitat at this location is broadly similar to that of the Lesser Xiangling. Throughout the region, the valley floors are quite dry, and cultivation extends far up the sides of the mountains. The pandas mainly frequent stands of bamboo above the

TABLE 9.2
Distribution by County of Giant Pandas in Different Mountain Ranges

MOUNTAIN RANGE	COUNTY[1]
Qinling	Shaanxi Province: Foping*, Yangxian*, Zhouzhi*, Taibai*, Ningshan, Liuba, Ningqiang, Chenggu
Minshan	Sichuan Province: Songpan*, Pingwu*, Beichuan*, Maoxian*, Ruoergai, Qingchuan*, Anxian, Shifang, Pengzhou, Dujiangyan, Mianzhu, Jiuzhaigou
	Gansu Province: Wenxian*, Diebu, Wudu, Zhouqu
Qionglai	Sichuan Province: Wenchuan*, Baoxing*, Tianquan*, Lushan, Chongzhou, Dayi, Luding, Kangding, Qionglai, Dujiangyan
Xiangling: Greater	Sichuan Province: Hongya*, Yingjing
Xiangling: Lesser	Sichuan Province: Mianning*, Shimian*, Jiulong, Yuexi
Liangshan	Sichuan Province: Mabian*, Meigu*, Ebian*, Ganluo, Yuexi, Leibo, Jinhekou

[1] Asterisk denotes primary area of occupancy in the indicated range.

cultivated areas. The Xiangling range is estimated to hold only about fifty pandas, the lowest number of any of the regions described here.

The southernmost distribution of giant pandas in China today occurs in the Liangshan Mountains. This range is located to the south of the Xiangling and extends for about 500 km from north to south. Here, one finds mainly broadleaf evergreen forest, whereas upper shadow conifer forests dominate in this range. Pandas are found in seven counties, but mainly in Mabian, Meigu, and Ebian (table 9.2).

BAMBOO HABITATS

It is well established that giant pandas feed mainly on bamboo. From all five ranges combined, feeding on over sixty bamboo species has been tabulated, and of these, thirty-five make up the main food source for the animals (table 9.3). Of the seven species occurring in the Qinling Mountains, giant panda feed mainly on *Bashania fargesii, Fargesia qinlingensis,* and *F. dracocephalai,* but occasionally on *Phyllostachys sulpurea* and *Indocalamus latifolius* (Pan et al. 1988; Ren 1998; Long et al., chapter 6). In the Minshan region, pandas forage mainly on *F. rufa, F. denudata, F. nitida,* and *F. obliqua,* although thirteen bamboo species in all may be utilized. In the Qionglai Mountains, pandas are more likely to feed on *Bashania faberi, Yushannia brevipaniculata,* and *F. robusta,* but will also forage on another ten species. In the Greater Xiangling, they feed on eight species, of which *Y. brevipaniculata* and *Chimonobambusa szechuanesis* are primary. In the Lesser Xiangling, of eleven available species *Y. tineloata, Y. cava,* and *B. spanostachya* are preferred. In Liangshan, although pandas forage on seventeen species of bamboo, they show a preference for *Y. brevipaniculata, Y. ailuropodina, Qiongzhuea macrophylla,* and *Y. tineloata.*

Although 99% of the pandas' diet is bamboo, other plants and animals in the different mountain ranges are occasionally found in their droppings. In autumn, they may visit farmlands to feed on bamboo core stems (Wenchuan and Tianquan, Qionglai range; the Yuexi region of Lesser Xiangling; and the Qingchuan region of the Minshan range), pumpkin, and kidney beans (Mabian, Liangshan region). In the winter, pandas forage on wheat (Wenxian, Minshan range) or even domestic pig food (Mabian and Meigu, Liangshan range; Tianquan, Qionglai range). At times, they may sample the fruits of such plants as *Actinida* spp. (Meigu, in Liangshan), *Uex franchetiana*, *Corylus ferox* (Tianquan), *Rubus* spp. (Yuexi, Meigu; Pingwu, Qionglai range), or eat such plants as *Trachyspermum scaberulum*, *Heracleum candicans*, *Houttuynia cordata*, and *Angelica* spp. During a period of bamboo die-off in the Minshan range, giant pandas were found to eat such plants as *Equisetum hiemale*, *Ligusticum sinensis*, *Allium* spp., *Notopterygium forbesii*, and the bark of *Abies* spp. In Baoxing, Wenchuan, and Pingwu counties, pandas have been observed feeding on carrion (e.g., musk deer, tufted deer, wild boar, serow; see also the diet review in Long et al. [chapter 6]). In the 1960s, a panda was captured in Baoxing County as it fed on carrion.

FEEDING BEHAVIOR AND DIET

Giant pandas are known to spend over half of each day in foraging and food consumption (Schaller et al. 1985). Due to differences in types and features of bamboo that are found throughout their present distribution, it is to be expected that the populations of these regions will utilize variant strategies and display variant feeding behaviors within the broad context of their adaptation to a bamboo niche.

From November to the following May, pandas in the Qinling Mountains range among the stands of *Bashania* bamboo and feed primarily on its leaves. In May, they begin selecting *Bashania* shoots. In June, they shift upward to about 2300 m to stands of *F. qinlingensis*. During the summer months, feeding is mainly on the middle section of young stems and culms, but during September and October, new branches and young leaves are preferred (Pan et al. 1988, 2001; Yong et al. 1993, 1994, chapter 10).

In the Minshan range, for most of the year pandas prefer the subtropical upland conifer regions, but some move to lower elevations with the onset of winter (November) to feed on *F. scabrida*. Because most of the habitat in these lower reaches has been converted to agriculture, the bulk of the panda population stays in its summer range to feed on old shoots and young stems or the fairly scanty remains of young leaves. In the spring, the entire population converges on freshly sprouted bamboo shoots, preferring those of robust size. Later, the animals shift their attention to the middle part of the stems, but by September, the diet is composed mainly of new leaves and branches.

In the Qionglai range, the pandas forage between 2500 and 3300 m during the winter months, although a small number migrates downslope at this time to feed on *Y. chungii* and *F. robusta*. These same two bamboo species provide the main source of stems and new leaves for summer and fall feeding. While feeding on leaves, pandas in the Qionglai sometimes bend stems over to forage on them. This behavior has not been seen in other mountain ranges.

In the Xiangling range, giant pandas feed mainly on the leaves of *B. spanostachya* in the winter, bamboo shoots in the spring, and stems in the summer.

A diversity of bamboos occurs in the Liangshan Mountains. The leaves of *Qiongzhuea* spp. are the main winter food, whereas in the spring, the shoots of both *Qiongzhuea* and *Yushania* spp. are utilized. The pandas feed on the stem of these two bamboos in the summer and on shoots and the new leaves of *Chimonobambus* spp. in the fall.

As a carnivore specialized to feed almost exclusively on bamboo (although occasionally they intake animal matter and a few other plants), how does the giant panda cope with a food source that is high in cellulose and low in protein? Research conducted in the different mountain ranges indicates a food-selection strategy that maximizes nutrient intake. First, pandas select the most nutritious species of bamboo available to them. For example, of the three species occurring on the Mabian Dafengding Reserve, they prefer *Q. macrophylla*, which is higher in crude protein and lower in cellulose and lignin than the other two species. Similar patterns are seen in Wolong, where the pandas prefer the more nutritious *B. faberi*, and in the Qinling region,

TABLE 9.3
Bamboo Species Consumed by Giant Pandas in Different Mountain Ranges

	MOUNTAIN RANGE[1]							MONTHS DURING WHICH SHOOTS GERMINATE
				XIANGLING				
BAMBOO SPECIES	QINLING	MINSHAN	QIONGLAI	GREATER	LESSER	LIANGSHAN	ELEVATION	
Bashania faberi		++	+++	++	+	++	1400–3500	June–July
B. fargesii	++	+					1200–2500	April–May
B. spanostachya					+++		3200–3600	
Chimonobambusa pachstachys						+	900–2200	September–October
C. neopurpurea			++			+	1000–2050	September–October
C. szechuanesis			++	+++		+	1100–3000	September–October
Fargesia angustissima		++	+				1570–2700	April–May
F. denudata		+++	+				1600–3500	June–July
F. dracocephala	++	+					1200–2300	April–May
F. dulcicula					++		2700–3500	
F. ferax		+	+		++	+	1300–3500	May–June
F. nitida		+++	++				1600–3400	August–September
F. obliqua		+++					1900–3300	June–July
F. qinlingensis	++						1000–3100	June–July
F. robusta		+	+++				1300–3300	May–June
F. rufa		+++					1500–3200	June–July
F. scabrida		++					1600–2600	June–July
Indocalamus latifolius	+						1200–1800	April–May

Species				Elevation	Period
I. longiauritus		+		800–2600	May–June
Phyllostachys nidularia		+	+	900–2000	April–May
P. sulpurea	+			700–1500	April–May
Qiongzhuea macrophylla			+++	2100–2900	May–June
Q. opienensis		++	+	1300–2900	April–May
Q. rigidula			+	1300–2200	August–September
Q. tumidinoda			++	1400–3000	April–May
Yushannia ailuropodina			+++	1800–3100	
Y. brevipaniculata	++	+++	+++	1400–3500	
Y. cava				3200–3800	May–June
Y. chungii		+	+++	1200–2800	June–August
Y. dafengdingensis			++	1700–3300	
Y. glauca			++	1900–3100	
Y. mabianensis			++	1600–3000	June–August
Y. maculata		+	++	1700–3400	
Y. tineloata		++	+++	1800–3800	May–June
Y. violascens			++	2600–3200	
Total number of species	5	14	13	8	11	17

[+], Scarce food; ++, main food; +++, most preferred food.

where *B. fargesii,* which contains higher protein and a higher proportion of amino acids, is preferred (Pan et al. 1988). In the Lesser Xiangling region of the Yele Reserve, the more nutritious *B. spanostachya* is the preferred food.

Second, in all five regions, pandas show a preference for leaves, shown to have higher crude protein and lower cellulose and lignin content than do stems and branches. Johnson et al. (1988) found, additionally, that the newly formed and preferred leaves of *B. faberi* contained 17.5% crude protein, in contrast with withering leaves, with a content of only 9%. Similarly, Pan et al. (1988) found the protein content of new leaves of *B. fargesii* to be higher (15.2%) than that of old leaves (10.2%)

Third, pandas prefer young and tender shoots of different diameters in different mountain ranges. Bamboo shoots are not only nutritionally similar to leaves, but also are digested more easily. For instance, shoots of *B. spanostachya* contained 12.9% protein, similar to leaves (15.0%), in the Lesser Xiangling, and shoots of *Q. macrophylla* contained 13.9% protein, also similar to leaves (15.5%), in the Liangshan region. Preferred bamboos in the Qinling, Minshan, Qionglai, and Xiangling ranges have the same characteristics. In addition, pandas prefer robust shoots, containing more shoot cores, over slender ones. For instance, we found that in the Lesser Xiangling, they selected the taller and more robust shoots of *B. spanostachya* (height >30 cm, basal diameter >10 mm) rather than those that were shorter and of smaller diameter. In other mountain ranges, a similar preference was found (Qinling: shoots of *B. fargesii* >10 mm [Pan et al. 1988]; Minshan: shoots of *F. scabrida* >7–8 mm and of *B. fargesii* >9 mm; Qionglai: shoots of *F. robusta* >10 mm; and Liangshan: shoots of *Q. macrophylla* >18 mm and of *Y. glauca* >16 mm). Consumption of robust shoots may provide greater nutritional intake per unit of feeding time than consumption of slender ones.

Finally, giant pandas preferred young and robust bamboo culms, for example, culms of *B. spanostachya* with basal diameter between 8 and 14 mm in the Xiangling Mountains, and culms of *Q. macrophylla* larger than 10 mm in the Liangshan Mountains. In addition, they preferred old shoots or young stems that are not only relatively more tender and easier to digest, but also contain higher crude protein than old stems. For instance, old shoots of *B. spanostachya* contain more crude protein than the old culms in the Xiangling Mountains and young culms of *Q. macrophylla* and *F. glauca* contain more protein than do old stems in the Liangshan Mountains.

SEASONAL RANGING

Vertical migration in the Qinling region is readily apparent (Yong et al., chapter 10), covering an elevational shift of up to 2300 m; the animals even occasionally crossed the crest to the northern slopes. Seasonal effects on the size of ranges are evident, the largest extent being about 10 km^2 in the summer, then declining to about 6 km^2 in the autumn and to only 2–3 km^2 in the winter (Yong et al. 1994; Liu 2001; Pan et al. 2001).

Cultivation at lower levels limits migration in the Minshan range. Winter snows and cold temperatures may lead to a shortage of food resources in this season, resulting in increased horizontal ranging of the pandas, even to their extending into villages to feed on garbage.

In the Qionglai, summer migration upward to ranges between 3100 and 3500 m is typical.

Because of extensive cultivation in the Xiangling region, pandas stay at the higher elevations on a year-round basis, but do exploit the lowest reaches of their range in the winter, when snow cover is heavy. They also seek out sunny exposures (thus, free of snow) at higher elevations to search for food.

Pandas in the Lesser Liangshan feed on *Qiongzhuea* spp. at lower elevations in winter and early spring, but move to higher elevations in the summer to take advantage of *Yushania* spp. and *B. faberi*. Because of the prevalence of human communities, giant pandas in Greater Liangshan stay at elevations of 3200 m or higher on a year-round basis.

REPRODUCTION

Female sexual maturity, indicated by onset of breeding, occurs most often at 6–7 years of age, except in the Qinling, where an abundance of bamboo could be a factor in the start of reproduction a year earlier. Males normally reach sexual maturity at 7–8 years of age.

Availability of bamboo and the timing of appearance of critical plant parts seem to affect timing of the mating season in the different regions. For example, at the Wolong Reserve, the spring estrus was delayed for about a month after large-scale bamboo die-off. And in the Lesser Liangshan, bamboo is abundant in the lower parts and mating commences in March. However, it is delayed by about a month in the higher region of the Greater Liangshan, where the winter shortage of bamboo is more persistent. Only in the Minshan area has the occasional occurrence of a fall estrus and spring birth been documented. Litters of twins are known to occur in the Minshan and Qionglai areas, but reliable data on litter size are not available.

In hollow tree bases or under large roots are the most frequent den sites in all areas except Qinling, where most of the large trees have been harvested. At this location, the pandas use natural limestone caves. Caves are also used occasionally in the Xiangling and Liangshan regions.

PREDATION ON GIANT PANDAS

There are no hard data on predation. However, the presence of predatory mammals according to region infers the possibility of predation pressure on the giant panda population, and occasionally episodes have been witnessed by local staff. It is believed that infants in dens and juveniles are the main victims.

The common leopard *(Panthera pardus)* occurs in all five regions, whereas dholes *(Cuon alpinus)* are found only in the Qinling and Minshan ranges. The yellow-throated marten *(Martes flavigula)* occurs in the Minshan, Qionglai, and Liangshan regions. The Asian golden cat *(Felis temmincki)* is believed to be an occasional predator in the Liangshan and Qionglai regions, and the clouded leopard *(Neofelis nebulosa)* may also prey on pandas in the Liangshan. Only in the Xiangling region is predation believed to be negligible because of the availability of alternate prey.

CURRENT POPULATION STATUS

In the Qinling range, giant pandas are found mainly in Foping and Yangxian counties. The relatively small population in this area lives in five different habitat fragments, but there are connecting corridors of *F. qinlingensis* along ridges. Genetic data reported by Fang et al. (1998) suggest inbreeding could become a problem for the pandas in this area, and, although seemingly stable at the moment, there is concern that the Qinling population may enter a state of slow decline in the future.

The largest single population of pandas in China is found in the middle region of the Minshan range, particularly in Pingwu, Songpan, and Beichuan counties. In other locations, fragmentation divides the pandas into nine small populations. Altogether, the genetic diversity of the Minshan pandas appears to be second only to Qionglai among the five regions (Fang et al. 1998). Like the Qinling region, the prospects for stability in the short run appear to be good, but overall, the population could continue the slow decline of the past.

In the Qionglai Mountains, the largest concentration of pandas is found on the southeastern slopes of the middle part of the range (i.e., in Wolong and Baoxing Reserves), but small, isolated populations occur here as well. The limited data available suggest that these pandas are genetically the most healthy of all the five regional populations (Fang et al. 1998; Pan et al., chapter 5). A decline in numbers was documented during the flowering of *F. faberi* during the 1980s, but the population appears to be stable at present. With added protection, the pandas in the Qionglai could be made relatively secure.

The great increase in the human population in recent years separated the pandas in Qionglai from those in Xiangling. Even within the Xiangling Mountains, tens of kilometers separate the Greater from the Lesser sections. In Lesser Xianglai, the population is divided into four groups of no more than ten individuals in each. Each of the two populations in Greater Xiangling numbers less than ten pandas. Not surprisingly, the genetic parameters (insofar as they are known) indicate the highest allele frequency and the lowest percentage of heterozygous alleles of any of the five regions, and in this region, the threat of extinction is the most severe (Fang et al. 1998).

The southernmost distribution of giant pandas in China today occurs in the Liangshan range. The numbers here are small and the habitat is heavily degraded. The population is divided into four groups that do have connecting links along ridges, but the genetic information available nevertheless indicates the Liangshan pandas are only slightly better off than those in Xiangling (Fang et al. 1998). We perceive the population as being in a fairly rapid state of decline.

CONSERVATION

The highest priority for giant panda conservation in China is protection of the remaining habitat. This is a long-term process and requires that conservation needs be balanced with those of community development for the human population. Another essential activity is to actually expand existing habitat through reforestation (e.g., as in China's "Grain to Green" program), particularly in areas where fragmented populations can be reconnected (Zhu and Ouyang, panel report 11.1). Increasing public awareness of the plight of China's pandas (Bexell et al., brief report 17.1) and improved management of the population (Yu and Deng, panel report 13.1) are additional steps that are necessary.

The total range of giant pandas in China today comes to slightly more than 21,000 km² (see table 9.1). Less than half this area is found in designated reserves. Expansion of the reserve system and upgrading infrastructure and training of reserve personnel are recognized as important by both provincial and central governments, and are being implemented. Bringing more area under effective protection will help to secure the future of the golden monkey (*Rhinopithecus* spp.), the takin *(Budorcas taxicolor)*, the dove tree *(Tetracentron sinense)*, and other rare and endangered endemics. This action will also provide protection for the ecological system of the upper Yangtze River and water resources for the lower Yangtze area. Giant pandas and human beings can co-exist in China, and this is a propitious time for reversing the long-term decline in the country's truly unique flora and fauna.

ACKNOWLEDGMENTS

This project was supported by the National Science Fund for Distinguished Young Scholars (30125006), and the National Natural Science Foundation of China (30230080, 30170148, 397700124). This is a long-term field study, and it was implemented with the strong support of different organizations and researchers. Special thanks are given to the State Forestry Administration, Sichuan Forestry Bureau, all the giant panda reserves, and to other colleagues in our research group, including Yi Wu, Chonggui Yuan, Chengming Huang, Guang Yang, Ping Tang, and Zejun Zhang.

REFERENCES

Fang, S. G., W. Feng, A. Zhang, H. Chen, J. Yu, G. He, X. Huang, X. Li, Y. Song, Z. Zhang, Q. Wang, R. Hou, and N. Fujihare. 1998. The research on genetic diversity of the giant panda. In *Proceedings of the international symposium on the protection of the giant panda, Chengdu, China, 1997*, edited by A. Zhang and G. He, pp. 141–53. (In Chinese.) Chengdu: Sichuan Publishing House of Science and Technology.

He, L., F. Wei, Z. Wang, Z. Feng, A. Zhou, P. Tang, and J. Hu. 2000. Nutritive and energetic strategy of giant pandas in Xiangling Mountains. (In Chinese.) *Acta Ecol Sin* 20:177–83.

Hu, J. C. 1985 Geographic and ecological distribution of the giant panda. (In Chinese.) *J Sichuan Normal College* 6:7–15.

Hu, J. C. 1986a. The giant panda in Qionglai. (In Chinese.) *J Sichuan Normal College* 7:21–28.

Hu, J. C. 1986b. The giant panda in Minshan. (In Chinese.) *Sichuan J Zool* 5:25–26.

Hu, J. C. 2001. *Research on the giant panda.* (In Chinese.) Shanghai: Shanghai Publishing House of Science and Technology.

Hu, J., G. B. Schaller, W. Pan, and J. Zhu. 1985. *The giant pandas of Wolong.* (In Chinese.) Chengdu: Sichuan Publishing House of Science and Technology.

Hu, J., G. B. Schaller, and K. Johnson. 1990a. Feeding ecology of giant pandas in Tangjiahe Reserve. (In Chinese.) *J Sichuan Normal College* 11:1–13.

Hu, J., F. Wei, C. Yuan, and Y. Wu. 1990b. *Research and progress in biology of the giant panda.* (In Chinese.) Chengdu: Sichuan Publishing House of Science and Technology.

Hu, J., F. Wei, C. Yuan, W. Deng, Y. Huang, and Y. Guo. 1990c. Herd dynamics of the giant panda before and after die-off of *Bashania faberi* in Wolong Reserve. (In Chinese.) *J Sichuan Normal College* 11:14–21.

Johnson, K. G., G. B. Schaller, and J. Hu. 1988a. Comparative behaviour of red and giant pandas in the Wolong Reserve, China. *J Mammal* 69:552–64.

Johnson, K. G., G. B. Schaller, and J. Hu. 1988b. Response of giant pandas to bamboo die-off. *Nat Geogr Res* 4:161–77.

Liu, X. 2001. *Mapping and modeling the habitat of giant pandas in Foping Nature Reserve, China.* Enschede, The Netherlands: Febodruk.

Pan, W., Z. Gao, and Z. Lü. 1988. *The giant panda's natural refuge in Qinling Mountains.* (In Chinese.) Beijing: Peking University Press.

Pan, W., Z. Lü, X. Zhu, D. Wang, H. Wang, Y. Long, D. Fu, and X. Zhou. 2001. *A chance for lasting survival.* (In Chinese.) Beijing: Peking University Press.

Reid, D. G., and J. Hu. 1991. Giant panda selection between *Bashania faberi* bamboo habitats in Wolong Reserve, Sichuan, China. *J Appl Ecol* 28:28–43.

Reid, D. G., J. Hu, S. Dong, W. Wang, and Y. Huang. 1989. Giant panda *Ailuropoda melanoleuca* behaviour and carrying capacity following a bamboo die-off. *Biol Conservation* 49:85–104.

Reid, D. G., J. C. Taylor, J. Hu, and Z. Qin. 1991. Environmental influences on *Bashania faberi* bamboo growth and implication for giant panda conservation. *J Appl Ecol* 28:185–201.

Ren, Y. 1998. *Plants of the giant panda's habitat of Qinling Mountains.* (In Chinese.) Xi'an: Shaanxi Publishing House of Science and Technology.

Schaller, G. S., J. Hu, W. Pan, and J. Zhu. 1985. *The giant pandas of Wolong.* Chicago: University of Chicago Press.

Schaller, G. B., W. Deng, K. Johnson, X. Wang, H. Shen, and J. Hu. 1989. Feeding ecology of giant pandas and Asiatic black bears in the Tangjiahe Reserve, China. In *Carnivore behavior, ecology and evolution*, edited by J. L. Gittleman, pp. 212–41. Ithaca: Cornell University Press.

Taylor, A. H., and Z. Qin. 1988a, Regeneration patterns in old-growth *Abies-Betula* forests in the Wolong Natural Reserve, Sichuan, China. *J Ecol* 76:1204–18.

Taylor, A. H., and Z. Qin. 1988b. Tree replacement patterns in subalpine *Abies-Betula* forests, Wolong Natural Reserve, China. *Vegetation* 78:141–49.

Taylor, A. H., and Z. Qin. 1988c. Regeneration from seed of *Sinarundinaria faberi*, a bamboo, in the Wolong Giant Panda Reserve, Sichuan, China. *Am J Bot* 75:1065–73.

Taylor, A. H., and Z. Qin. 1989. Structure and composition of selectively cut and uncut *Abies-Tsuga* forest in Wolong Natural Reserve, and implications for panda conservation in China. *Biol Conserv* 47:83–108.

Wei, F., and J. Hu. 1994. Studies on the reproduction of wild giant panda in Wolong Natural Reserve. (In Chinese.) *Acta Theriol Sin* 14:243–48.

Wei, F., J. Hu, C. Yuan, and Y. Wu. 1990. Body size difference of giant pandas in Minshan and Qionglai mountains. (In Chinese.) *Acta Theriol Sin* 10:243–47.

Wei, F., W. Wang, A. Zhou, J. Hu, and Y. Wei. 1995. Preliminary study on food selection and feeding strategy of red pandas. (In Chinese.) *Acta Theriol Sin* 15:259–66.

Wei, F., C. Zhou, J. Hu, G. Yang, and W. Wang. 1996a. Bamboo resources and food selection of giant pandas in Mabian Dafengding Natural Reserve. (In Chinese.) *Acta Theriol Sin* 16:171–75.

Wei, F., A. Zhou, J. Hu, G. Yang, and W. Wang. 1996b. Habitat selection by giant pandas in Mabian Dafengding Reserve. (In Chinese.) *Acta Theriol Sin* 16:241–45.

Wei, F., Z. Feng, and J. Hu. 1997. Population viability analysis computer model of giant panda population in Wuyipeng, Wolong Natural Reserve, China. *Int Conf Bear Res Manag* 9:19–23.

Wei, F., Z. Feng, and Z. Wang. 1999a. Habitat selection by giant and red pandas in Xiangling Mountains. (In Chinese.) *Acta Zool Sin* 45:57–63.

Wei, F., Z. Feng, Z. Wang, A. Zhou, and M. Li. 1999b. Feeding strategy and resource partitioning between giant and red pandas. *Mammalia* 63:417–30.

Wei, F., Z. Feng, Z. Wang, A. Zhou, and J. Hu. 1999c. Nutrient and energy requirements of red panda (*Ailurus fulgens*) during lactation. *Mammalia* 63:3–10.

Wei, F., Z. Feng, Z. Wang, A. Zhou, and J. Hu. 1999d. Use of the nutrients in bamboo by the red panda (*Ailurus fulgens*). *J Zool Lond* 248:535–41.

Wei, F., Z. Feng, Z. Wang, and J. Liu. 1999e. Association between environmental factors and growth of bamboo species *Bashania spanostachya*, the food of giant and red pandas. (In Chinese.) *Acta Ecol Sin* 19:710–14.

Wei, F., Z. Feng, Z. Wang, and J. Hu. 2000a. Habitat use and separation between the giant panda and the red panda. *J Mammal* 81:448–55.

Wei, F., Z. Wang, and Z. Feng. 2000b. Energy flow through populations of giant pandas and red pandas in Yele Natural Reserves. (In Chinese.) *Acta Zool Sin* 46:287–94.

Yang, G., J. Hu, F. Wei, and W. Wang. 1994. Population dynamics of giant panda in Mabian Dafending Natural Reserve. (In Chinese.) *J Sichuan Normal College* 15:114–18.

Yang, G., J. Hu, F. Wei, and W. Wang. 1998. The spatial distribution pattern and seasonal vertical movement of the giant panda in Dafengding Natural Reserve, Mabian, Sichuan. In *Resource and conservation of vertebrates,* edited by J. Hu, Y. Wu, W. Hou, F. Wei, and Q. Yang, pp. 8–14. (In Chinese.) Chengdu: Sichuan Publishing House of Science and Technology.

Yong, Y., J. Zhang, and S. Zhang. 1993. Distribution and population of the giant panda in Foping. (In Chinese.) *Acta Theriol Sin* 13:245–50.

Yong, Y., K. Wang, and T. Wang. 1994. Movement characteristics of the giant panda in Foping. (In Chinese.) *Acta Theriol Sin* 14:9–14.

Zhou, A., F. Wei, P. Tang, and J. Zhang. 1997. A preliminary study on food nutrient of red pandas. (In Chinese.) *Acta Theriol Sin* 17:266–71.

PANEL REPORT 9.1

Assessing the Habitat and Distribution of the Giant Panda

METHODS AND ISSUES

Colby Loucks and Hao Wang

THE CONSERVATION of pandas in the wild requires the protection of adequate habitat to support viable populations. Road construction and logging have fragmented panda habitat, dissecting historically large areas into much smaller ones. An essential first step in developing a conservation plan for wild pandas is to establish a snapshot of where remaining habitat and wild panda populations still persist today. We provide an overview of past efforts, summarize the important points derived from panel discussions, and recommend future actions to identify remaining habitat and quantify panda distributions.

The historical range of the giant panda *(Ailuropoda melanoleuca)* once extended from northern Vietnam northward to the vicinity of Beijing and eastward as far as Fujian. As humans intensified their activities in these fertile areas, the panda was forced to retreat from the subtropical forests to the temperate forests east of the Tibetan Plateau. The populations remaining today are restricted to several mountain ranges in Sichuan, Gansu, and Shaanxi Provinces (see figure 9.1 in Hu and Wei, chapter 9). Pandas live in the mixed deciduous-coniferous temperate forests of these mountain ranges. Although there are approximately forty nature reserves dedicated to conserving the panda and its habitat, they fall short of providing secure long-term protection of this endangered species. Approximately one-half of the remaining panda habitat remains unprotected. Current nature reserves are underfunded, and once-continuous habitat, stretching across mountain ranges, is being fragmented as human settlements, roads, and logging and hydrologic development destroy or convert habitat in the river valleys. An accurate assessment of the distribution of remaining habitat and panda populations is of immediate importance.

With continuous improvements in the analysis and application of such remote-sensing technology as satellite imagery and aerial photography, mapping the remaining habitat has become a cost-effective tool for conservation planning. De Wulf et al. (1988) conducted the first analysis of panda habitat using satellite imagery. Using LANDSAT MSS imagery (80-m resolution), they were able to produce coarse forest habitat maps for the Min and Qionglai Mountains. Since this initial analysis, the Qinling Mountains and the Qionglai Mountains surrounding Wolong Nature Reserve have received most of the recent efforts at remote-sensing analysis.

Loucks et al. (2003) used SPOT satellite imagery (20-m resolution) to assess remaining forest-nonforest habitat in the Qinling Mountains. This effort built on initial fieldwork by Pan et al. (1988). Liu et al. (chapter 11) have completed a habitat assessment of Foping Nature Reserve in the Qinling Mountains, using an integrated expert system and artificial neural network system, which incorporated LANDSAT TM images (30-m resolution), ancillary terrain data, and animal radiotracking locations. This method has shown great potential, because it can extract detailed information from georeferenced points and further extrapolate the information to other areas. D. Wang (pers. comm.) recently completed a fragmentation analysis of panda habitat for the northern Min Shan region. This analysis identified road construction as the primary cause of habitat fragmentation.

Wolong Nature Reserve, as the centerpiece of giant panda protection, has received the most detailed remote-sensing analyses. Liu et al. (1999, chapter 14) have developed a framework and methods to analyze potential habitat (without human impacts) and realized habitat (with human impacts) over time by integrating ecology with human demography and socioeconomics. Liu et al. (2001) have used LANDSAT TM and MSS data, as well as Corona data (declassified U.S. photo reconnaissance imagery), to analyze panda habitat changes from before the reserve was established to more than 30 years afterward.

Additional new mapping approaches are also making sophisticated analyses possible. Linderman et al. (in press) have developed methods to map bamboo distribution using satellite imagery, neural networks, ground truth data, and global positioning systems (GPS). These methods are able to achieve a high accuracy, dramatically reduce the amount of time required for field surveys, and make it feasible to detect bamboo distributions in topographically challenging areas. The methods and results from this study are not only critical for panda conservation, but also essential for regional and global assessments of understory biodiversity, evaluation of habitat availability and fragmentation, and resource management (e.g., reserve design and ecological restoration).

The World Wildlife Fund (WWF) and China's Ministry of Forestry (MoF) have collaborated to fund the third national survey of panda populations and habitat distribution (Yu and Liu, panel report 15.1). The field work was completed in 2001. Using topographic maps in conjunction with GPS, the survey compiled critical information on the approximate distribution of pandas, their remaining habitat, and bamboo species, as well as catalogs of threats and other environmental data (see survey form, figure PR.9.1).

The national survey provided the first systematic on-the-ground analysis of panda habitat in over ten years. The data collected were compiled into a geographic information system (GIS) database. Although this database is limited to coarse habitat assessments and approximate panda distributions, its use in conjunction with remotely sensed habitat and bamboo distribution information will enable policymakers to propose new protected areas, identify areas in which to restore habitat or develop corridors, and limit potential fragmentation threats.

In addition to habitat data, the survey amassed information on the presence and number of giant pandas. Survey teams, using local reserve and community staff, conducted extensive field surveys of all fifty-four counties that are known to contain pandas. There are nine counties in the Qinling Mountains of Shaanxi Province, nine counties in Gansu Province, and thirty-six counties in Sichuan Province. The national survey started with Pingwu County, Sichuan Province, in the summer of 1998. The last county was surveyed in the late fall of 2001.

The survey methodology for identifying and counting pandas was constrained by steep and difficult terrain, accessibility, cost, and time. The combination of these factors precluded the use of the traditional line-transect techniques typically employed to survey mammals (Anderson

Location Information

ID			Date		Weather	
Location	County/City		Township		Place	
Surveyor No.		ID of RTEs			Surveyor name	
Altitude (m)		GPS	N ° "		E ° "	

Landscape Information

Landform	Ridge	Up Slope	Mid-Slope	Low Slope	Valley	Flat	Shape	Even	Concave	Convex	Mixed	None
Aspect	N	NE		E	SE		S	SW	W	NW	None	
Slope	0–5°		6–20°		21–30°		31–40°		>41°			

Panda Information

Panda Sign	Panda	Feces	Eating Site	Footprint		Den Site	Cave Site	Scratch Marks	Scent Marks	Body		
Panda	Number		Size			Gender			Behavior			
Feces	Number	1	2	3–10	11–20	>20	Age (days)	1–3	4–15	>15		
	Mastication	Fine		Mid	Coarse	Composition	Stem	Stem and Leaves	Leaves	Shoot		Other
ID of Sample								Small feces	Number			
Size of Feces	(cm)				(cm)				Size	(cm)		

Habitat Information

Habitat Type	Evergreen Broadleaf	Evergreen-Deciduous		Deciduous Broadleaf	Deciduous and Conifer	Conifer	Shrub	Grassland	Farmland
Forest Type	Original Forest				Secondary Forest		Plantation Forest		
Canopy Structure	Overstory Height (m)		5–9		10–19		20–29		>30
	Coverage		0–0.24		0.25–0.49		0.5–0.74		0.75–1
	Ave. Diameter (cm)		0–10		11–20	21–30		31–50	>50
Shrub	Height (m)	0–1	1–2	2–3	3–4	4–5	Cover	0–24%	25–49%	50–74%	75–100%

FIGURE PR.9.1. Survey form used for the Third National Survey to Collect Information on Pandas and Their Habitat.

(continued)

Bamboo Information

Bamboo species:										
Height (m)	0–1	1–2	2–3	3–5	>5	Cover	0–24%	25–49%	50–74%	75–100%
Type	Dispersed	Cluster	Mixed		Condition	Good	Average	Bad	Water Source	Yes	No

Threat Information

Disturbance		Human								Nature			Other	
		Road	Log	Poacher's Stand	Trap	Herd	Bamboo Cut	Herb	Farm	Tour	Bamboo Flowering	Fire	Slope Slide		
Stage	Finished														
	Ongoing														
	Started														
	Planned														
Intention	None														
	Minimal														
	Average														
	Max														
Comment															

FIGURE PR.9.1. *(continued)*

et al. 1979; Buckland et al. 1993). Instead, each survey team member was allocated a section of habitat limited to 2 km² to survey each day. Typically, a survey team established a base camp in a river valley and partitioned the valley into survey blocks. The typical survey team consisted of forty people; a team could therefore cover approximately 80 km² of habitat daily. A total of approximately 22,700 km² of habitat was surveyed.

Each team member was free to decide the best route to traverse from river to ridgeline. Team members searched for signs of panda activity, including feces, tree scent marks, feeding sites, den sites, or visual sightings. A total of 16,950 person-days of activity was needed to complete the survey in Sichuan. During that time, only twenty-one live wild pandas were seen. Panda feces was the most common sign, and when a feces specimen was found, the surveyor filled out the survey form (see figure BR9.1), noting both signs and habitat information.

The technique used to count pandas was the "bite size technique" (see Pan et al., chapter 5). Pandas digest very little of the plant material they consume and pass the majority of their bamboo intake as waste. Giant pandas eat for more than 12 hours every day and do so in a repetitive manner (Schaller et al. 1985). While eating the bamboo stalk, the panda will push the stalk into its mouth and bite down. When this bite has been cleared of its mouth, it will bite down again and continue eating in this manner for hours at a time. Chinese investigators have determined that the average bite size of an individual panda is not likely to vary by more than a few mm, and that individual pandas have different average bite sizes. Therefore, it was possible to collect and measure the average bite size of bamboo pieces in feces collected during the survey, and from this information, determine the number of pandas that produced the collected feces.

Because individual pandas have different average bite sizes, this information, when combined with average home range sizes, allows the surveyor to determine the minimum number of pandas in a given habitat block. The national panda survey that took place during the 1980s also used this technique, and therefore it is possible to compare the results of the two surveys.

One of the limitations of this analysis is the lack of home range and basic natural history information for the giant panda across its range. Schaller et al. (1985) provided early natural history information on the giant pandas of the Wolong and Tangjiahe Nature Reserves. Pan et al. (1988, 2001) provided additional natural history information for the pandas of the Qinling Mountains, and conducted the only long-term home range analysis of the species. Lacking any other long-term studies of panda home ranges, the national survey has applied the results of the home range analysis for Qinling pandas to populations in other mountain ranges.

To determine the home range size of wild pandas in the Qinling Mountains, Pan et al. (1988, 2001) used radiotelemetry to augment direct observation of wild pandas in Foping and Changqing Nature Reserves. By capturing and attaching radiocollars on thirty pandas over thirteen years, they were able to develop an accurate database of panda movement during different times of the year. Radiotelemetry or GPS-telemetry methodology is the most accurate and efficient way to determine movement patterns, eating preferences, habitat preferences, and the reproductive success of pandas in the wild. Unfortunately, the use of radiotelemetry or GPS-telemetry methodology on giant pandas has not gained acceptance from the national or provincial governments in China, severely limiting the collection of valuable information in other locations outside Wolong, Tangjiahe, Changqing, and Foping Nature Reserves.

RECOMMENDED ACTIONS

The third national survey provides an important snapshot of the status of the giant panda in the wild. However, its application to monitoring and evaluating the health and status of wild populations is limited. To be able to make proper management decisions, we need to learn more about the natural history of the panda. We offer several recommendations that are urgently needed to save this endangered species:

- Apply remote sensing techniques to identify potential habitat in the Min, Xiangling, and Liang Mountains, and that portion of the Qionglai Mountains lying outside Wolong Reserve.

- Apply remote sensing techniques in all ranges to identify potential linkage areas between habitat blocks that may serve as corridors for panda immigration/emigration.

- Utilize direct methods, including radio-telemetry and GPS telemetry, to gather information on the natural history of giant pandas.

- Quantify threats to the integrity of remaining habitat throughout the panda's range with the help of remote sensing techniques.

- Apply artificial intelligence techniques to delineate bamboo distributions throughout the panda's range.

- Disseminate habitat maps derived from the national survey to policymakers and all national, provincial, county, and reserve staff in the region.

- Establish a GIS-based panda and habitat monitoring and evaluation system in all nature reserves, and a central clearinghouse

- of GIS/remote sensing information for public use.
- Encourage technology transfer and capacity building through workshops on GIS and radio-GPS telemetry applications for use by provincial and reserve staff.
- Incorporate human demographic and socioeconomic data into habitat analyses, and use the data to predict the temporal and spatial dynamics of future habitat changes.

Past studies of panda habitat have taken place in the Qinling Mountains and Wolong Reserve (Qionglai Mountains). Future efforts should be focused on other parts of their range. The national survey provides critical information on the distribution of pandas, bamboo, habitat, and threats to the panda's survival. This database, in conjunction with GIS, will assist in efforts to develop new areas for protection, habitat restoration, corridors, and infrastructure development. These tools may also be used to revise China's management plan for pandas and their habitat.

Direct methods for gathering information on the natural history of pandas must be used more frequently. The need to employ radiotelemetry and GPS telemetry is urgent. The best information to date on the location and activities of pandas has been acquired through the use of radiocollars, especially in the Qinling Mountains. GIS technology is essential to the development of management plans (applicable to individual nature reserves and to the entire habitat range) for pandas and their habitat; for instance, in projects to map bamboo and study the interactions between humans and their environment. Use of these technologies will increase our prospects for saving the wild population of pandas in China.

ACKNOWLEDGMENTS

We thank Xuehua Liu, Bill McShea, Jianguo Liu, and Dajun Wang for their significant contributions to the discussions that resulted in this report.

REFERENCES

Anderson, D. R., J. L. Laake, B. R. Crain, and K. P. Burnham. 1979. Guidelines for line transect sampling of biological populations. *J Wildlife Manage* 43:70–78.

Buckland, S. T., D. R. Anderson, K. P. Burnham, and J. L. Laake. 1993. *Distance sampling: Estimating abundance of animal populations.* London: Chapman and Hall.

De Wulf, R. R., R. E. Goossens, J. R. MacKinnon, and W. Shen-cai. 1988. Remote sensing for wildlife management: Giant panda habitat mapping from LANDSAT MSS images. *Geocarto Int* 1:41–50.

Linderman, M., J. Liu, J. Qi, L. An, Z. Ouyang, J. Yang, and Y. Tan. In press. Using artificial neural networks to map the spatial distribution of understory bamboo from remote sensing data. *Int J Remote Sensing*.

Liu, J., Z. Ouyang, W. Taylor, R. Groop, Y. Tan, and H. Zhang. 1999. A framework for evaluating effects of human factors on wildlife habitat: The case of the giant pandas. *Cons Biol* 13:1360–70.

Liu, J., M. Linderman, Z. Ouyang, L. An, J. Yang, and H. Zhang. 2001. Ecological degradation in protected areas: The case of Wolong Nature Reserve for giant pandas. *Science* 292:98–101.

Loucks, C., L. Zhi, E. Dinerstein, W. Dajun, F. Dali, and W. Hao. 2003. The giant pandas of the Qinling Mountains, China: A case study in designing conservation landscapes for elevational migrants. *Cons Biol* 17:558–65.

Pan, W., Z. Gao, and Z. Lü. 1988. *The giant panda's natural refuge in the Qinling Mountains.* (In Chinese.) Beijing: Peking University Press.

Pan, W., Z. Lü, X. Zhu, D. Wang, H. Wang, Y. Long, D. Fu, and X. Zhou. 2001. *A chance for lasting survival.* (In Chinese.) Beijing: Peking University Press.

Schaller, G. B., J. Hu, W. Pan, and J. Zhu. 1985. *The giant pandas of Wolong.* Chicago: University of Chicago Press.

BRIEF REPORT 9.1

Using DNA from Panda Fecal Matter to Study Wild-Living Populations

*Yunwu Zhang, Hemin Zhang, Guiquan Zhang,
Oliver A. Ryder, and Yaping Zhang*

THE GIANT PANDA *(Ailuropoda melanoleuca)* is a rare and native animal of China. Its origin can be dated to 8zo million years ago, and it was once distributed in many areas in East Asia (Hu 1991). However, due to a low rate of reproduction, the continued shrinking and isolation of its habitat as a result of human activities, and the disruption of genetic exchange between populations, the number of giant pandas in the wild has decreased rapidly. Today, pandas can be found only in scattered forests in several mountain ranges in western China (Hu and Wei, chapter 9; Loucks and Wang, panel report 9.1), and the species is on the verge of extinction. Estimates from population surveys in the mid-1980s suggested that the number of pandas was only about one thousand (Hu 1991), leading to the classification of its status as "endangered" by the International Union for the Conservation of Nature (Zhou 1996).

Genetic study of the giant panda is essential to the development of a viable conservation strategy. However, the acquisition of genetic information from wild-living pandas is limited by the difficulty of acquiring blood and tissue samples, and this in turn limits the application of molecular technology to the assessment of genetic status. Noninvasive methods of extracting DNA from feces, hair, and scent marks have been developed and used for genetic study of the great apes (Morin et al. 1994; Garner and Ryder 1996), bears (Taberlet et al. 1997), seals (Reed et al. 1997), as well as giant pandas (Zhang et al. 1994; Ding et al. 1998). To expand the feasibility of applying simple and efficient methods for noninvasive DNA preparation, we present results obtained from different methods of extracting DNA from giant panda feces.

DNA EXTRACTION

Our objective was to test several different methods for the extraction of DNA from giant panda feces. Blood and fecal samples were collected from captive individuals at the Wolong Breeding Center, China, and stored at –70°C. DNA extractions from four blood samples were performed as described in Maniatis et al. (1982) and used as a positive control. We used different methods for fecal DNA extraction.

METHOD 1

Materials used were Qiagen Loading Buffer (QLB: 500 mM Tris-HCl, 16 mM EDTA, 100 mM NaCl, pH 6.0) plus the DNA Isolation and Purification kit (Shanghai Watson Biotechnology, Inc., Shanghai). We first added 600 μl of QLB to 200 mg of feces, vortexed the mixture briefly, then centrifuged it at 13,000 rpm for 6 min. Four hundred μl of supernatant was then transferred to a new tube and 400 μl of DP buffer (provided in the kit) was added. The final step was to follow the instructions of the small tissue/cell genomic DNA extraction kit provided in the company's handbook (Shanghai Watson Biotechnology, Inc., Shanghai).

METHOD 2

Our second method was based on the work of Taberlet et al. (1997), but with slight modifications. First, we added 1 ml of L6 extraction buffer (10 M GuSCN, 0.1 M Tris-HCl pH 6.4, 0.02 M EDTA pH 8.0, 1.3% Triton X-100) to 100 mg of feces, and incubated overnight. We then transferred 500 μl of the liquid phase to a new tube, and added 500 μl of fresh L6 extraction buffer and 40 μl of Ultra Bind (UltraClean™ 15, Mo Bio Laboratories, Inc., Solana Beach, California). The mixture was incubated at room temperature for 10 min, with constant agitation. After centrifugation at 8000 rpm for 1 min, the pellet was washed three times with 500 μl of L2 buffer (10 M GuSCN, 0.1 M Tris-HCl pH 6.4), once with 1 ml of 100% ethanol, and once with 1 ml of acetone. The pellet was dried at 55°C, and redissolved in 200 μl of TE (pH 8.0) for 5 min. It was then centrifuged again, and the supernatant was gently transferred to another tube without disturbing silica particles. The supernatant was then ready for polymerase chain reaction (PCR).

METHOD 3A

Method 3a was based on the work of Machiels et al. (2000), but with slight modifications. Four hundred μl of PBS (115 mM NaCl, 8 mM KCl, 1.6 mM KH_2PO_4 pH 7.4), along with 0.9% SDS and 18 mM EDTA, was added to 100 mg of feces. We then added 400 μl of phenol/chloroform/isoamyl alcohol (25:24:1) buffer at pH 8.0, and vortexed vigorously. This mixture was heated to 65°C for 10 min, then rapidly cooled to −20°C. It was then centrifuged at 8000 rpm for 20 min, after which the water phase was transferred to a new tube.

Four hundred μl of phenol/chloroform/isoamyl alcohol (25:24:1) was added to the extract twice, followed by the addition of 10% cetyltrimethylammonium bromide 170 μl and 0.7 M NaCl 170 μl to the water phase, and then by mixing. We then added 400 μl of chloroform/isoamyl alcohol (24:1) to the extract twice. Ten mM NH_4Ac 60 μl and isopropyl alcohol 700 μl were added and mixed, then kept at −20°C for at least 2 h. This was centrifuged at 13,000 rpm for 2 min, and the DNA pellet was washed with 70% ethanol and dried at 55°C. The pellet was redissolved with 100 μl TE (pH 8.0).

METHOD 3B

Because different phosphate buffered saline (PBS) preparation methods are available, we again followed the method of Machiels et al. (2000), but with another PBS buffer prepared according to Maniatis et al. (1982) to extract DNA from fecal samples.

In the present study, we used methods 1, 2, and 3a for DNA extraction from fecal sample gp1-1. Methods 1 and 2 were used for fecal sample gp1-2. Methods 3a and 3b were used for fecal samples gp1-3 and gp1-4.

PCR AMPLIFICATION AND SEQUENCING

A pair of primers—L15513 and H15915 (Anderson et al. 1981; Ding et al. 1998)—was used to amplify the fragments of the partial mtDNA cytochrome *b* gene and the $tRNA^{Thr}$ gene. Amplification was performed in 30-μl reactions. Each reaction contained 5 μl of prepared DNA, 3 μl of GeneAmp® 10XPCR buffer (Perkin Elmer, Boston), 0.6 μl of 10-μg/ml BSA, 0.6 μl of 10-mM dNTPs (Invitrogen, Carlsbad, California), 0.6 μl of each 10-μM primer, and 0.18 μl of 5-U/μl AmpliTaq® Gold DNA polymerase (Perkin Elmer, Boston). Thermal cycling conditions were 40 cycles at 94°C for 40 s, 53°C for 40 s, and 72°C for 40 s, with 94°C for 10 min at the beginning and 72°C for 10 min at the end. After PCR, DNA products were purified with the QIAquick™ PCR purification kit (QIAGEN, Valencia, California). Both DNA strands were sequenced using the Dye Terminator Cycle DNA Sequencing kit and ABI 373 automated sequencer (PE Applied Biosystems, Foster City, California). Sequence alignments were performed by using Sequencher 3.1 (Gene Codes Corporation, Ann Arbor, Michigan). For sequence comparison, we used a 372-bp fragment at the 3'-end of the cytochrome *b* gene (nucleotide sites 769–1140, counted from the start codon). Two giant panda sequences from GenBank, U23552 (Talbot and Shields 1996) and X94918 (Ledje and Arnason 1996), were also retrieved and used here for comparison.

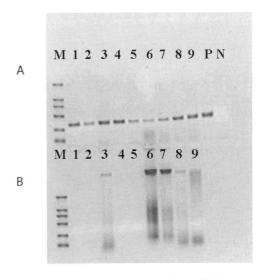

FIGURE BR.9.1. Results of DNA extraction and PCR amplification from giant panda feces. (A) mtDNA PCR amplification results. M: 100-bp ladder PCR marker. Bands 1–3: results of gp1-1 using methods 1, 2, and 3a, respectively. Bands 4 and 5: results of gp1-2 using methods 1 and 2, respectively. Bands 6 and 7: results of gp1-3 using methods 3a and 3b, respectively. Bands 8 and 9: results of gp1-4 using methods 3a and 3b, respectively. P: positive control. N: negative control. Fecal DNA extraction results. (B) M and bands 1–9 are the same as described in (A).

RESULTS

Different methods were used to extract DNA from four different giant panda fecal samples, gp1-1, gp1-2, gp1-3, and gp1-4. The yields of DNA from methods 1 and 2 were low, whereas methods 3a and 3b resulted in much higher DNA yields (figure BR.9.1B).

A 449-bp PCR product was successfully amplified from DNA isolated from blood and feces using all four methods (figure BR.9.1A). To verify that the PCR products were actually derived from panda DNA, all PCR products were sequenced. For each fecal sample, the PCR products from DNA extracted by different methods always have the same sequence. Sequence comparison between different fecal samples showed that there were two haplotypes: samples gp1-1, gp1-2, the four positive controls, and X94918 shared one haplotype (haplotype 1). Samples gp1-3, gp1-4, and U23552 shared another haplotype (haplotype 2). There were two nucleotide substitutions between haplotype 1 and haplotype 2 at site 852 (C-T) and site 1014 (C-T).

DISCUSSION

Our sequencing results indicate that all PCR products were derived from the giant panda templates. Therefore, all extraction methods successfully provided target DNA that would be usable for genetic analyses of giant pandas. Thus we demonstrated the feasibility of routine mtDNA amplification using DNA extracted from fecal samples.

It appears that different PBS preparation methods do not affect DNA yields from fecal samples. However, although both method 3a and method 3b gave much higher DNA yields than the other two methods, most of the DNA is probably not from the pandas themselves, but from their food source. As PCR can be inhibited by an excess of total DNA, a high ratio of background DNA:target DNA may affect the efficiency of target DNA amplification. We did not observe an affect on mtDNA amplification, but when nuclear genes are studied, because they require more template DNA, an effect may become evident.

ACKNOWLEDGMENTS

This work was supported by the State Key Basic Research and Development Plan (grant G20000468), NSFC, and the Chinese Academy of Sciences.

REFERENCES

Anderson, S., A. T. Bankier, B. G. Barrell, M. H. de Bruijn, A. R. Coulson, J. Drouin, I. C. Eperon, D. P. Nierlich, B. A. Roe, F. Sanger, P. H. Schreier, A. J. Smith, R. Staden, and I. G. Young. 1981. Sequence and organization of the human mitochondrial genome. *Nature* 290:457–65.

Ding, B., Y. P. Zhang, and O. A. Ryder. 1998. Extraction, PCR amplification, and sequencing of mitochondrial DNA from scent mark and feces in the giant panda. *Zoo Biol* 17:499–504.

Garner, K. J., and O. A. Ryder. 1996. Mitochondrial DNA diversity in gorillas. *Mol Phylogenet Evol* 6:39–48.

Hu, J. 1991. *Advances in the study of the biology of the giant panda.* (In Chinese.) Chengdu: Sichuan Science and Technology Press.

Ledje, C., and U. Arnason. 1996. Phylogenetic analyses of complete cytochrome *b* genes of the order Carnivora with particular emphasis on the Caniformia. *J Mol Evol* 42:135–44.

Machiels, B. M., T. Ruers, M. Lindhout, K. Hardy, T. Hlavaty, D. D. Bang, V. A. M. C. Somers, C. Baeten, M. von Meyenfeldt, and F. B. J. M. Thunnissen. 2000. New protocol for DNA extraction of stool. *Biotechniques* 28:286–90.

Maniatis, T., E. F. Fritsch, and J. Sambrook. 1982. *Molecular cloning: A laboratory manual.* New York: Cold Spring Harbor Laboratory.

Morin, P. A., J. J. Moore, R. Chakaraborty, L. Jin, J. Goodall, and D. S. Woodruff. 1994. Kin selection, social structure, gene flow, and the evolution of chimpanzees. *Science* 265:1193–201.

Reed, J. Z., D. J. Tollit, P. M. Thompson, and W. Amos. 1997. Molecular scatology: The use of molecular genetic analysis to assign species, sex and individual identity to seal faeces. *Mol Ecol* 6:225–34.

Taberlet, P., J. J. Camarra, S. Griffin, E. Uhres, O. Hanotte, L. P. Waits, C. Dubois-Paganon, T. Burk, and J. Bouvet. 1997. Noninvasive genetic tracking of the endangered Pyrenean brown bear population. *Mol Ecol* 6:869–76.

Talbot, S. L., and G. F. Shields. 1996. A phylogeny of the bears (Ursidae) inferred from complete sequences of three mitochondrial genes. *Mol Phylogenet Evol* 5:567–75.

Zhang, Y. P., O. A. Ryder, Q. Zhao, Z. Fan, G. He, A. Zhang, H. Zhang, T. He, and C. Yucun. 1994. Non-invasive giant panda paternity exclusion. *Zoo Biol* 13:569–73.

Zhou, M. Y. 1996. INF project stirs debate over how to preserve pandas. *Science* 272:1580–81.

10

Giant Panda Migration and Habitat Utilization in Foping Nature Reserve, China

Yange Yong, Xuehua Liu, Tiejun Wang,
Andrew K. Skidmore, and Herbert H. Prins

IT IS WELL ESTABLISHED that the giant panda *(Ailuropoda melanoleuca)* is an endangered species. Partial surveys from the mid-1980s estimated that about nine hundred to one thousand pandas exist in the wild, and are found only in the western part of China (figure 10.1). The giant panda is a solitary mammal (Schaller et al. 1985). It is quite difficult to locate individuals in remote mountain areas covered with dense vegetation. Radiotracking is an effective way to locate pandas to observe their behavior and how they relate to their habitat. This technology also permits a quantitative assessment of activity patterns, habitat preferences, and requirements for survival, particularly when combined with other analytical technologies, such as the geographic information system (GIS). Both technologies have been widely used in the study of many different species (White and Garrott 1990).

Prior work on panda migrations has been done mainly in Wolong Nature Reserve (WNR) in the Qionglai Mountains and Changqing Nature Reserve (CNR) in the Qinling Mountains. Radiotracking data have revealed that pandas in WNR spend most of the year in the higher elevations, where they feed mainly on arrow bamboo *(Bashania fangiana)*. However, the animals migrate to lower elevations during May and June to feed on the shoots of umbrella bamboo *(Fargesia robusta)* (Hu 1985; Schaller et al. 1985; Hu and Wei, chapter 9). The work of Pan et al. (1988) in CNR revealed a reverse movement pattern from that at WNR: pandas spent most of the year at the lower elevations to feed on *B. fargesii*, but only about 3 months of the year (June–August) at higher locations, where *F. spathacea* predominates.

It now appears that the migratory patterns of giant pandas differ according to the mountain region in which they occur (see Hu and Wei, chapter 9). Giant pandas occur at a relatively high density in Foping Nature Reserve (FNR), a part of the Qinling Mountains distribution; this population is one of the few for which radio-telemetry has been used in tracking their movements. However, the proximate factors responsible for these migrations have not previously been studied (Pan et al. 1988). Our objective was to quantitatively confirm the existence of seasonal ranges at FNR and the characteristics of the habitat in each range, as well as determine the temporal pattern of movements between them.

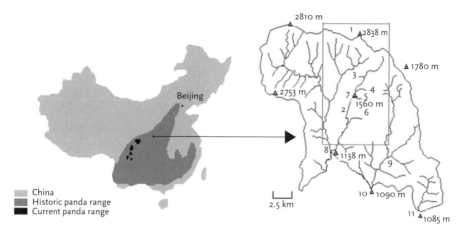

FIGURE 10.1. Historic changes in the geographic distribution of giant pandas in China (left), and radiotracking locations at Foping Nature Reserve (right) used in this study. 1, GuangTouShan Mountain; 2, DongHe River; 3, ZiChangGou Valley; 4, PaiFangGou Valley; 5, WaFangGou Valley; 6, ShuiJingGou Valley; 7, SanGuanMiao Village; 8, DaGuPing Village; 9, DaChengHao Village; 10, YueBa Village; 11, LongTanZi Village.

The first survey of giant pandas was carried out at FNR in 1973 (Shaanxi Biological Resources Survey 1976). In addition, initial ecological observations of the species were carried out at this location (Yong 1981, 1989; Wu 1981, 1986; Ruan 1983). Research during the 1990s built on these earlier studies. Notably, Yong et al. (1993, 1994) published their observations on panda demography, distribution, and movement patterns. Li et al. (1997) have reported the results of a population viability analysis (PVA) for the FNR pandas, and studies of the relationship of pandas to their habitats have appeared in recent years (Yang et al. 1997, 1998; Ren et al. 1998; Yang and Yong 1998).

Previous work has shown that, in general, the Qinling pandas move annually between winter and summer ranges (Pan et al. 1988, 1989; Yong et al. 1994). Over a 5-year period (1991–1995), radiotelemetry was used to track FNR pandas, but data from this effort have been published only in China, and they lacked statistical analysis or the additional benefit of being combined with GIS analysis.

GIS analysis provides a flexible tool for managing resources and understanding and predicting complex and changing environmental systems (Peuquet et al. 1993). Our study combines GIS information with radiotracking data for six pandas, and utilizes statistical analyses to demonstrate spatial distributions quantitatively and graphically. We also address relationships between the movements of radiocollared pandas and a number of habitat parameters, such as vegetation type, elevation, slope, and aspect.

STUDY AREA

Foping Nature Reserve is located on the southern slopes of the Qinling Mountains at 32°32′–33°45′N latitude, and 107°40′–107°55′E longitude (see figure 10.1). It covers an area of about 293 km², and the elevation ranges from 1000 to about 2910 m. From west to east, four river drainage systems are found, the XiHe, DongHe, JinShuiHe and LongTanZi. Annual rainfall averages about 900 mm. The lowest temperature occurs on average in January and is about –3°C, and the highest in July, about 15°C.

The vegetation at FNR consists predominantly of coniferous forests, mixed coniferous and broadleaf forests, deciduous broadleaf forests, shrub, and meadow (Ren et al. 1998). The two bamboos that are the pandas' main food are *B. fargesii*, found primarily below 1900 m, and *F. spathacea*, which occurs mainly above 1900 m.

TABLE 10.1
Profiles of the Six Radiocollared Pandas Used in This Study

ID NUMBER	SEX	START DATE	AGE AT START (YEARS)	END DATE	TRACKING TIME (MONTHS)
127	Male	May 1991	10	May 1995	34
043	Female	July 1991	12	Aug. 1992	9
065	Male	Feb. 1992	0.7	Dec. 1995	34
045	Female	May 1992	6	Dec. 1995	29
005	Male	Apr. 1994	15	Dec. 1995	20
083	Female	Jan. 1995	1.5	Aug. 1995	9

About sixty-four giant pandas were found at FNR in a survey conducted in 1990, occurring at an average density of one individual per 5 km² (Yong et al. 1993). The human population residing in the reserve numbered about two hundred, concentrated in the three villages of SanGuanMiao, DaGuPing, and DaChengHao.

USING RADIOTELEMETRY TO TRACK GIANT PANDA MOVEMENTS

Because preliminary evidence indicated that, like giant pandas in other locations, those at Foping were seasonally migratory, we placed fifty-nine fixed towers primarily along the ridge top known as GuangTouShan and in the SanGuanMiao valley for use in tracking panda movements over successive annual cycles. The radiotracking equipment used in our work consisted of a MOD-500 telemetry radiocollar, a TR-2 receiver, and an RA-2AK hand-held H-style antenna (U.S. Telonics, Mesa, Arizona). Six pandas (three females, three males) were collared and tracked at different times during the study period (table 10.1).

According to White and Garrott (1990), location can be estimated by triangulation to obtain the coordinates of a radiocollared animal on the basis of two bearings received at two different towers, or more directional bearings from additional receiving towers. Although three or more bearings undoubtedly yield location data that are more accurate, it was usually possible to obtain only two readings on an individual, given the mountainous terrain of FNR. In this study, all panda locations were estimated from the cross point of two bearings received at two towers.

APPLICATION OF GIS TO HABITAT ASSESSMENT

A survey of giant panda habitats at FNR was carried out during the summer (July–August) of 1999. Line transects for sampling were established and the global positioning system (GPS) was used to record the geolocations of all sample plots and important landmarks. A total of 110 plots measuring 10 × 10 m² were sampled every 100 m along a transect at a given elevation height to obtain detailed habitat information. Additional GPS points were randomly sampled along transects to collect general environmental data. Data on elevation, slope, aspect, and ground cover were collected at all GPS points. Mapping ground cover and classifying it according to habitat types was complemented by a knowledge-based expert classifier system (Liu 2001; Liu et al. 2002a). To obtain high accuracy in map production, nine data layers were used: LANDSAT TM bands 15 and 7, elevation, and slope and aspect models, all of which were digital images.

A total of 1756 raw location records from the latitude-longitude coordinate system were transformed into the Universal Transverse Mercator coordinate system of the GIS software (Integrated

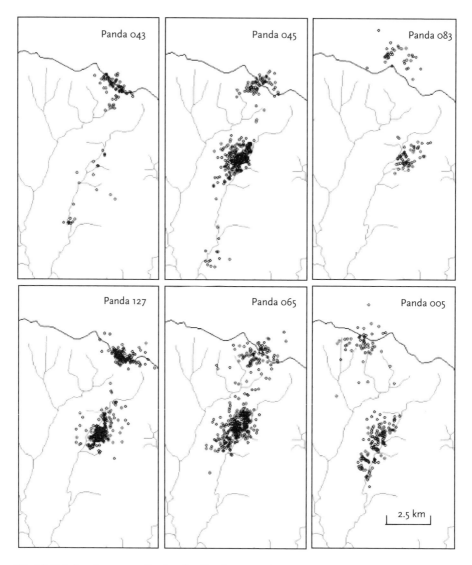

FIGURE 10.2. Ranging patterns for six radiocollared giant pandas in the SanGuanMiao region between 1991 and 1995. Upper cluster, summer range; lower cluster, winter range.

Land and Water Resources Information System, Enschede, The Netherlands). After checking carefully the entire tracking dataset taken from such distinct aspects as the calculated geocoordinate, distance of daily moves, and locations by season of the year, a final dataset of 1639 records was selected for analysis. Three kinds of raw tracking records were eliminated (Liu et al. 2002b): (1) no cross point could be found by two bearings, (2) the cross point obtained was located on the side of the mountain, or (3) only one bearing was available. Individual panda locations determined at different times were then placed on a background map on which drainage systems and reserve boundaries were georeferenced with the same coordinate system (figure 10.2). The linked habitat type of each tracking location was extracted from the classified map and was counted for winter and summer habitats, respectively. We also calculated the percentages of tracking locations in different habitat types over the two seasons.

SEASONAL PATTERNS OF SPATIAL DEPLOYMENT

The clustered radio tracking points obtained from this study clearly indicated two seasonally distinct ranging areas for each of the six radio-collared pandas. The northern cluster (figure 10.2) in each case was operationally designated as the location of high-elevation summer ranges, whereas the southern cluster was regarded as that of the winter range, farther down the slopes.

These tracking points were then used to demarcate the elevational boundaries of the two clusters. As explained earlier, we used data from October 6 to June 7 in the following year to define the elevation zone of the winter range, whereas the data from June 15 to September 1 were used to define the elevation zone of the summer range. Because elevation data were not normally distributed, we used the boxplot method (Moore and McCabe 1998) when performing our calculations. The first quarter (Q1), median, and third quarter (Q3) were used to determine the elevation zones of the summer and winter ranges as follows:

$$IQR = Q3 - Q1,$$
$$\text{Up-bound} = Q3 + 1.5 IQR,$$
$$\text{Low-bound} = Q1 - 1.5 IQR,$$

where IRQ is the interquartile range.

Elevation zones were then defined as extending from the low-bound to the up-bound limits. We hypothesized that there is a significant difference between the elevation medians of the winter and summer ranges. The Mann-Whitney U-test (SPSS 1997) was used to test our hypothesis.

Boxplots revealed that the northern cluster (summer range) fell between approximately 2150 and 2800 m, whereas the southern one (winter range) was located at about 1400–1950 m (figure 10.3). The median elevations for these two zones differ significantly (Mann-Whitney U-test, $df = 1$, $p < 0.01$). Our finding is broadly similar to those of Pan et al. (1988), who found

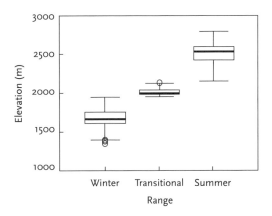

FIGURE 10.3. Boxplots of winter ($N = 1170$), transitional ($N = 22$), and summer ($N = 447$) ranges for giant pandas in the FNR. N is the number of samples.

winter ranges in the adjacent reserve (CNR) to be below 1900 m, whereas summer ranges were above 2300 m.

Each panda had its own core area within its seasonal range, indicated by the high density of points within the two clusters (see figure 10.2). However, there was overlap, even in core areas, for some individuals. Pandas 127 and 043 each had overlapping areas in their summer ranges, whereas pandas 065 and 045 overlapped in both their summer and winter ranges. The summer range of panda 083 during 1995 was on the northern slope of GuangTouShan, far away from the other individuals. Panda 065 is the subadult male offspring of panda 045, and they had nearly total range overlap. However, it appeared that panda 065 had begun to expand his ranging, perhaps to find his own core area. In their winter habitat, pandas 045, 065, and 005 confined themselves to the left side of the DongHe River, whereas pandas 043, 083, and 127 stayed on the right side.

The pandas spent relatively little migration time (June 7–15 and September 1–October 6) in the area between 1950 and 2150 m in elevation (located between the summer and winter ranges). We regarded this area as a transitional zone. We identified only twenty-two tracking points in this 200-m zone (figure 10.3).

ANNUAL MIGRATION PATTERN AND TIME SPENT IN SUMMER AND WINTER RANGES

The existence of elevationally distinct summer and winter ranges requires strategically timed migrations for their exploitation (Baker 1978). The ecological studies referenced earlier indicate that the approximate timing of migratory movements of pandas by season can be used to study migration patterns in detail. For our study, we analyzed data from the periods of May 20–July 3 and August 15–October 15 to document the temporal pattern of migrations more precisely. These periods were subjectively determined, based on the obvious changes in elevation shown in figure 10.4.

Our data reveal that the pandas ascended from winter to summer ranges over a period of about 8 days (June 7–15), but often in as little as 2–3 days. Descents in the fall typically lasted 36 days on average (September 1–October 6; figure 10.4). Excluding the average of 44 days in transition time annually, the FNR pandas spent 240 days in their winter habitat each year (October–May) and 78 days in the summer habitat (June–August).

These shifts occurred in each year of the study and may be regarded as a long-term pattern of exploitation of the FNR habitat. Data are similar for CNR pandas, except that the spring move upward occurs in May–June (Pan et al. 1988). According to Pan et al. (1988), pandas at WNR spend the greater part of the year above 2700 m to feed on *B. fangiana,* and some apparently remain there on a year-round basis. There is a relatively short period from roughly late April/early May to about mid-June, however, when the majority moves down to lower elevations to feed on the new shoots of *F. robusta.* Although it is sometimes concluded that the WNR pandas fail to show seasonal movements, we believe theirs is just a different pattern of migration than is found in other locations.

During our period of study, tracking data on some pandas were missed in the spring move because they departed their winter range too quickly to be tracked. Having good information in hand on this point (i.e., from this study) should make future tracking efforts more complete.

SIZE OF SEASONAL RANGES

Once zones (ranges) had been defined by elevation, the minimum convex polygon method (see White and Garrott 1990) was used to determine their size. The intermediate (transition) zone was excluded from the two seasonal ranges because it was used as a temporal corridor only for a short period of movement between them. For comparative purposes, the average sizes of winter and summer ranges for males and females were calculated separately (table 10.2). We hypothesized that males, based on research at WNR (Hu 1990), generally would have larger ranges than did females. We also hypothesized that winter ranges would be larger than summer ones for all FNR pandas. We used the Mann-Whitney U-test to confirm or reject these hypotheses.

For all radiocollared pandas, winter ranges were significantly larger than those used in the summer (table 10.3, Mann-Whitney U-test, $df = 1$, $p < 0.05$). However, when broken down by sex, only females had a significantly larger winter range ($p < 0.05$). The smaller size of summer ranges derives from the convergence of pandas on dense stands of *F. spathacea* near the top of GuangTouShan Mountain. No shortage in the availability of *F. spathacea* was evident during the months spent at this elevation (see also Hu 1990).

The total range size for each individual was larger when data from all 5 years were pooled than for any one of their annual ranges (see figure 10.2), indicating that each panda shifted its range somewhat from year to year. When range size for the two annual seasons are summed, the average size for males is 6.7 km² and that for females is 5.0 km². In WNR, males are reported to have a range of about 6–7 km², and females a smaller range of about 4–5 km² (Hu 1990). However, transitional ranges were included in determinations of seasonal ranges at WNR. It is possible that range size is to some

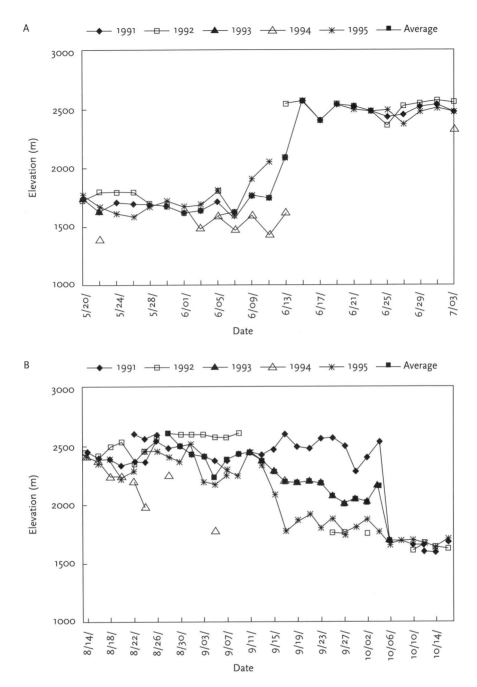

FIGURE 10.4. Migration between summer and winter ranges and time spent per annum in each. (A) Elevational change of panda movement from winter to summer activity ranges. (B) Elevational change of panda movement from summer to winter activity ranges.

TABLE 10.2
Comparison of the Size of Winter and Summer Ranges (km²)

ID NUMBER	SEX	1991 W[1]	1991 S	1992 W	1992 S	1993 W	1993 S	1994 W	1994 S	1995 W	1995 S	AVERAGE W	AVERAGE S
005	Male							2.2	1.6	6.0	2.3		
127	Male	3.1	4.3	5.2	2.1	1.1				4.2		3.8	2.9
065	Male			4.5	1.3	3.8		2.3	5.5	5.8	3.4		
045	Female			2.3	0.7	4.2		2.6		4.5	2.4		
043	Female		1.4	2.7	1.2							3.3	1.7
083	Female									3.2	3.0		

[1] w, Winter; s, summer.

TABLE 10.3
Significance Tests for Winter versus Summer and Male versus Female Range Size

GROUP	VARIABLE		SIGNIFICANCE[1]
Male versus female	Winter activity range	0.550	Not significant
	Summer activity range	0.167	Not significant
	Total activity range	0.089	Not significant
Winter versus summer activity ranges	Male panda's winter and summer activity range	0.261	Not significant
	Female panda's winter and summer activity range	0.045	Significant
	All pandas' winter and summer activity ranges	0.037	Significant

[1] Mann Whitney U-test at the level of $p = 0.05$.

extent related to sample size; that is, winter ranges are larger because pandas were in this range for a much longer period of time each year.

HABITAT UTILIZATION

The distribution of tracking points in relation to the type of ground cover, elevation, steepness of slope, and aspect for winter and summer ranges is shown in figure 10.5. There were many more tracking points to use when measuring wintertime preferences, because the pandas spend four times as much time in the winter range as in their summer range.

Each range can be summarized according to the highest preferences shown for each variable. Thus, in winter, the pandas preferred deciduous broadleaf forest at an elevation of 1600–1800 m, on slopes between 10° and 20° in steepness, and facing in a southward direction, possibly to increase exposure to the sun. In contrast, in their summer ranges, pandas were found nearly equally often in conifer and mixed conifer/broadleaf forest at an elevation of 2400–2600 m,

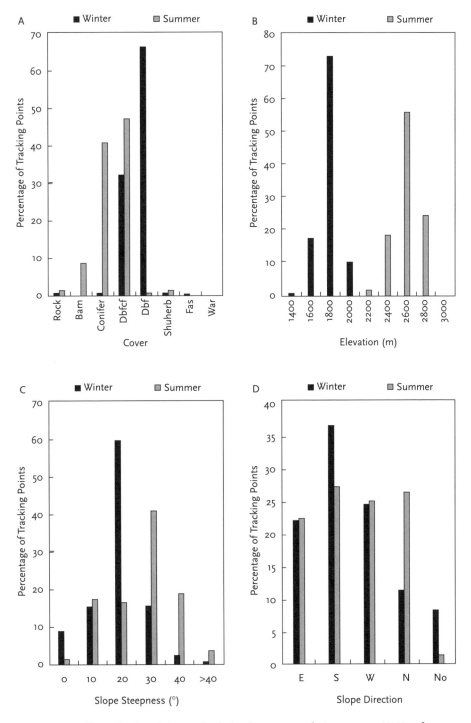

FIGURE 10.5. Habitat utilization of giant pandas during the summer and winter seasons. (A) Use of groundcover based habitat types: Rock, rock and bare soil; Bam, mixed bamboo grove and meadow; Conifer, conifer forest; Dbfcf, mixed conifer and broadleaf forest; Dbf, deciduous broadleaf forest; Shuherb, shrub and herb cover; Fas, dry arable land and settlements; War, water and rice land areas. (B) Use of elevation ranges. (C) Use of slope steepness ranges. (D) Use of slope direction: E, east (45°–135°); S, south (135°–225°); W, west (225°–315°); N, north (315°–360°; 0°–45°); No, no aspect.

on slopes between 20° and 30° in steepness, and with no clear preference for slope face. Note that some of the variables we measured are not independent, which partly accounts for the preferences shown.

In describing habitat usage, it is assumed that (1) a species will select areas that are best able to satisfy its life requirements; and (2) as a result, the species will make greater use of the higher-quality habitats (Verner et al. 1986). It has been shown for other species having seasonal ranges (e.g., African elephants [*Loxodonta africana*]) that food availability is the major factor in range selection (Babaasa 2000). In FNR, the vegetation types are vertically distributed, with deciduous broadleaf forest and an abundant understory of *B. fargesii* occurring between 1600 and 1800 m. This is the main food source for giant pandas in the winter season. At higher elevations (2400–2600 m), *F. spathacea* occurs with mixed conifer and deciduous broadleaf forests, providing a higher-altitude summer food source and necessitating a seasonal pattern of migration. In contrast, at approximately 1900–2200 m, bamboo is of limited availability and accounts for the fairly rapid migration upward in the spring.

Our findings confirm the belief of Chinese biologists that giant pandas prefer gentle over steep slopes. The somewhat steeper slope preference in the summer range is undoubtedly because gradients in this area are generally steeper than in the lower reaches, and the pandas show a preference for the gentlest slopes available to them. Our data on aspect do not suggest strong preferences, especially in the summer range.

SUMMARY

We combined GIS analysis with radiocollaring to explore the seasonal movements of giant pandas at the FNR and the habitat factors regulating those patterns. In all 5 years of the study, the entire population followed a migratory pattern between summer and winter ranges that was documented in detail for six radiocollared individuals. The pandas spent only about 3 months of the year higher elevations in the summer, when abundant *F. spathacea* became the primary food source. Migration from winter to summer ranges occurs in a fairly brief period of about 8 days, there being little bamboo available in the transition zone between the two ranges. Winter ranges, where the main food was *B. fargesii*, were significantly larger than summer ones, but this effect is probably due to the densely concentrated bamboo in the summer range. As was predicted, the FNR pandas preferred gentler slopes over steep ones, but showed only modest differences in their preferences for the direction of slope faces.

ACKNOWLEDGMENTS

Financial and facility support from several organizations was essential to the completion of this work. We thank the International Institute for Aerospace Survey and Earth Science in Enschede, The Netherlands, for scholarship and Ph.D. support for Dr. Xuehua Liu. Our thanks also go to the American Zoo and Aquarium Association, Silver Spring, Maryland, and China's State Forestry Administration, Beijing, whose assistance made possible the earlier stages of fieldwork. We appreciate the assistance of Dr. Xiaoming Shao, Chinese Agricultural University, Beijing, with our fieldwork in the summer of 1999. We thank an anonymous reviewer and the editors for comments on earlier drafts of this chapter.

REFERENCES

Babaasa, D. 2000. Habitat selection by elephants in Bwindi Impenetrable National Park, south-western Uganda. *Africa J Ecol* 38:116–22.

Baker, R. R. 1978. *The evolutionary ecology of animal migration*. London: Hodder and Stoughton.

Hu, J. 1985. *The giant panda in Wolong Nature Reserve*. (In Chinese.) Sichuan: Sichuan Science Press.

Hu, J. 1990. The biological study of the giant panda in Sichuan. In *Research and progress in the biology of the giant panda*. (In Chinese.) Sichuan: Public House of Science and Technology.

Li, X., Y. Yong, and J. Zhang. 1997. Report on the giant pandas' survival in Foping. (In Chinese.) *Acta Zool Sin* 43:285–93.

Liu, X. 2001. *Mapping and modeling the habitat of the giant panda in Foping Nature Reserve, China.* The Netherlands: ITC.

Liu, X., A. K. Skidmore, and H. Van Oosten. 2002a. Integration of classification methods for improvement of land-cover map accuracy. *ISPRS J Photogram Remote Sensing* 56:257–68.

Liu, X., A. K. Skidmore, T. Wang, Y. Yong, and H. H. T. Prins. 2002b. Giant panda movements in Foping Nature Reserve, China. *J Wildlife Manage* 66:1179–88.

Moore, D. S., and G. McCabe. 1998. *Introduction to the practice of statistics.* New York: W. H. Freeman.

Pan, W., Z. Gao, and Z. Lü. 1988. *Natural refuge of the giant panda in Qinling.* (In Chinese.) Beijing: Peking University Press.

Pan, W., Z. Lü, and G. Meng. 1989. *The seasonal movement behavior of the giant panda in Qingling.* (In Chinese.) Abstract, 50th annual academic conference, Zoological Society of China, pp. 474–75.

Peuquet, D., J. R. Davis, and S. Cuddy. 1993. Geographic information system and environmental modelling. In *Modelling change in environmental systems,* edited by A. J. Jakeman, M. B. Beck, and M. J. McAleer, pp. 543–56. New York: John Wiley and Sons.

Ren, Y., M. Wang, M. Yue, and Z. Li, eds. 1998. *Plants found in the giant panda's habitat in the Qinling Mountains.* (In Chinese.) Xian, Shaanxi: Sciences and Technology Press.

Ruan, S. 1983. Observations on feeding by giant pandas in the field. (In Chinese.) *Wild Anim Mag* 1:5.

Schaller, G. B., J. Hu, W. Pan, and J. Zhu. 1985. *The giant pandas of Wolong.* Chicago: University of Chicago Press.

Shaanxi Biological Resources Survey. 1976. Primary survey of the giant panda in Qinling Mountains, Shaanxi. (In Chinese.) *Biol Surv (Anim Part)* 3: 91–104.

SPSS 1997. *SPSS base 7.5 applications guide.* Chicago: SPSS.

Verner, J., M. L. Morrison, and C. J. Ralph. 1986. *Wildlife 2000—Modelling habitat relationships of terrestrial vertebrates.* Madison: University of Wisconsin Press.

White, G. C., and R. A. Garrott 1990. *Analysis of wildlife radio-tracking data.* San Diego: Academic.

Wu, J. 1981. Research history of the giant pandas of Qinling. (In Chinese.) *Wild Anim Mag* 4:8–10.

Wu, J. 1986. The giant panda in Qinling. (In Chinese.) *Acta Zool* 32:92.

Yang, X., and Y. Yong. 1998. Winter habitat selection of giant panda in Foping. In *Vertibrate animal resource and protection,* edited by Hu Jinchu, pp. 20–31. (In Chinese.) Sichuan: Science and Technology Press.

Yang, X., S. Meng, Y. Yong, T. Wang, and J. Zhang. 1997. Research on the giant pandas' environmental ecology in Foping (I)—Vegetation classification of summer habitat and environmental factors. (In Chinese.) *J Xibei Univ Nat Sci* 27:508–13.

Yang, X., S. Meng, Y. Yong, T. Wang, and J. Zhang. 1998. Research of the giant pandas' environmental ecology at Foping (II)—Summer habitat selection. (In Chinese.) *J Xibei Univ Nat Sci* 28: 348–53.

Yong, Y. 1981. Preliminary observations of the giant panda of Foping. (In Chinese.) *Wild Anim Mag* 3:10–16.

Yong, Y. 1989. Preliminary observations of panda breeding dens in Foping. (In Chinese.) *Zool Mag* 3:10–14.

Yong, Y., J. Zhang, and S. Zhang. 1993. The distribution and population of the giant panda in Foping. (In Chinese.) *Acta Theriol Sin* 13:245–50.

Yong, Y., K. Wang, and T. Wang. 1994. Movement habits of the giant pandas in Foping. (In Chinese.) *Acta Theriol Sin* 14:9–14.

BRIEF REPORT 10.1

Noninvasive Techniques for Monitoring Giant Panda Behavior, Habitat Use, and Demographics

Matthew E. Durnin, Jin Yan Huang, and Hemin Zhang

RELIABLE DATA on the distribution, abundance, and behavior of wild giant pandas (*Ailuropoda melanoleuca*) are often impractical or impossible to acquire due to their secretive and elusive natures and/or highly dispersed populations. Furthermore, researchers and wildlife managers have found it difficult to implement standardized sampling designs to measure population dynamics because of the mountainous and densely vegetated habitats that pandas inhabit. As part of a larger research effort focusing on giant panda reproductive behavior, demographics, and genetic sampling, we are devising, evaluating, and establishing practical monitoring protocols that are noninvasive and cost-effective on a variety of scales ranging from single forest stand to entire mountain range.

Although such noninvasive sampling techniques as remotely activated camera systems and barbed-wire hair snares have become widely utilized in bear population research and monitoring (Garshelis et al. 1993; Mace et al. 1994; Taberlet and Luikart 1999; Mills et al. 2000), prior to our work, they have not been applied to the study of giant pandas. We were interested in assessing the potential of remotely activated cameras and hair snares to monitor pandas' use of critical habitat (e.g., scent stations, dens), identify the presence of pandas in specific areas (e.g., those under consideration for protection), and monitor and estimate panda populations.

REMOTE ANIMAL-ACTIVATED CAMERAS AND PANDA BEHAVIOR

Studies of giant pandas in both the wild and captivity suggest that olfactory communication plays a critical role in reproduction, social structure, dispersion, territorial maintenance, and competition in giant panda populations (Kleiman 1983, 1985; Schaller et al. 1985; Swaisgood et al. 1999, chapter 7). Therefore, detailed study of scent-marking behavior in wild giant pandas can provide a better understanding of the role and importance of olfactory communication in reproduction, the formation of social bonds between adult and subadult animals, spacing patterns, and cross-species interactions, such as may occur between giant pandas and Asiatic black bears (*Ursus thibetanus*).

To determine the feasibility of using noninvasive techniques to study panda scent-marking behavior in the wild, we first needed to monitor the frequency of their visits to communal scent stations. Scent stations are usually found in prominent sites throughout the landscape (e.g., knolls, high ridges, low passes) and are located along routes frequented by both resident and transient pandas. Large conifer and deciduous trees make up the majority of scent stations, which are readily identified through their characteristic build up of secretions, smoothed bark, and claw marks (Schaller et al. 1985). During walks along randomly located transects, established trail systems, and game trails, we identified forty-one scent stations in the Wuyipeng study area (roughly 33 km²), lying in the Wolong Nature Reserve in Sichuan Province.

Eight scent stations were randomly selected and remote camera systems were set up at each station. Camera stations consisted of a Trail-Master® TM-500 passive infrared trail monitors combined with the TM-35-1 camera kit (Goodson and Associates, Lenexa, Kansas). The monitor detects the combination of heat and motion in the area and sends a signal to the camera, triggering its mechanism. The system was programmed to record an event and take photos at 1-min intervals for as long as an animal was in the area. A time and date stamp was imprinted on each photo taken.

The eight scent stations were monitored for a total of 648 camera days (1 day = a camera in place for 24 h), and produced 378 photos; five panda photographs (1.3%) and 107 (28.3%) photographs of other animals (table BR.10.1). Of the remaining photos, 36 (9.5%) were test photos and 230 (60.9%) were false triggers. The five panda visits translate to a photographic rate or sampling effort (defined as the number of camera days per panda photograph summed across all camera traps) of 129.6 camera days/panda photo.

Remotely activated camera systems were also deployed at seventeen barbed-wire hair snares at the Wuyipeng site and at a second location known as the Niu Tou Shan study site. Our objective was to photograph animals that would be genetically tagged through DNA analysis of hair samples, thereby setting up a mark-recapture system for estimating population size. Also, because baits were disappearing from the hair-snare sites, we wanted to identify the animals involved, so that adjustments could be made in trap design to reduce bait loss.

The seventeen hair-snare stations were monitored for 445 camera days and produced 236 photographs: one panda photograph (0.4%) and twenty-one (8.9%) photographs of other animals (table BR.10.2). Of the remaining photographs, fifty-three (22.5%) were test photos and 161 (68.2%) were false triggers. The single panda visit translates to a photographic rate or sampling effort of 445 camera days/panda photo.

Two small to medium-sized carnivores, the Siberian weasel *(Mustela sibirica)* and yellow throated marten *(Martes flavigula)*, were the most common visitors to the hair snares. Of the twenty-two photographs of animals, eighteen (82%) were either weasels or martens. Their small size and climbing ability likely enabled them to negotiate the bait cables at each trap with relative ease. We concluded that they were the major cause of bait loss.

BARBED-WIRE HAIR SNARES AND POPULATION MONITORING

Estimating abundance and population trends is critical to wildlife conservation. Unfortunately, as is too often the case for such rare and elusive species as the giant panda, estimates of population size or survival rates have been plagued by small sample size and have been limited to raw counts or indices with unknown relationships to true population size. For many species, the lack of reliable estimates has led wildlife biologists to consider other indirect measures, such as the photographic "capture" technique just described, or, more recently, noninvasive sampling of DNA to estimate population size (Garshelis et al. 1993; Mace et al. 1994; Cutler and Swann 1999; Woods et al. 1999). In this study, we wished to determine the feasibility and efficacy of using barbed-wire snares to collect hair for molecular analysis. If successful, this would provide a

TABLE BR.10.1
Photographic Records from Eight Randomly Selected Scent Stations

	STATION ID (SCENT STATION NUMBER)									
	ps1 (ss1)	ps3 (ss15)	ps4 (ss16)	ps6 (ss13)	ps7 (ss27)	ps8 (ss28)	ps9 (ss29)	ps10 (ss31)	TOTALS	ESTIMATED NUMBER OF INDIVIDUALS
Trap days	200	184	52	69	42	90	50	51	738	
Giant panda	—[1]	3/2[2]	—	—	—	—	1/1	1/1	5	4
Asiatic black bear	—	—	—	—	—	—	1/1	—	1	1
Red panda (*Ailurus fulgens*)	3/2	4/2	—	—	—	—	—	—	7	4
Forest musk deer (*Muschus berezovskii*)	3/3	—	—	—	—	—	2/2	—	5	5
Sambar deer (*Cervus unicolor*)	2/2	—	—	—	3/2	—	7/3	—	12	7
Takin (*Budorcus taxicolor*)	1/1	—	—	—	—	—	—	—	1	1
Wild pig (*Sus scrofa*)	—	—	—	—	—	—	12/3	—	12	3
Tragopan (*Tragopan caboti*)	2/2	63/6	—	—	—	—	1/1	—	66	9
Flying squirrel (*Petaurista petaurista*)	—	1/1	—	—	—	—	—	—	1	1
Dog (*Canis familiaris*)	—	—	—	—	1/1	—	—	—	1	1
Porcupine (*Hystrix hodgsoni*)	—	—	—	—	1/1	—	—	—	1	1

NOTES: Days monitored varied for several reasons. In some cases equipment malfunctioned (e.g., battery failure) and in some other cases, an entire roll of film was shot on a single animal. In these instances, the number of days monitored stopped on the date of the last photo taken during a session.

[1] —, No data.

[2] Total photos/estimated number of individuals; that is, the first number listed is the number of photos taken, second number is the estimated number of different individuals in the photos. So, for example, three photos of pandas were taken at station 3 but represent only two individuals.

TABLE BR.10.2
Photographic Records from Seventeen Barbed-Wire Hair-Snare Stations

	STUDY AREA		TOTAL NUMBER OF PHOTOGRAPHS	ESTIMATED NUMBER OF INDIVIDUALS
	WUYIPENG	NIU TOU SHAN		
Trap days	269	176		
Giant panda	—[1]	1/1[2]	1	1
Red panda	1/1	2/1	3	2
Siberian weasel (*Mustela sibirica*)	5/5	10/3	15	8
Yellow throated marten (*Martes flavigula*)	3/2	—	3	2

[1] —, No data.
[2] Total photos/estimated number of individuals; that is, the first number listed is the number of photos taken, second number is the estimated number of different individuals in the photos.

unique genetic profile of the immediate population and allow us to estimate population size using a mark-recapture technique.

Utilizing methods described by Zielinski and Kucera (1995), we divided two areas (Wuyipeng and Niu Tou Shan), totaling approximately 56 km², into twelve sampling units, each measuring about 4 km² in size. Two barbed-wire hair snares were placed in each 4-km² sampling unit and located a minimum of 1 km apart. Trap locations within sampling units were chosen to be representative of the elevation gradients and habitat types available to pandas in each area. Six sampling units (twelve traps) were monitored in the Wuyipeng area for 1 month in the spring and again for a month in the fall of 1999. Six sampling units (eleven traps) in the Niu Tou Shan area were monitored for 1 month in the spring of 1999.

Hair snares consisted of a bait (smoked chicken or goat meat) strung between two trees out of reach of a panda standing on its hind legs (approximately 3–4 m high) and a perimeter fence of barbed wire running around three or more trees at about 50 cm above the ground (Woods et al. 1999). Traps were baited at the beginning of each month and again with fresh bait at mid-month. We checked traps every 7 days.

Hair samples from single barbs were placed in 4-oz Whirl-Pak® sterile sample bags (Nasco Industries, Fort Atkinson, Wisconsin) filled with silica desiccant. Samples were put in a −20°C freezer within 24 h of collection.

The twenty-three hair snares were monitored for 830 trap days (1 trap day = a trap in place for 24 h), and produced twenty-seven hair samples (30.7 trap days/hair sample). Each trap was operational for 23 days on average. Nine of them (39.1%) captured more than one hair sample and four (17.4%) captured more than one giant panda hair sample over the entire sampling period. Samples were visually inspected under a dissecting microscope and given an initial classification as panda, other animal, or undetermined. Four samples were determined to be from pandas (14.8%), fourteen (51.9%) from other animals, and nine (33.3%) were undetermined (table BR.10.3). The four panda samples translate to a capture rate (defined as the number of 24-h trap days per hair sample summed across all barbed-wire hair snares) of 207.5 trap days/sample capture.

DISCUSSION

Our experience with remote animal-activated cameras and barbed-wire hair snares suggests

TABLE BR.10.3
Number of Hair Samples Collected by Study Area and Hair Trap

	WUYIPENG STUDY AREA		NIU TOU SHAN STUDY AREA	
	SAMPLING SESSION			SAMPLING SESSION
TRAP NUMBER	APRIL 1–22, 1999	SEPTEMBER 17– OCTOBER 15, 1999	TRAP NUMBER	OCTOBER 20– NOVEMBER 11, 1999
HT1	0	0	HT14	0
HT2	2	5	HT15	1
HT3	0	0	HT16	0
HT4	0	0	HT17	0
HT5	3	0	HT18	0
HT6	0	0	HT19	0
HT7	3	1	HT20	0
HT8	0	0	HT21	0
HT9	0	5	HT22	2
HT10	0	0	HT23	0
HT11	0	3	HT24	1
HT12	1	0		

that the usefulness of these techniques is of limited value in giant panda studies, due to the low rate of return per sampling effort. Further refinement of techniques, such as testing different baits, may improve the rate and quality of return and thus their usefulness.

Our initial objective was to study the scent marking-behavior of pandas, but procurement of photographs of only five visits over 648 camera days suggests that, at least for this activity, such techniques as radiocollaring or ear tagging might be required. Scent station monitoring did confirm that four stations were in fact being used for marking and at least four different pandas were in the area. As these and other results indicate, the remote cameras may be useful in presence/absence studies. We were able to confirm, for example, the presence of two species afforded Level I protected status under Chinese law and four Level II protected species. By increasing sample unit size, camera stations could be deployed on a reserve-wide scale as a means of creating coarse distribution maps.

Additionally, even though remote cameras were inexpensive and easy to use, they required significant labor and were subject to loss or damage by humans. The high number of false triggers were most likely attributable to weather variables (e.g., cold and wet conditions) or even the sensitivity of the passive infrared units to small changes in background heat signatures, as when bamboo is warmed by the sun and then blown by the wind. More frequent visits to camera stations to remedy weather-caused problems and use of infrared triggers that require a break in the sensor beam might improve their usefulness.

Baited barbed-wire hair snares experienced much greater visitation by carnivores other than pandas, (e.g., weasels, martens). Although bamboo comprises more than 99% of its diet, the panda's taste for meat is quite well documented (Schaller et al. 1985). At this time, we do not know if pandas were less attracted to meat baits than were other carnivores in the area. Those panda hair samples that were collected could be the result of chance encounters with the traps while foraging for bamboo.

Noninvasive sampling of DNA would seem to be a very promising approach to estimating population size and monitoring panda populations. However, barbed-wire hair snares may not be the best method for the collection of samples for genetic tagging. For example, in the same 3 months that hair snares were employed in the Wuyipeng and Niu Tou Shan study areas, we collected forty-eight hair and fifty-six fecal samples by walking established trail systems, game trails, and randomly selected strip transects. Thus, we collected not only a much larger sample but probably a more representative one of the resident pandas. Furthermore, even though deploying and monitoring hair snares require fewer person-hours than traditional transect methods, the method provides a lower return per unit of effort. For example, during our first monitoring session, three people walked transects every day for 21 days (63 person-days) and collected seventeen hair and twenty-nine fecal samples. This translates into an effort of 1.4 person days of effort per sample collected. By contrast, the twelve hair snares set up in the Wuyipeng area yielded only two panda hair samples during the same 21-day session. When deployment and monitoring efforts were tabulated, we found that this approach required 21 person-days per sample.

Hair snares can be deployed in remote locations and require less investment of time on average than transect methods. They would therefore seem to be the preferred method for a census and monitoring program at larger scales (e.g., reserves, mountain ranges). However, our results indicate that it is the return per unit of effort that should guide decisions on which sampling methods to use. A reduction of effort might be realized through improved trap design, more strategic location of traps, or more novel baits. In work with grizzly bears *(Ursus arctos)*, Mowat and Strobeck (2000), for example, realized a higher return of samples at sites where vegetation concealed the wire snares. The same authors report that results improved when more novel baits, such as rotting fish oil and beef formulas, were used.

REFERENCES

Cutler, T. L., and D. E. Swann. 1999. Using remote photography in wildlife ecology: A review. *Wildlife Soc Bull* 27:571–81.

Garshelis, D. L., P. L. Coy, and B. D. Kontio. 1993. Applications of remote animal-activated cameras in bear research. *Int Union Game Biol Congr* 21: 315–22.

Kleiman, D. G. 1983. Ethology and reproduction of captive giant pandas *(Ailuropoda melanoleuca)*. *Z Tierpsychol* 62:1–46.

Kleiman, D. G. 1985. Social and reproductive behavior of the giant panda *(Ailuropoda melanoleuca)*. *Bongo (Berlin)* 10:45–58.

Mace, R. D., S. C. Minta, T. L. Manley, and K. E. Aune. 1994. Estimating grizzly bear population size using camera sightings. *Wildlife Soc Bull* 22:74–83.

Mills, L. S., J. J. Citta, K. P. Lair, M. K. Schwartz, and D. A. Tallmon. 2000. Estimating animal abundance using noninvasive DNA sampling: Promise and pitfalls. *Ecol Appl* 10:283–94.

Mowat, G., and C. Strobeck. 2000. Estimating population size of grizzly bears using hair capture, DNA profiling, and mark-recapture analysis. *J Wildlife Manag* 64:183–93.

Schaller, G. B., J. Hu, W. Pan, and J. Zhu. 1985. *The giant pandas of Wolong*. Chicago: University of Chicago Press.

Swaisgood, R. R., D. G. Lindburg, and X. Zhou. 1999. Giant pandas discriminate individual differences in conspecific scent. *Anim Behav* 57:1045–53.

Taberlet, P., and G. Luikart. 1999. Non-invasive genetic sampling and individual identification. *Biol J Linn Soc* 68:41–55.

Woods, J. G., D. Paetkau, D. Lewis, B. N. McLellan, M. Proctor, and C. Strobeck. 1999. Genetic tagging of free-ranging black and brown bears. *Wildlife Soc Bull* 27:616–27.

Zielinski, W. J., and T. E. Kucera. 1995. *American marten, fisher, lynx, and wolverine: Survey methods for their detection*. Albany, Calif.: Pacific Southwest Research Station, U.S. Department of Agriculture Forest Service.

11

Mapping Habitat Suitability for Giant Pandas in Foping Nature Reserve, China

*Xuehua Liu, M. C. Bronsveld, Andrew K. Skidmore,
Tiejun Wang, Gaodi Dang, and Yange Yong*

THE GIANT PANDA *(Ailuropoda melanoleuca)* survives today in five mountain regions in China (see Hu and Wei, chapter 9). With the passage of time, its forest habitat has been reduced in extent and has also become fragmented. Although the shrinking of the panda's range is partially the result of climatic changes during the Pleistocene epoch, in recent times, it has been caused primarily by human activity (Schaller et al. 1985; WNR et al. 1987; Schaller 1993). China's population explosion and economic development are the main factors responsible for this shrinkage.

The detailed mapping of forest cover from visual interpretation of satellite images reveals that the area of potential panda habitat in Sichuan Province shrank from 20,000 km² in 1974 to only 10,000 km² in 1989. The situation in Gansu and Shaanxi Provinces is similar (Chinese Ministry of Forestry and World Wildlife Fund 1989). It is essential to know the extent of pandas' habitat and its spatial and temporal changes for conservation to be effective. Restoration of lost habitat may not always be possible, but remaining habitat may be maintained and protected.

In assessing habitat through mapping, De Wulf et al. (1988) emphasized that, in the long term, the creation of a digital database and a monitoring system would prove useful to efficient conservation management. Remote sensing (RS) and geographic information system (GIS) utilization are also valuable techniques for analyzing, monitoring, and managing resources (Al-Garni 1996). The need for relatively quick and potentially more economical ways to compile habitat information has led to the use of high-resolution satellite data (Ormsby and Lunetta 1987), and, with the aid of GIS, to the depiction of relatively complex spatial relationships (Wheeler and Reid 1984). The importance of integrating RS and GIS data has been confirmed by many investigators in evaluating habitat (Scepan et al. 1987; Tappan et al. 1991; Roy et al. 1995; Al-Garni 1996; Amuyunzu and Bijl 1999).

The use of RS techniques in surveys of wildlife habitat is still a developing field (Li 1990), and its application to assessment of giant panda habitat is limited. During the past two national censuses of pandas (1974–1975, 1985–1988), habitat mapping was done manually on topographic maps. Although RS data have been applied to panda habitat assessment in a few reserves, the assessments were based mostly on visual interpretation (Morain 1986; De Wulf et al. 1988, 1990; Ren 1989; Chui and Zhang 1990; Li 1990; Ren et al. 1993; MacKinnon and De Wulf 1994). Application of the GIS approach to mapping and evaluating panda habitat started with Ren et al. (1993) and has been developed by Ouyang et al. (1996), Liu et al. (1997, 1998) and Chen et al. (1999).

Our study builds on a mapping approach that integrates an expert system and a neural network system into our method of assessing habitat suitability. Use of an artificial intelligence (AI) system could have a notable impact on future evaluations, because learning procedures can adapt GIS to the imprecise and voluminous nature of geographically based data as the data are acquired (Peuquet et al. 1993). The aim of the entire system is to effectively integrate RS data, other digital environmental data, ground data, and expert knowledge to extract and map panda habitat suitability with a high degree of accuracy.

METHODS

STUDY AREA

This study was carried out on Foping Nature Reserve (FNR) in Shaanxi Province. (For a description of FNR, see Yong et al., chapter 10.)

FIELD SURVEY AND RADIOTRACKING

A survey of FNR habitats was carried out in July–August 1999, and was combined with LANDSAT TM data acquired in July 1997. The line transect sampling method was adopted. Eight ground-cover-based habitat types (conifer forest, mixed conifer and broadleaf forest, deciduous broadleaf forest, bamboo mixed with meadow, shrub-grass-herb land, farmlands with settlements, rock and bare land, water and rice land) were sampled for 160 survey points (Liu 2001). Sightings of panda sign were also recorded for each survey point. Sample plots with measured habitat parameters or observations, following Doering and Armijo (1986), were assumed to reliably reflect habitat conditions.

Radiotracking data were assumed to reflect the pandas' selection of habitat and were extrapolated to the entire reserve. Because of governmental restrictions on radiocollaring of giant pandas, we were fortunate to obtain tracking data from six individuals over a period of 5 years (see Yong et al., chapter 10) and to use such data in an assessment of their habitat. The core areas of these individuals were identified for their summer and winter ranges and the spring mating season (figure 11.1). As a first effort of this kind, our methodology is somewhat exploratory in nature. However, because we used ten digital data layers in our model, we believe we were able to reduce the biases of small sample size and can generalize from the FNR pandas to the larger population.

ESNNC ALGORITHM

The expert system classifier (ESC), the neural network classifier (NNC), and the integrated system derived from these two components (ESNNC) were described by Liu et al. (1999). The NNC learns from the samples and therefore depends on the accuracy of the information the sample sets provide. However, the ESC extracts expert knowledge from impressions of data distribution in different patterns and the field survey experience. Therefore, the ESC is a sample-free method. The ESNNC approach integrates ESC and NNC, and trains the whole system to reach the targets through learning from known samples. Useful information from the ESC results is introduced into the NNC before and after running the system.

The initial stage of ESNNC (i.e., inputting the output of the ESC into the NNC as an additional information layer) is based on the principle that the neural network system is very sensitive to subtle changes in the input data. The system is then trained by different, randomly selected sample sets and produces several output maps. A frequency-checking program is used to compare all output maps to obtain the majority class for a single pixel and assigns that pixel to the majority class. In this way, the combined suitability-based habitat map is formed.

The second stage of ESNNC is to use the output of the ESC through its producer accuracy

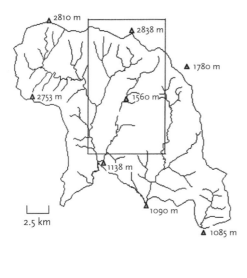
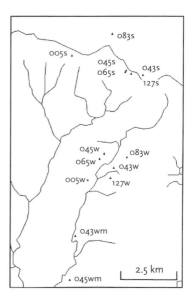

FIGURE 11.1. The center of individual pandas' winter, summer, and mating activity ranges. Numbers refer to individuals; w, winter range; s, summer range; m, mating range.

and some newly built-in rules based on expert knowledge to correct the output of the initial stage of the ESNNC. For example, the winter habitat should not occur in high-elevation areas and slope steepness should not be greater than 35°, based on the definitions of the classes. Figure 11.2 shows the schematic of this mapping approach. For comparison, the traditional maximum likelihood classifier (MLC) is applied to classification of habitat types, using the same data.

MAPPING SUITABILITY-BASED PANDA HABITAT TYPES

Ten digital data layers (figure 11.2) were used as the initial information source of the classification system; that is, RS data (LANDSAT TM bands 1–5 and 7, acquired in July 1997), terrain data (elevation, slope steepness, slope aspect), and social data (distance to nearest area of human activity). Survey points (160) and nonoverlapping radiotracking points (1425) were used for mapping. All the data have the same coordinate system—Universal Transverse Mercator (UTM)—and all digital maps were georeferenced with the same pixel size of 10×10 m². For compari-

son, the ten data layers and a grand total of 1585 points were used for mapping suitability-based habitat types by use of the ESC, NNC, and ESNNC, as well as the traditional MLC systems.

Eight classes of habitat suitability were defined: very suitable summer habitat (vss), suitable summer habitat (ss), transitional habitat (tr), suitable winter habitat (sw), very suitable winter habitat (vsw), marginal habitat (ms), unsuitable habitat (us), and water-rich areas (w). Multiple criteria were used to define suitability (table 11.1) to render the habitat model more accurate and reasonable.

A stratified random sampling strategy was applied for all 1585 points in order to obtain samples for each class. Initially, seven hundred samples were selected for use as a test set. The remaining 885 points were used to select the training samples (table 11.2). The classification of suitability-based habitat types was carried out fifteen times on the initial-stage ESNNC with fifteen different training sets. By checking the resulting maps against a frequency-checking program, a final output map from the ESNNC was produced.

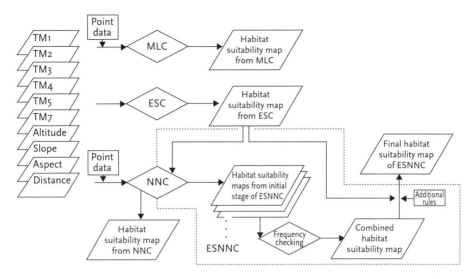

FIGURE 11.2. Integrated expert system and neural network classifier (ESNNC) for mapping suitable panda habitat. TM1–5 and 7 represent LANDSAT TM image bands. The input map "Distance" represents the distance to areas of human activity. MLC, maximum likelihood classifier; ESC, expert system classifier; NNC, neural network classifier; ESNNC, integrated expert system and neural network classifier.

The other three classifiers (NNC, ESC, and MLC) were carried out only once, using one training sample set. Four output maps from the ESNNC, NNC, ESC, and MLC were tested by seven hundred independent testing points, and the mapping accuracies of the integrated ESNNC and the other three classifiers were compared using Z statistics.

DEFINING HABITAT SUITABILITY

Eight suitability-based habitat types were identified and their spatial distribution mapped from ESNNC (figure 11.3). Suitable and very suitable summer habitats were found in the north, northwest, and northeast boundary areas of FNR (figure 11.4). The very suitable summer habitat constituted only a small part of the total summer habitat and was distributed mainly in the GuangTouShan area. The suitable and very suitable winter habitats were mainly in the central and southern areas (figure 11.5), together with small human activity areas. The transition zone between the two seasonal habitats was identified, and was found to be fairly wide in the northeastern part.

Marginal habitats with slope steeper than 35° were scattered mainly on the southern slopes of GuangTouShan, XiHe, JinShuiHe, and LongTangZi Rivers. Only a small part of the reserve is made up of rock and bare land, farming and settlements, shrub-herb-grass land, and water and rice land. Unsuitable areas (i.e., rock and bare land, farming and settlements, shrub-herb-grass land) were found at the mountain crests or in the river valleys.

The availability of our eight suitability-based habitat types are shown in table 11.3. More than 50% of the area at FNR was winter habitat, and nearly half of this area was classified as very suitable. Summer habitat made up less than 20% of the reserve, and only about 6% was classified as very suitable. Nearly one-fifth of the reserve was used as a transition zone. Just under 13% of the reserve was classified as marginal habitat (e.g., steep slopes, barren areas).

The integrated mapping approach was compared with the three other methods. Table 11.4 shows four main aspects in mapping suitability-based habitats; namely, the number of classes identified, overall mapping accuracy, kappa value,

TABLE 11.1
Criteria Used in Assigning Sample Data to Habitat Suitability Classes

CRITERION[1]		VSW	SW	TR	VSS	SS	MS	US	W
Elevation (m)		1949	1949	1949–2158	2158	2158			
Panda signs		Many	Present		Many	Present			
Slope (°)							>35		
Groundcover-based habitat types								fas rab shgr	war
Distance to the center of summer activity ranges (m)					1000	>1000			
Distance to the center of winter activity ranges (m)	Panda 005 and panda 043 Panda 127 and panda 065 Panda 045 and panda 083	1500 1300 1000	>1500 >1300 >1000						
Distance to the center of mating activity ranges (m)	Panda 043 Panda 045	500 1000	>500 >1000						

ABBREVIATIONS: fas, farmland and settlement; ms, marginal habitat; rab, rock and bare land; shgr, shrub-herb-grass land; ss, suitable summer habitat; sw, suitable winter habitat; tr, transition zone; us, unsuitable habitat; vss, very suitable summer habitat; vsw, very suitable winter habitat; w, water-rich area; war, water and rice land.

TABLE 11.2
Stratified Random Sampling for Suitability-Based Habitat Types through 1585 Samples

CLASS NAME	TOTAL NUMBER OF COLLECTED SAMPLES	SELECTED TESTING SAMPLES	SELECTED TRAINING SAMPLES	REMAINING SAMPLES
Very suitable summer habitat	328	150	150	28
Suitable summer habitat	73	30	30	13
Transition zone	30	14	14	2
Suitable winter habitat	183	80	80	23
Very suitable winter habitat	853	376	377	100
Marginal habitat	60	25	25	10
Unsuitable habitat	47	20	20	7
Water-rich area	11	5	4	2
Total	1585	700	700	185

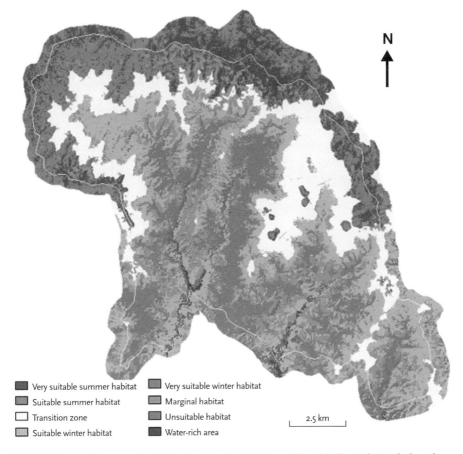

FIGURE 11.3. Spatial pattern of suitability-based panda habitat types. The white line indicates the boundary of FNR.

FIGURE 11.4. Example of *Fargesia spathacea* bamboo habitat at FNR utilized by giant pandas during the summer season. Photograph by Liu Xuehua.

FIGURE 11.5. Example of *Bashania fargesii* bamboo habitat at FNR utilized by giant pandas during the winter season. Photograph by Liu Xuehua.

TABLE 11.3
Availability of Different Panda Habitat Types Based on the ESNNC Approach

SUITABILITY-BASED HABITAT TYPES	AREA (km^2)	PERCENTAGE OF ENTIRE AREA
Very suitable summer habitat	16.6	5.7
Suitable summer habitat	29.3	10.0
Transition zone	57.4	19.6
Suitable winter habitat	88.4	30.1
Very suitable winter habitat	64.9	22.1
Marginal habitat	31.4	10.7
Unsuitable habitat	4.6	1.7
Water-rich area	0.7	0.2

TABLE 11.4
Accuracy Assessment in Mapping Suitability-Based Habitat Types from Comparisons between ESNNC and NNC, ESC and MLC

CLASSIFIER	NUMBER OF CLASSES IDENTIFIED	OVERALL ACCURACY (%)	KAPPA VALUE	KAPPA VARIANCE	Z STATISTIC
ESNNC	8	83	0.742	0.0004	
NNC	8	76	0.640	0.0005	3.25[1]
ESC	8	48	0.358	0.0005	12.72[1]
MLC	7	—	—	—	—

NOTE: —, Not necessary to check accuracy.

[1] Significant difference at the 95% confidence interval. The Z-statistical result indicates that ESNNC produced a habitat suitability map with significantly higher accuracy than NNC or ESC.

and kappa variance. The ESNNC classified habitat types with the highest overall mapping accuracy (83%), followed by NNC (76%) and ESC (48%). The traditional statistical methods (MLC) did not yield satisfactory classification results in this case, because it recognized only seven classes. An unidentified class derived from small sample size, precluding the statistical robustness needed for MLC to run the classification. Results from pairwise comparisons between ESNN, NNC, and ESC systems, using Z statistics, are shown in table 11.4. The ESNNC produced a significantly higher accuracy than did the other two.

ISSUES IN HABITAT IDENTIFICATION AND MAPPING

Our study, for the first time, mapped and assessed panda habitats in FNR using multitype data (i.e., RS, radiotracking, ground survey, and human influence data) in a GIS environment. Wildlife ecologists and managers often map potential animal habitats based on vegetation or land cover and use. However, such an approach may not tell us where the habitats actually preferred by animals are located and, consequently, where they spend most of their time. Mapping of the different habitats derived from our study provided an objective measure of habitat quality. Suitability assessments revealed that the two drainage systems of the DongHe and XiHe Rivers make up the best seasonal habitats, with an easy connection between them. The transition zone in these regions is of moderate width, adding to our conclusions on suitability. The DongHe area is further enhanced by having almost no habitat with a slope steepness greater than 35°.

Surveys in the 1990s revealed a panda population in the DongHe and XiHe river basins of twenty-six and twenty-three individuals, respectively. The third-highest area in terms of numbers of pandas was the DaChengHao area with seven (Yong 1993). The wider portions of transition zones at that location are comparatively flat and require more time for pandas to pass through, particularly when environmental conditions are less favorable. As the transition zone normally lacks robust bamboo, we believe the pandas select those locations with gentle slopes and suitable width in moving between seasonal ranges. Accordingly, pandas in the LongTanZi and DaChengHao areas had the longest transition times.

Summer habitat within the reserve boundary is about 46 km², or only 16% of the entire reserve (see table 11.3), and is probably not sufficient to meet the requirements of the sixty-four pandas counted in the 1990 survey, given an average summer range for individuals of about 2.5 km² (Liu et al. 2002). Adjoining areas outside the boundary undoubtedly form an important part of the summer range. The six radio-collared pandas in the SanGuanMiao area at times moved along the ridges (known as GuangTouShan) and foraged on both slopes (Liu et al. 2002), lending support to this supposition. Because pandas spend less time in their summer ranges than in those occupied during the winter,

it does not necessarily mean that duration of stay is correlated to range size.

The mapping approach used in our study is applicable to the detection and monitoring of changes in panda habitat. As shown in figure 11.3, suitable and very suitable winter habitats occur primarily in the center and southern areas of FNR, coinciding with areas of human activity. Competition between pandas and the local people for forest resources is especially pronounced at lower elevations. For example, the pandas use the understory bamboo as their staple food and the canopy forest as shelter, whereas the local people clear away the groves of bamboo and cut down the trees to grow mushrooms (Liu 2002). This activity rapidly converts forest into other types of land cover. The centrally located SanGuanMiao area is surrounded by both suitable and very suitable panda winter habitats, and relocation of the human population in this area would obviously enhance habitat quality.

Maintaining the continuity of the pandas' summer habitat would also be beneficial. For example, the local government is planning to establish a site in the northeastern corner of the transition area for tourists to use in visiting SanGuanMiao. A path used by the local people in moving between SanGuanMiao and points outside the reserve transects very suitable winter habitat, and with increased usage, will have a negative effect in the form of added noise, garbage, and the like. Moreover, a narrow strip along the mountain ridge, used as a corridor for transition between GuangTouShan and the LongTanZi and DaChengHao areas, would be adversely affected.

CONCLUSIONS

In this chapter, we describe an approach for mapping panda habitat based on digital data (ESNNC). Compared with NNC, ESC, and MLC, ESNNC obtained the highest accuracy in mapping suitability-based panda habitat types and produced needed information for reserve management and conservation. Ours is a practical mapping approach, relying on limited sampling in an area with very difficult terrain. The results show that FNR maintains good quality panda habitat. An area of about 200 km² (70% of the reserve) is suitable for pandas, and 30% (82 km²) is very suitable habitat. The DongHe and XiHe areas are very good areas for pandas and should be maintained as high-quality habitat.

We suggest that farming and other human activities that have a negative impact on pandas in the central SanGuanMiao area should cease. The area in the northeastern corner along the boundary should not be developed as a tourist site, because it would degrade an already limited summer range. Mushroom production, which entails the removal of understory bamboo and canopy trees in the pandas' winter range, should be forbidden, especially in the SanGuanMiao area.

ACKNOWLEDGMENTS

We thank Dr. Xiaoming Shao, Chinese Agricultural University, for his great contribution to the field survey in 1999. We thank Dr. Lalit Kumar for critical comments on an earlier draft of this chapter. We also thank the Chinese State Forestry Administration and FNR for granting us permission to carry out the field survey. Thanks to the Memphis Zoo for sponsoring the first author's attendance at the *Panda 2000* International Conference in San Diego.

REFERENCES

Al-Garni, A. M. 1996. A system with predictive least-squares mathematical models for monitoring wildlife conservation sites using GIS and remotely-sensed data. *Int J Remote Sensing* 17:2479–503.

Amuyunzu, C. L., and B. C. Bijl. 1999. Integration of remote sensing and GIS for management decision support in protected areas: Evaluating and monitoring of wildlife habitats. In *Geoinformation Technology Application for Resource and Environmental Management in Africa*, edited by P. O. Adeniyi, pp. 113–121. Harare, Zimbabwe: The Environment and Remote Sensing Institute.

Chen, L., X. Liu, and B. Fu. 1999. Evaluation on giant panda habitat fragmentation in Wolong Nature Reserve. (In Chinese.) *Acta Ecol Sin* 19:291–97.

Chinese Ministry of Forestry and World Wide Fund for Nature. 1989. *National conservation management plan for the giant panda and its habitat*. Hong Kong: Chinese Ministry of Forestry and World Wide Fund for Nature.

Chui, H. T., and M. D. Zhang. 1990. Giant panda habitat analysis in Xinglongling, Qinling Mountain, using remote sensing. In *Research of biogeography and soil geography*, edited by the Committee of Chinese Geography, pp. 64–70. (In Chinese.) Beijing: Science Press.

De Wulf, R. R., R. E. Goossens, J. R. MacKinnon, and S. Wu. 1988. Remote sensing for wildlife management: Giant panda habitat mapping from LANDSAT MSS images. *Geocarto Int* 1:41–50.

De Wulf, R. R., R. E. Goossens, F. C. Borry, B. P. De Roover, and J. R. MacKinnon. 1990. Monitoring deforestation for nature conservation management purposes in Sichuan and Yunnan provinces, People's Republic of China, using multitemporal Landsat MSS and SPOT-1 HRV multispectral data. *Proceedings, the ninth EARSeL symposium, 1989, Espoo, Finland, June 27–July 1, 1989*, pp. 317–23.

Doering, III J. P., and M. B. Armijo. 1986. Habitat evaluation procedures as a method for assessing timber-sale impacts. *Proceedings, Wildlife 2000 on modeling habitat relationships of terrestrial vertebrates. Stanford Sierra Camp, Fallen Leaf Lake, California, 7–11 October 1984*, edited by J. Verner, M. L. Morrison, and C. J. Ralph, pp. 407–10. Madison: University of Wisconsin Press.

Li, Z. 1990. Investigation for the giant panda habitat using remote sensing. (In Chinese.) *Remote Sensing Environ* (China) 5:94–101.

Liu, X. 2001. Mapping and modeling the habitat of giant pandas in Foping Nature Reserve, China. Ph.D. dissertation, International Institute for Aerospace Survey and Earth Science, Enschede, The Netherlands.

Liu, X. 2002. People, environment and giant pandas in Foping Nature Reserve, China: Mushroom production and panda habitat. (In Chinese.) *Biol Div* 4:63–71.

Liu, X., M. C. Bronsveld, A. G. Toxopeus, M. S. Kreijns. 1997. GIS application in research of wildlife habitat change: A case study of the giant panda in Wolong Nature Reserve. *J Chinese Geogr* 7:51–60.

Liu, X., M. C. Bronsveld, A. G. Toxopeus, M. S. Kreijns, H. Zhang, Y. Tan, C. Tang, J. Yang, and M. Liu. 1998. Application of digital terrain model (DTM) in the habitat research of the endangered animal species. (In Chinese.) *Prog Geogr* 17:50–58.

Liu, X., A. K. Skidmore, H. Van Oosten, and H. Heine. 1999. Integrated classification systems. *Proceedings of the second international symposium on operationalization of remote sensing.* (CD-ROM). Enschede, The Netherlands, 16–20 August.

Liu, X., A. K. Skidmore, T. Wang, Y. Yong, and H. H. T. Prins. 2002. Giant panda movements in Foping Nature Reserve, China. *J Wildlife Manage* 66:1179–88.

MacKinnon, J., and R. R. De Wulf. 1994. Designing protected areas for giant pandas in China. In *Mapping the diversity of nature,* edited by R. I. Miller, pp. 127–42. London: Chapman and Hall.

Morain, S. A. 1986. Surveying China's agricultural resources: Patterns and progress from space. *Geocarto Int* 1:15–24.

Ormsby, J. P., and R. S. Lunetta. 1987. White-tail deer food availability maps from thematic mapper data. *Photogramm Engin Remote Sensing* 53:1081–85.

Ouyang, Z., Z. Yang, Y. Tan, and H. Zhang. 1996. Application of geographical information system in the study and management in Wolong Biosphere Reserve. *MaB Cjoma's Biosphere Reserve,* special issue:47–55.

Peuquet, D., J. R. Davis, and S. Cuddy. 1993. Geographic information system and environmental modeling. In *Modeling change in environmental systems,* edited by W. Sussex, A. J. Jakeman, M. B. Beck, and M. J. McAleer, pp. 543–56. London: John Wiley and Sons.

Ren, G. 1989. Investigation of giant panda bamboo resources using remote sensing. (In Chinese.) *Remote Sensing Environ* (China) 2:34–35.

Ren, G., G. Yu, and M. Yan. 1993. Approach for surveying and management of giant panda bamboo resources by means of GIS. (In Chinese.) *Southwest China J Agr Sci* 6:33–39.

Roy, P. S., S. A. Ravan, N. Rajadnya, K. K. Das, A. Jain, and S. Singh. 1995. Habitat suitability analysis of *Nemorhaedus goral:* A remote sensing and geographic information system approach. *Curr Sci* 69:685–91.

Scepan, J., F. Davis, and L. L. Blum. 1987. A geographic information system for managing California condor habitat. In *Proceedings of the 2nd Geographic Information Systems Conference and Workshops, San Francisco, October 26–30, 1987,* pp. 476–86. Falls Church, Virginia: American Society of Photogrammetry and Remote Sensing.

Schaller, G. B. 1993. *The last panda.* Chicago: University of Chicago Press.

Schaller, G. B., J. Hu, W. Pan, and J. Zhu. 1985. *The giant pandas of Wolong.* Chicago: University of Chicago Press.

Tappan, G. G., D. G. Moore, and W. I. Knausenberger. 1991. Monitoring grasshopper and locust habitats in Sahelian Africa using GIS and remote sensing technology. *Int J Geogr Inf Sys* 5:123–35.

Wheeler, D. J., and M. K. Reid. 1984. A geographical information system for resource managers based on multi-level remote sensing data. Center for remote sensing and cartography report 84-1. Salt Lake City: Utah University.

WNR, STUB, and FBOSP (Wolong Nature Reserve, Sichuan Teachers' University [Biological Department], Forest Bureau Of Sichuan Province). 1987. *The vegetation and resource plants in Wolong.* (In Chinese.) Chengdu: Sichuan Science and Technology Publishing House.

Yong, Y. 1993. The distribution and population of the giant panda in Foping. (In Chinese). *Acta Theriol Sin* 13:245–50.

PANEL REPORT 11.1

Restoring Giant Panda Habitat

Chunquan Zhu and Zhiyun Ouyang

Owing to road construction, widespread logging, and the expansion of agriculture, the habitat of the giant panda *(Ailuropoda melanoleuca)* has been significantly reduced and seriously fragmented. Satellite survey maps indicate that the forest vegetation in the panda's range has decreased by about 50% over the past thirty years. The resulting isolation of some populations has prevented gene exchange between groups. Habitat degradation and decline therefore represents a major threat to the long-term survival of the giant panda (figure PR.11.1).

Data from surveys carried out during 1985–1988 (Chinese Ministry of Forestry and World Wildlife Fund 1989) indicated a low population density for giant pandas compared with other mammals in the same range. However, if given sufficient high-quality forested habitat, panda populations will increase in size and density. This reinforces the contention that habitat protection and restoration are the highest priority in conservation of the giant panda.

To evaluate critical habitat changes and to develop effective restoration programs, the World Wildlife Fund—China initiated a study of habitat restoration in June 2000. The objectives were to (1) reveal the main causes of forest degradation, (2) develop and test a set of criteria for assessing forest status in terms of both degradation and potential for restoration, and (3) determine the habitat needed by pandas in relation to existing and potentially restorable areas.

A first step was to determine the rates of decline in both area and quality of existing habitat. Of particular importance is the loss of suitable stands of bamboo needed by pandas for food. An in-depth analysis of the causes of habitat degradation and possible countermeasures that will curb this trend must also be identified and developed. Remedial measures that hold promise include the following:

- Limiting human access to mountainous regions in order for natural forests to recover;
- Increased protection of the main bamboo feeding areas;
- Establishment of corridors that will facilitate giant panda migration and dispersal;
- Large-scale planting of appropriate trees and bamboos; and
- Enhanced protection of the forested areas lying outside of giant panda reserves.

China's burgeoning human population and the resultant encroachment on panda habitat underscore the need for community development and habitat rehabilitation to proceed hand

FIGURE PR.11.1. Example of degraded habitat in Foping Nature Reserve, lacking understory bamboo and suitable for restoration. Photograph by Liu Xuehua.

in hand. This requirement appears to be recognized by the Chinese central government, as evidenced by China's "Grain to Green" policy for restoring forests in marginal farmland areas. Other factors, such as judicious infrastructure construction, will place a restraint on further habitat fragmentation, and the development of sustainable use of nonforest products by forest-dependent citizens will alleviate degradation. Providing benefits for the local people to conserve and protect panda habitat is now recognized by the authorities in China as a necessary policy.

ACKNOWLEDGMENTS

We thank Eric Dinerstein, Changqing Yu, and Junqing Li for their contributions to this chapter.

REFERENCE

Chinese Minisitry of Forestry and World Wildlife Fund. 1989. *The national conservation management plan of the giant panda and its habitat*. Beijing: Ministry of Forestry for China and World Wildlife Fund.

12

Sympatry of Giant and Red Pandas on Yele Natural Reserve, China

Fuwen Wei, Ming Li, Zuojian Feng, Zuwang Wang, and Jinchu Hu

THE GIANT PANDA *(Ailuropoda melanoleuca)* and red panda *(Ailurus fulgens)* are endemic to the Himalayan-Hengduan Mountains. The giant panda is today found only in Sichuan, Shaanxi, and Gansu Provinces of China (Schaller et al. 1985; Hu et al. 1990). The red panda, in contrast, has a larger range, extending from central Nepal eastward along the Himalayas through Bhutan, India, and Myanmar into China (Roberts and Gittleman 1984; Glatston 1994; Wei et al. 1999a). In China, both species are sympatric in the Qionglai, Minshan, Xiangling, and Liangshan Mountains of Sichuan Province (Schaller et al. 1985; Hu et al. 1990; Wei et al. 1999a).

Giant and red pandas are rare animals and are listed as Category I and Category II species, respectively, in the National Protected Animal List in China. Both species are classified as carnivores and share a number of anatomical and ecological characteristics. They represent not only monotypic genera, but are also the sole representatives of the subfamilies Ailuropodinae and Ailurinae (Glatston 1989). Second, both species specialize in feeding on bamboo diets and have shared the same bamboo species in regions of sympatry (Schaller et al. 1985; Johnson et al. 1988; Reid et al. 1991a; Wei et al. 1995, 1996a). However, they retain the short, relatively simple digestive tracts typical of other carnivores and cannot digest cellulose (Dierenfeld et al. 1982; Schaller et al. 1985; Warnell et al. 1989; Wei et al. 1999b,c). Finally, both species are confronted by the same environmental pressures, such as habitat loss, population isolation, and human interference, and suffer high mortality in the wild (Schaller et al. 1985; Yonzon 1989; Hu et al. 1990; Glatston 1994; Wei and Hu 1994; Wei et al. 1997, 1999a).

Because giant and red pandas inhabit the same areas and feed on the same bamboo species, understanding the mechanism of coexistence is an interesting scientific question. Our objective in this study was to determine what resource and habitat characteristics were used by the pandas and whether differential resource and habitat use provided significant separation between the species.

FIELD RESEARCH IN CHINA

Fieldwork was conducted in Yele Natural Reserve, Mianning County, southwestern Sichuan Province, China (101°58′–102°15′E, 28°50′–29°02′N).

The reserve is situated in the western Lesser Xiangling Mountains and southeastern Daxueshan Mountains and includes about 242 km² of rugged ridges and narrow valleys at elevations of 2600–5000 m. Our research base was located at 3100 m, and a concentrated study area of about 25 km² was established in the upper Shihuiyao Valley.

The vegetation showed characteristic vertical zonation. Mixed coniferous and deciduous broadleaf forest occurred below 2800 m. The original vegetation was dominated by *Tsuga chinensis, Betula platyphylla, B. utilis,* and *Acer* spp. However, most of the forest has been degraded—because of cultivation, firewood cutting, and logging—to shrubland and meadow; that is, habitats unsuitable for pandas. Between 2800 and 3700 m, a subalpine coniferous forest is predominant. Dominant conifers were *Abies fabri* and *Sabina pingii*. *Betula, Acer,* and *Prunus* were the most common deciduous broadleafed trees. Dominant shrubs were *Bashania spanostachya, Rhododendron, Lonicera, Sorbus,* and *Rosa*. This forest was the main habitat of both pandas. Alpine shrub and meadow extended from 3700 to 4400 m.

Five species of bamboo *(Bashania spanostachya, Fargesia dulcicula, F. exposita, F. ferax,* and *Yushania tineloata)* occurred in the reserve, but *B. spanostachya* was dominant. This species, which covered whole hillsides of our study areas, was the main food resource of both pandas. Because the other four species occurred at low elevations and were disturbed extensively by human activities, giant and red pandas seldom fed on them.

COLLECTING DATA ON FOOD HABITS

Giant and red pandas are rarely observed in the field. Therefore, feeding signs and droppings left by both species of pandas were used to gather information on feeding behaviors and food habits. Three droppings were collected from dropping clusters (multiple defecation) or one dropping was collected from a single defecation encountered along the survey trails in different places.

At least thirty droppings of giant pandas and sixty droppings of red pandas were collected monthly for food-habit analysis. The collected droppings were oven dried and different items, such as stems, leaves, shoots, and fruits, were separated and weighed. The proportion of each food item in the diet was estimated by weight of dry matter.

METHOD FOR DETERMINING FOOD SELECTION

A total of 143 1-m² bamboo plots were set up randomly along different hillsides from 3000 to 4000 m in the study area. Each bamboo culm was assigned to one of four age classes: new shoot, old shoot, young culm, and old culm, as described by Schaller et al. (1985). The total number of culms of each age class in each plot was counted, and the basal diameter and height of each culm were measured to estimate the frequency distribution of culm size in the environment.

At feeding sites of giant pandas, basal diameters of all culm stumps of different age classes were measured to estimate the proportion of different sized culms eaten by giant pandas. When giant pandas fed on new shoots taller than 50 cm, basal diameters of remaining shoot stumps were measured. When feeding on new shoots shorter than 50 cm, no shoot stumps were left. Under this circumstance, we measured shoots not eaten by giant pandas at feeding sites and shoots in nearby places where a panda had not foraged, using the method outlined by Schaller et al. (1985). This method was also applied to red pandas, as they did not leave shoot stumps at feeding sites when feeding on new shoots.

The Vanderploeg and Scavia selectivity index (E_i) was used to quantify preferences for bamboo culms and shoots (Lechowicz 1982; Schaller et al. 1985; Reid et al. 1991b; Wei et al. 1996a,b). The value of E_i was calculated by:

$$E_i = (W_i - 1/n)/(W_i + 1/n),$$
$$W_i = (r_i/p_i)/\sum(r_i/p_i),$$

where W_i is the Vanderploeg and Scavia selectivity coefficient, i is the shoot size class, n is the total number of shoot size classes defined, p_i is the number of shoots in the ith class in the environment, and r_i is the number of shoots in the ith class eaten or not eaten by giant or red pandas at the feeding sites. The value of the index is scaled between -1 and $+1$.

INDEX OF NICHE BREADTH AND OVERLAP OF RESOURCES

The Shannon-Wiener index of niche breadth was calculated using the following formula (Sun 1992; Krebs 1999):

$$B_i = \frac{\log \sum N_{ij} - (1/\sum N_{ij})(\sum N_{ij} \log N_{ij})}{\log r},$$

where B_i is the niche breadth of the ith species, r is the number of resource divisions, and N_{ij} is the proportion of the ith species that uses the jth resource division.

An index of niche overlap was calculated using the formula (Colwell and Futuyma 1971; Sun 1992):

$$C_{ih} = 1 - \frac{1}{2} \sum \left| \frac{N_{ij}}{N_i} - \frac{N_{hj}}{N_h} \right|.$$

where C_{ih} is the index of overlap between ith species and hth species, N_{ij} is the proportion of the ith species using the jth resource division, N_i is the total proportion of ith species using all resources divisions, N_{hj} is the proportion of the hth species using jth resource division, and N_h is the total proportion of hth species that uses all resources divisions.

SAMPLING THE HABITAT

Both giant and red pandas usually leave multiple defecations at feeding sites. Number of droppings in each vary significantly. When feeding and resting for a short time, the giant panda often leaves one to four droppings in a group, infrequently five to ten. During long rests, it leaves more than ten droppings (Schaller et al. 1985; Reid and Hu 1991). The number of droppings in a single red panda defecation is usually around eight to fifteen or, less commonly, fifteen to thirty, but sometimes more than one hundred in repeatedly used sites (called "latrines") (Yonzon 1989; Reid et al. 1991a; Wei et al. 1995). Field observations indicated that the longer the animals spent at the feeding sites, the greater the number of droppings that were left. Therefore, there was a positive linear relationship between total time spent in feeding sites and number of droppings deposited (Reid and Hu 1991; Wei et al. 1996b). Because of the difficulty of observing the activities of either panda in the field, droppings were selected as an indirect index for quantifying habitat use of both pandas.

A modified habitat sampling method created by Dueser and Shugart (1978) was applied to measure nineteen habitat variables (see appendix). Whenever a fresh dropping cluster was encountered, three sampling units were centered on its location. The sampling units included one 1.0-m^2 bamboo plot at the center, two 20-m^2 rectangular transects (each 2 × 10 m^2), and one 400-m^2 square plot. At the center of each 100-m^2 quadrant within the 400-m^2 square plot, an additional 1.0-m^2 bamboo plot also was sampled. These five 1.0-m^2 bamboo plots supplied a detailed measure of such bamboo parameters as density, height, basal diameter, and proportion of old shoots. Two 20-m^2 rectangular transects and the 400-m^2 square plot provided information about canopy, slope, aspect, numbers of trees and shrubs, diameter at breast height (DBH), distance from dropping cluster to the nearest trees and shrubs, numbers of fallen logs and tree stumps, and diameter of the nearest fallen logs and tree stumps. Although more than one dropping cluster may have occurred in some 20 × 20-m^2 plots, only one was used as the plot center to measure habitat.

The Bartlett-Box test was used to evaluate the homogeneity of variance for each variable between giant and red pandas. The Mann-Whitney

U-test was used to test whether habitat variables of both pandas differed. Stepwise discriminant function analysis (DFA) was applied in examining habitat separation between species, because DFA can be used as an exploratory tool to identify predictor variables from potentially useful parameters (Marnell 1998). We selected the significantly different variables between both species after processing the parametric and non-parametric tests to enter the stepwise approach. All tests were performed in SPSS for Windows (Norusis 1994).

SEASONAL CHANGES IN FOOD HABITS

Giant pandas fed only on *B. spanostachya* bamboo all year round. The red panda, however, foraged not only on this bamboo, but also on different kinds of arboreal fruits (table 12.1). Giant pandas showed a quite different pattern of utilization for different parts of bamboo among seasons. From March to April, they mainly fed on stems, which accounted for 87.5–91.3% of the monthly diet. Shoots of *B. spanostachya* emerged from mid-April through July, and giant pandas began to forage on these in May. In June, shoots made up the highest proportion of their monthly diet (64.7%). From July to October, they shifted to feeding mainly on leaves. Leaves counted for the highest proportion of the monthly diet in September (92.8%). From December to next March, as bamboo leaves withered and dropped from the branches, giant pandas shifted again to foraging mainly on stems and old shoots.

The leaves of *B. spanostachya* are the main food resource of red pandas, accounting for 89.9% of their annual diet (table 12.1). From November to April, leaves were the only food of red pandas. Like giant pandas, red pandas also preferred shoots during their early growth. The proportion of shoots eaten increased from 17.8% in May to 42.5% in June, and dropped to 9.4% in August, a time when finding new shoots is difficult. From late June to October, red pandas commonly fed on a variety of arboreal fruits besides bamboo leaves and shoots. These fruits included *Prunus pilosiuscula*, *Prunus* spp., *Ribes meyerri*, *R. tenue*, *R. longiracemoaum*, *Rosa sericea*, *Rubus foliolosus*, *R. innominatus*, *R. flosculosus*, and *Sorbus koehneana*, of which *Prunus* spp. and *Rubus* spp. are most dominant. Surprisingly, uncrushed seeds of *Sabina squamata* were also found in red panda droppings. We do not know why red pandas feed on these seeds, as they are hard to digest.

SELECTING SHOOTS AS FOOD

The selectivity index (E_i) indicates that both pandas selectively foraged on different kinds of new shoots. When feeding on shoots shorter than 50 cm, giant pandas preferred taller and robust ones (height >30 cm, basal diameter >10 mm) rather than shorter shoots (height <30 cm) of any diameter (table 12.2). When foraging on shoots taller than 50 cm, they also preferred robust ones (diameter >10 mm) and avoided slender shoots (diameter <8 mm) (table 12.3). Red pandas, by contrast, did not feed on shoots taller than 50 cm, possibly because they are more difficult for them to digest. They preferred robust shoots shorter than 50 cm with wider basal diameter (>10 mm, especially those >16 mm; table 12.4).

SELECTING LEAVES AS FOOD

Giant pandas fed on bamboo leaves in all seasons (see table 12.1). While feeding, they bit off stems and then held the stems with their paws to bite off leaves and branch tips (see Long et al., chapter 6). Red pandas were found to walk over fallen logs or branches of such shrubs as rhododendrons to reach leaves. Unlike giant pandas, red pandas nipped off leaves one by one, eating only laminae. They often ate only a portion of each lamina, leaving the remainders on the leafstalk. We counted all leaves eaten by red pandas in fifteen feeding sites and found that the proportion of partially eaten laminae was very high, making up 67.4% of the total ($n = 7316$). We also found that red pandas appeared to prefer newly formed leaves in late spring and early summer, and occasionally foraged on furled new leaves.

TABLE 12.1
Proportion of Dry Matter of B. spanostachya in Giant and Red Panda Droppings from Yele Natural Reserve

SPECIES	FOOD	JANUARY	FEBRUARY	MARCH	APRIL	MAY	JUNE	JULY	AUGUST	SEPTEMBER	OCTOBER	NOVEMBER	DECEMBER	ANNUAL AVERAGE
Giant panda	Stems	85.3	82.1	91.3	87.5	46.5	35.3	38.9	22.4	7.2	28.5	56.5	75.8	54.8
	Leaves	14.7	17.9	8.7	12.5	6.1		46.8	77.6	92.8	71.5	43.5	24.2	34.7
	Shoots					47.4	64.7	14.3						10.5
Red panda	Leaves	100	100	100	100	82.2	56.9	61.7	86.1	95.6	98.8	100	100	89.94
	Shoots					17.8	42.5	37.2	9.4					8.90
	Sabina sp. fruit						0.6	1.1						0.14
	Prunus spp. fruit								1.7	1.5	1.2			0.37
	Rubus spp. fruit								2.1	1.4	1.6			0.43
	Sorbus spp. fruit								0.7	0.9	0.4			0.17
	Ribes spp. fruit									0.6				0.05

NOTE: All values are percentages of dry matter by weight. A blank entry signifies that the food is not consumed in the indicated month.

TABLE 12.2
Selection of New Shoots of B. spanostachya (Height <50 cm) by Giant Pandas

SHOOT HEIGHT (cm)	DIAMETER CLASS (i) (mm)	RANDOM SAMPLE (p_i)	NOT EATEN BY PANDA (r_i)	W_i	E_i	PREFERENCE STATUS
<30	≤8.0	35	72	0.107	0.126	Not preferred
	8.1–10.0	75	152	0.106	0.118	Not preferred
	10.1–12.0	63	130	0.108	0.127	Not preferred
	12.1–14.0	59	125	0.111	0.140	Not preferred
	14.1–16.0	54	107	0.103	0.107	Not preferred
	>16.0	30	71	0.123	0.194	Not preferred
>30	≤8	31	79	0.133	0.229	Not preferred
	8.1–10.0	72	162	0.117	0.169	Not preferred
	10.1–12.0	84	41	0.025	−0.532	Preferred
	12.1–14.0	90	37	0.021	−0.591	Preferred
	14.1–16.0	56	25	0.023	−0.563	Preferred
	>16.0	26	11	0.022	−0.581	Preferred

NOTE: E_i is the selectivity index; p_i is the number of shoots with ith class in the environment; r_i is the number of shoots with ith class not eaten; W_i is the Vanderploeg and Scavia selectivity coefficient.

TABLE 12.3
Selection of New Shoots of B. spanostachya (Height >50 cm) by Giant Pandas

DIAMETER CLASS (i)	RANDOM SAMPLE (p_i)	EATEN BY PANDA (r_i)	W_i	E_i	PREFERENCE STATUS
≤8.0	45	2	0.008	−0.905	Avoided
8.1–10.0	87	35	0.075	−0.380	Not preferred
10.1–12.0	104	114	0.204	0.101	Preferred
12.1–14.0	88	98	0.207	0.109	Preferred
14.1–16.0	62	83	0.249	0.199	Preferred
>16.0	32	44	0.256	0.211	Preferred

NOTE: E_i is the selectivity index; p_i is the number of shoots with ith class in the environment; r_i is the number of shoots with ith class eaten; W_i is the Vanderploeg and Scavia selectivity coefficient.

NICHE BREADTH AND OVERLAP

The breadth and overlap of giant and red panda food resource niches (table 12.5) were calculated from data in table 12.1. Niche breadth of food resources was found to be 0.5797 and 0.2250, respectively, for giant and red pandas. Niche overlap between the two species was 0.3470. Although there was some overlap for food resource between the two, the overlap was small.

HABITAT USE

The mean and standard deviation of the nineteen habitat variables demonstrated some differences between the two species (table 12.6).

TABLE 12.4
Selection of New Shoots of B. spanostachya (Height <50 cm) by Red Pandas

DIAMETER CLASS (i) (mm)	RANDOM SAMPLE (p_i)	NOT EATEN BY PANDAS (r_i)	W_i	E_i	PREFERENCE STATUS
≤8.0	29	51	0.385	0.396	Not preferred
8.1–10.0	72	97	0.295	0.278	Not preferred
10.1–12.0	88	48	0.120	–0.165	Preferred
12.1–14.0	91	39	0.094	–0.279	Preferred
14.1–16.0	57	21	0.081	–0.347	Preferred
>16.0	26	3	0.025	–0.737	Preferred

NOTE: E_i is the selectivity index; p_i is the number of shoots with *i*th class in the environment; r_i is the number of shoots with *i*th class not eaten; W_i is the Vanderploeg and Scavia selectivity coefficient.

TABLE 12.5
Index of Resource Niche Breadth and Overlap for Giant and Red Pandas

SPECIES	LEAVES (%)	STEMS (%)	NEW SHOOTS (%) TALL AND ROBUST	NEW SHOOTS (%) SHORT AND ROBUST	FRUITS (%)	NICHE INDEX BREADTH	NICHE INDEX OVERLAP
Giant panda	34.7	54.8	10.5	0	0	0.5797	0.3470
Red panda	89.94	0	0	8.90	1.16	0.2250	

Bartlett-Box univariate homogeneity of variance tests indicated that variances of eleven of nineteen variables were equal or homogeneous, whereas eight were unequal.

Because the data obtained had nonnormal distributions, parametric and nonparametric tests were applied to compare the results. The Mann-Whitney U-test detected that twelve of the nineteen variables differed significantly between species ($p < 0.05$). The results of parametric and nonparametric tests were almost the same with respect to variables and level of probability, revealing that both species used different microhabitats (table 12.6).

The DFA of the two species was significant (eigenvalue = 3.577, Wilks' λ = 0.219, χ^2 = 253.999, $df = 8$, $p < 0.001$), which suggests that the two species exhibited different patterns in habitat use. The DFA correctly classified 96.5% (167 of 173 samples) of the habitat samples according to species, 97.5% (79 of 81 samples) for giant pandas and 95.7% (88 of 92 samples) for red pandas. Although the Mann-Whitney U-test detected twelve potential variables in identifying sites of giant and red pandas, the stepwise approach identified only eight predictor variables that appeared to be most significant in discriminating sites of both species (table 12.7).

Standardized canonical discriminant function coefficients and correlations between discriminating variables and canonical discriminant functions can be used to judge the relative contribution to the power of the discriminant function. Larger absolute values of correlations or coefficients indicate a more significant contribution to the power of the function (Cooley and Lohnes

TABLE 12.6
Mean, Standard Deviation, ANOVA, and Mann-Whitney U-Test for Nineteen Habitat Variables of Giant and Red Pandas

HABITAT VARIABLE[1]	GIANT PANDA (N = 81)		RED PANDA (N = 92)		ANOVA (df = 1/172)		MANN-WHITNEY U-TEST	
	MEAN	STANDARD DEVIATION	MEAN	STANDARD DEVIATION	F	p	U	p
Canopy	3.11	0.69	3.27	0.63	2.561	0.111	3274.5	0.128
Slope	1.79	0.75	2.73	0.68	73.971	<0.00	528.0	<0.00
Aspect	1.91	0.73	2.27	0.74	10.204	0.002	2760.0	0.002
Bamboo culm density	27.40	3.59	33.45	6.41	56.648	<0.00	1490.0	<0.00
Bamboo basal diameter (mm)	11.11	0.79	10.23	1.01	39.984	<0.00	1842.0	<0.00
Bamboo culm height (cm)	233.83	33.79	210.73	35.37	19.160	<0.00	2432.5	<0.00
Old shoot proportion (%)	0.14	0.10	0.11	0.06	5.174	0.024	3237.0	0.137
Tree density	1.26	0.50	1.14	0.50	2.399	0.123	3256.5	0.136
Tree size	42.91	7.55	43.12	8.20	0.030	0.863	3690.5	0.914
Tree dispersion	2.44	0.55	4.02	1.40	90.484	<0.00	1049.0	<0.00
Shrub density	1.11	0.87	2.79	1.38	89.276	<0.00	1084.0	<0.00
Shrub size	21.61	4.77	22.08	4.39	0.449	0.504	3369.5	0.278
Shrub dispersion	4.03	1.81	1.95	0.96	92.156	<0.00	1329.5	<0.00
Fallen-log density	0.90	0.53	1.76	0.69	83.082	<0.00	996.5	<0.00
Fallen-log size	30.61	6.69	31.31	6.78	0.468	0.495	3507.5	0.506
Fallen-log dispersion	4.18	1.44	1.98	0.97	140.659	<0.00	537.0	<0.00
Tree-stump density	0.66	0.57	0.88	0.59	6.114	0.014	2680.5	0.001
Tree-stump size	29.81	5.71	30.73	6.82	0.911	0.341	3530.0	0.551
Tree-stump dispersion	4.30	1.20	2.49	1.11	106.717	<0.00	931.0	<0.00

[1] Density variables without units are in number per sample plot. See text and appendix for details.

TABLE 12.7
*Stepwise Approach of Discriminant Functional Analysis for the
Significantly Different Habitat Variables for Giant and Red Pandas*

HABITAT VARIABLE	STANDARDIZED CANONICAL DISCRIMINANT FUNCTION COEFFICIENT	CORRELATION BETWEEN DISCRIMINATING VARIABLES AND CANONICAL DISCRIMINANT FUNCTIONS
Fallen-log dispersion	−0.407	−0.480
Shrub density	0.366	0.382
Slope	0.353	0.348
Fallen-log density	0.350	0.369
Shrub dispersion	−0.336	−0.388
Bamboo culm density	0.322	0.304
Tree-stump dispersion	−0.243	−0.418
Tree dispersion	0.218	0.385

NOTES: Maximum significance of F to enter = 0.05; minimum significance of F to remove = 0.1.

1971). Correlations of the eight indicator variables with the discriminant function fell within a narrow range of absolute values (between 0.304 and 0.480; table 12.7). Fallen-log dispersion contributed most to the power of the discriminant function, and bamboo density contributed least. Standardized coefficients of the eight selected variables also fell within a narrow range (between 0.218 and 0.407). Fallen-log dispersion contributed most to the power of the function, but tree dispersion contributed least. Although results of the two analyses differed in some variables, others were ranked the same. Because correlations and coefficients were similar, those eight variables appeared to contribute almost equally to the power of the discriminant function and could be treated as indicators in identifying sites of giant and red pandas.

In sum, the giant panda occurred at sites on gentle slopes with a lower density of fallen logs, shrubs, and bamboo culms. The sites also were close to trees and far from fallen logs, shrubs, and tree stumps. Conversely, the red panda occurred at sites on steeper slopes with a higher density of fallen logs, shrubs, and bamboo culms. The sites were close to fallen logs, shrubs, and tree stumps.

DISCUSSION

Resource partitioning between sympatric species may reduce interspecific competition and is often considered as enabling multispecies coexistence (Lack 1945; Brown and Lieberman 1973; Schoener 1974; Mushinsky and Hebrard 1977; Giller and McNeill 1981; Van Horne, 1982). Our results clearly demonstrate that the two coexisting panda species do have significantly different patterns of resource utilization in the area of sympatry, although their resource niches partially overlap (overlap index, 0.347). Such differences may be crucial to the coexistence of the two species, for the following reasons.

First, both pandas differ somewhat in food habits, although they feed mainly on the same bamboo species. Giant pandas might sample other animals and plants; however, their annual diet consists of over 99% bamboo in the Xiangling (Wei et al. 1999d), Liangshan (Wei et al. 1996a), Qionglai (Schaller et al. 1985), and

Minshan Mountains (Schaller et al. 1989). By contrast, the annual diet of red pandas consists of about 98% bamboo in sympatric areas (Johnson et al. 1988; Reid et al. 1991a; Hu and Wei 1992; Wei et al. 1995). In addition, red pandas prefer different kinds of fruits in the summer and autumn in the Xiangling, Qionglai (Reid et al. 1991a), and Liangshan Mountains (Wei et al. 1995).

Second, giant pandas utilize almost every part of the bamboo plant (leaves, shoots, branches, and culms). Red pandas, however, only use leaves and shoots. Bamboo leaves make up 89.9% of the red panda's annual diet in Xiangling, 91.4% in Liangshan (Wei et al. 1995), and over 90% in Qionglai Mountains (Reid et al. 1991a). By contrast, leaves make up only 34.7% of the annual diet of giant pandas in the Xiangling, 38.8% in Liangshan (Wei et al. 1996a), and 40.7% in Qionglai Mountains (Schaller et al. 1985).

Third, giant and red pandas select different kinds of early growth bamboo shoots. Giant pandas prefer tall and robust shoots. They select shoots of *B. spanostachya* taller than 30 cm and larger than 10 mm in basal diameter in the Xiangling Mountains, and shoots of *F. robusta* taller than 25 cm and larger than 10 mm in the Qionglai Mountains (Schaller et al. 1985). Red pandas, however, prefer the short and robust shoots, which giant pandas rarely consume. For instance, red pandas select shoots of *B. spanostachya* shorter than 50 cm and with a basal diameter larger than 10 mm.

Finally, both pandas exhibit quite different patterns of microhabitat utilization. Giant pandas, having larger body size, use sites with lower densities of shrubs, fallen logs, and bamboo culms. Feeding and moving in this more open microhabitat could reduce energy expenditures. Energy conservation could be important for the giant panda, because its daily energy intake exceeds expenditures by only a small margin (Schaller et al. 1985; Wei et al. 1997). In contrast, the smaller red panda usually walks on the fallen logs, branches of shrubs (especially rhododendrons), and tree stumps, which give the animal ready access to bamboo leaves (Johnson et al. 1988; Reid et al. 1991a; Wei et al. 1995; Wang et al. 1998). This explains why red pandas use microhabitats with high densities of shrubs, tree stumps, and fallen logs. Giant pandas use bamboo stands on relatively gentle slopes because of suitability for feeding (Schaller et al. 1985; Hu et al. 1990; Reid and Hu 1991; Wei et al. 1996b). Previous reports revealed that the giant panda uses slopes of less than 20° (especially slopes <10°) in the Qionglai and Liangshan Mountains (Reid and Hu 1991; Wei et al. 1996b). This study and others (Wei et al. 1995; Wang et al. 1998) indicate that red pandas use relatively steep slopes. Use of steeper slopes by red pandas may be correlated with density of the branches of shrubs (particularly rhododendron) and density of fallen logs within the leaf stratum of the bamboo. On steeper slopes, rhododendrons have longer branches that extend into the leaves of bamboo, and fallen logs are also more likely to intersect the leaf layer. Therefore, the smaller red panda chooses places with good access to bamboo leaves, which happen to be associated with steeper slopes.

APPENDIX: DESCRIPTION OF NINETEEN HABITAT VARIABLES

Canopy (%): Canopy of overstory vegetation in 400-m² plot, five categories: <20, 20–40, 40–60, 60–80, and >80

Slope (°): Slope of 400-m² plot, four categories: <10°, 10–20°, 20–30°, and >30°

Aspect: Aspect of 400-m² plot, four categories: eastern slope (45–135°), southern slope (135–225°), western slope (225–315°), and northern slope (315–45°)

Bamboo culm density (culms/m²): Average number of culms in five 1.0-m² bamboo plots

Bamboo basal diameter (mm): Average basal diameter of culms in five 1.0-m² bamboo plots (five culms were randomly measured at each plot)

Bamboo culm height (cm): Average height of culms in five 1.0-m² bamboo plots (five culms were randomly measured at each plot)

Old shoot proportion (%): Average proportion of old shoots in five 1.0-m² bamboo plots

Tree density: Average number of trees in two 20-m² rectangular transects

Tree size (cm): Average diameter at breast height of nearest tree from the center in each 100-m² square plot

Tree dispersion (m): Average distance to nearest tree in each 100-m² square plot

Shrub density: Average number shrubs in two 20-m² rectangular transects

Shrub size (cm): Average diameter at breast height of nearest shrub in each 100-m² square plot

Shrub dispersion (m): Average distance to nearest shrub in each 100-m² square plot

Fallen-log density: Average number of fallen logs >15 cm in diameter in each 100-m² square plot

Fallen-log size (cm): Average diameter of nearest fallen logs >15 cm in diameter in each 100-m² square plot

Fallen-log dispersion (m): Average distance to nearest fallen logs >15 cm in diameter in each 100-m² square plot

Tree-stump density: Average number of tree stumps >15 cm in diameter in each 100-m² square plot

Tree-stump size (cm): Average diameter of nearest tree stumps >15 cm in diameter in each 100-m² square plot

Tree-stump dispersion (m): Average distance to nearest tree stumps >15 cm in diameter in each 100-m² square plot.

ACKNOWLEDGMENTS

This project was supported by the National Science Foundation of China (grant 30230080) and the National Science Fund for Distinguished Young Scholars (grant 30125006). Ping Tang, Eren Gutie, and Wenqing Lu participated in parts of the fieldwork. The Sichuan Forest Bureau and the Mianning Forest Bureau gave us assistance during the research. Previous versions of the manuscript benefited from comments by D. G. Reid and Don Lindburg.

REFERENCES

Brown, J. H., and G. A. Lieberman. 1973. Resource utilization and coexistence of seed-eating desert rodents in sand dune habitats. *Ecology* 54:788–97.

Colwell, R. K., and D. J. Futuyma. 1971. On the measurement of niche breadth and overlap. *Ecology* 52:567–76.

Cooley, W. W., and P. R. Lohnes. 1971. *Multivariate data analysis*. New York: John Wiley and Sons.

Dierenfeld, E., H. Hintz, J. Robertson, P. Van Soest, and O. T. Oftedal. 1982. Utilization of bamboo by the giant panda. *J Nutr* 112:636–41.

Dueser, R. D., and H. H. Shugart Jr. 1978. Microhabitats in forest-floor small mammal fauna. *Ecology* 59:89–98.

Giller, P. S., and S. McNeill. 1981. Predation strategies, resource partitioning and habitat selection in Notonecta (Hemiptera/Heteroptera). *J Anim Ecol* 50:789–808.

Glatston, A. R. 1989. *Red panda biology*. The Hague: SPB Academic.

Glatston, A. R. 1994. *Status survey and conservation action plan for Procyonids and Ailurids: The red panda, olingos, coatis, raccoons and their relatives*. Gland, Switzerland: IUCN.

Hu, J., and F. Wei. 1992. Feeding ecology of red pandas. (In Chinese.) *J Sichuan Normal College* 13:83–87.

Hu, J., F. Wei, C. Yuan, and Y. Wu. 1990. *Research and progress in biology of the giant panda*. (In Chinese.) Chengdu: Sichuan Publishing House of Science and Technology.

Johnson, K. G., G. B. Schaller, and J. Hu. 1988. Comparative behavior of red and giant pandas in the Wolong Reserve, China. *J Mammal* 69:552–64.

Krebs, C. J. 1999. *Ecological methodology*. Second edition. Menlo Park: Addison-Wesley.

Lack, D. L. 1945. Ecology of closely related species with special reference to cormorant (*Phalacrocorax carbo*) and shag (*P. aristootelis*). *J Anim Ecol* 14:12–16.

Lechowicz, M. J. 1982. The sampling characteristics of selectivity indices. *Oecologia* 52:22–30.

Marnell, F. 1998. Discriminant analysis of the terrestrial and aquatic habitat determinations of the smooth newt (*Triturus vulgaris*) and the common frog (*Rana temporaria*) in Ireland. *J Zool Lond* 244:1–6.

Mushinsky, H. R., and J. J. Hebrard. 1977. Food partitioning by five species of water snakes. *Herpetologica* 33:162–66.

Norusis, M. J. 1994. *SPSS professional statistics 6.1*. Chicago: SPSS.

Reid, D. G., and J. Hu. 1991. Giant panda selection between *Bashania fangiana* bamboo habitats in

Wolong Reserve, Sichuan, China. *J Appl Ecol* 28: 28–43.

Reid, D. G., J. Hu, and Y. Huang. 1991a. Ecology of the red panda in the Wolong Reserve, China. *J Zool Lond* 225:347–64.

Reid, D. G., M. Jiang, Q. Deng, Z. Qin, and J. Hu. 1991b. Ecology of the Asiatic black bear *(Ursus thibetanus)* in Sichuan, China. *Mammalia* 55: 221–37.

Roberts, M. S, and J. L. Gittleman. 1984. *Ailurus fulgens. Mammal Species* 222:1–8.

Schaller, G. B., J. Hu, W. Pan, and J. Zhu. 1985. *The giant pandas of Wolong.* Chicago: University of Chicago Press.

Schaller, G. B., Q. Deng, K. Johnson, X. Wang, H. Shen, and J. Hu. 1989. Feeding ecology of giant pandas and Asiatic black bears in the Tangjiahe Reserve, China. In *Carnivore behavior, ecology and evolution,* edited by J. L. Gittleman, pp. 212–41. Ithaca: Cornell University Press.

Schoener, T. W. 1974. Resource partitioning in ecological communities. *Science* 185:27–39.

Sun, R. 1992. *The principle of animal ecology.* (In Chinese.) Second edition. Beijing: Beijing Normal University Press.

Van Horne, B. 1982. Niches of adult and juvenile deer mice *(Peromyscus maniculatus)* in seral stages of coniferous forests. *Ecology* 63:992–1003.

Wang, W., F. W. Wei, J. C. Hu, Z. J. Feng, and G. Yang. 1998. Habitat selection by red panda in Mabian Dafengding Reserve. *Acta Theriol Sin* 18:15–20.

Warnell, K. J., S. D. Crissey, and O. T. Oftedal. 1989. Utilization of bamboo and other fiber sources in red panda diets. In *Red Panda Biology,* edited by A. R. Glatston, pp. 51–56. The Hague: SPB Academic.

Wei, F., and J. Hu. 1994. Studies on the reproduction of wild giant panda in Wolong Natural Reserve. (In Chinese.) *Acta Theriol Sin* 14:243–48.

Wei, F., W. Wang, A. Zhou, J. Hu, and Y. Wei. 1995. Preliminary study on food selection and feeding strategy of red pandas. (In Chinese.) *Acta Theriol Sin* 15:259–66.

Wei, F., C. Zhou, J. Hu, G. Yang, and W. Wang. 1996a. Bamboo resources and food selection of giant pandas in Mabian Dafengding Natural Reserve. (In Chinese.) *Acta Theriol Sin* 16:171–75.

Wei, F., A. Zhou, J. Hu, G. Yang, and W. Wang. 1996b. Habitat selection by giant pandas in Mabian Dafengding Reserve. (In Chinese.) *Acta Theriol Sini* 16:241–45.

Wei, F., Z. Feng, and J. Hu. 1997. Population viability analysis computer model of giant panda population in Wuyipeng, Wolong Natural Reserve, China. *Int Conf Bear Res Man* 9:19–23.

Wei, F., Z. Feng, Z. Wang, and J. Hu. 1999a. Current distribution, status and conservation of wild red pandas *Ailurus fulgens* in China. *Biol Conserv* 89: 285–91.

Wei, F., Z. Feng, Z. Wang, A. Zhou, and J. Hu. 1999b. Nutrient and energy requirements of red panda *(Ailurus fulgens)* during lactation. *Mammalia* 63:3–10.

Wei, F., Z. Feng, Z. Wang, A. Zhou, and J. Hu. 1999c. Use of the nutrients in bamboo by the red panda *(Ailurus fulgens). J Zool Lond* 248:535–41.

Wei, F., Z. Feng, Z. Wang, A. Zhou, and M. Li. 1999d. Feeding strategy and resource partitioning between red and giant pandas. *Mammalia* 63:417–30.

Yonzon, P. B. 1989. Ecology and conservation of the red panda in Nepal Himalayans. Ph.D. dissertation. University of Maine, Orono.

13

Balancing Panda and Human Needs for Bamboo Shoots in Mabian Nature Reserve, China

PREDICTIONS FROM A LOGISTIC-LIKE MODEL

Fuwen Wei, Guang Yang, Jinchu Hu, and Stephen Stringham

THE SCARCITY of the endemic giant panda (*Ailuropoda melanoleuca*) in China is a consequence of population decline and isolation, habitat degradation, and human interference (Schaller et al. 1985; Hu et al. 1990; Hu 2001). Mortality seems high even for an ursid: about 57% over the first year of life for cubs in the Qionglai Mountains (Wei and Hu 1994; Wei et al. 1997a). Conservation challenges and the unique biology of giant pandas have attracted worldwide attention. Their ecology has been studied in a number of habitats (Schaller et al. 1985, 1989; Pan et al. 1988; Hu et al. 1990; Reid and Hu 1991; Johnson et al. 1988; Wei et al. 1996a,b, 1997a,b, 1999a,b,c, 2000a,b; see also Long et al. [chapter 6], Yong et al. [chapter 10], and Liu et al. [chapter 11]).

To date, the only quantitative assessment of interactions between the panda and its food supply (bamboo) is that by Yuan et al. (1990) in the Qionglai Mountains. They modified Smith's (1979) dual logistic predation model to account for two classes of predators and prey (bamboo). Our study further revises Smith's model to account for human harvest, so as to better understand the dynamic relationship between bamboo, pandas, and human harvest, and to provide an improved basis for conserving each species. Vital rates for bamboo in Mabian Dafenging Natural Reserve, in the Liangshan Mountains, were ascertained by a 1-year field survey.

STUDY AREA

The Liangshan Mountains are one of the main habitats of pandas in China (Hu and Wei, chapter 9). In 1978, to protect core habitats and their panda populations, the Chinese government established two adjacent refuges: Mabian Dafengding and Meigu Dafengding Nature Reserves. The fieldwork of our study was performed mainly in Mabian.

Mabian encompasses several habitat types that are separated altitudinally (table 13.1). Six

TABLE 13.1
Altitudinal Ranges of Vegetation Communities in Mabian Natural Reserve

ELEVATION (m)	HABITAT TYPE
<2000	Evergreen broad-leaved forest
2000–2400	Mixed broad-leaved and deciduous forest
2400–2800	Mixed coniferous and broad-leaved forest
2800–3700	Coniferous forest
>3700	Subalpine shrubs and alpine meadows

bamboo species *(Qiongzhuea macrophylla, Indocalamus longiauritus, Bashania fangiana, Chimonobambus pachstachys, C. szechuanesis,* and *Yushania mabianensis)* occur in the reserve. Of these, *Q. macrophylla* (1900–2450 m) and *I. longiauritus* (2400–2900) constituted the main food of pandas. We analyzed the dynamic relationship between pandas and these two bamboo species.

METHODS AND MODELS

The methods used in this fieldwork have been described by Yuan et al. (1990) and Yang et al. (1994).

PREDATOR-PREY MODELS

Smith (1979) used a pair of logistic-like models to assess the dynamics of a single species of predator and prey with a constant per capita rate of predation:

Prey: $dB/dt = (r_B/B_{max})(B_{max} - B)B - cP$; (13.1)

Predator: $dP/dt = (r_P/P_K)(P_K - P)P$, (13.2)

where:

r = the density-independent (DI) rate of increase for each species: $r = (b_{DI} - d_{DI})$;

b_{DI} = maximum per capita birth rate (%);

d_{DI} = minimum per capita mortality rate (%);

B and P = densities per km² of prey (tons of bamboo) and predators (number of pandas);

K = equilibrium density (of bamboo or pandas), the density where reproduction is exactly balanced by mortality;

B_{max} = bamboo density so high that bamboo reproduction ceases (at K, reproduction $\neq 0$ unless mortality = 0);

P_K = equilibrium size for a predator population governed solely by prey supply; and

c = mean consumption rate of prey per predator (e.g., tons of bamboo/year/panda).

Additional terms can be added to the logistic-like submodels to account for multiple predator or prey species. For example, Yuan et al. (1990) used the cP term for pandas and added a second term i to represent a constant percentage (i) loss of bamboo to predation by insects and rats.

Prey: $dB/dt = [(r_B/B_{max})(B_{max} - B)B]$
$- [iB + cP]$, (13.3)

where the term $[(r_B/B_{max})(B_{max} - B)B]$ represents predation-free dynamics of the prey and the term $[iB + cP]$ represents the impact of predation.

This revised predation model was applied to data from Wuyipeng, Wolong Nature Reserve, to explore the dynamic relationship between pandas and two local bamboo species, *Fargesia robusta* and *B. fangiana*, whose vital rates were determined by fieldwork (Yuan et al. 1990).

To apply this model to Mabian Natural Reserve, further revision was needed to account for the harvest of bamboo shoots by local residents. Shoot harvesting is permitted by the local gov-

TABLE 13.2
Statistical Results from Bamboo Plots in Mabian Natural Reserve

PARAMETERS	QIONGZHUEA MACROPHYLLA		INDOCALAMUS LONGIAURITUS	
	AVERAGE	STANDARD DEVIATION	AVERAGE	STANDARD DEVIATION
A: Density before increasing (culms/m^2)	18.35	10.28	16.55	8.93
B: Number of shoots (culms/m^2)	4.37	3.48	4.02	3.90
C: Density after increasing (culms/m^2) = (A + B)	22.72	13.32	20.57	12.35
D: Shooting rate = (B/A)	0.238	0.215	0.243	0.223
E: Loss by insects and bamboo rats (culms/m^2)	2.29	0.943	2.21	0.824
F: Loss rate (%) by insects and bamboo rats = (E/C)	0.101	0.078	0.108	0.065

ernment and the Reserve administration. Historically, the harvest quota has been set by the local government without consideration of the population densities of bamboo or pandas. To investigate impacts of shoot harvesting on the bamboo growth and the local panda population, a new term h, representing a constant annual amount (tons) of bamboo harvesting, was added to the model of Yuan et al. (1990):

$$\text{Prey: } dB/dt = [(r_B/B_{max})(B_{max} - B)B] - [iB + cP + h]. \quad (13.4)$$

EQUILIBRIUM DENSITY OF PANDAS

All of the above equations are variants of the Verhulst-Pearl logistic. In a logistic model, there are two possible equilibria: 0 and K (P_K in the case of pandas and B_K in the case of bamboo). Our long-term goal is to determine P_K for Mabian pandas and the corresponding bamboo densities. We make the simplifying assumption that P_K is determined only by the supply of food (bamboo). The value of P_K would be no greater than the highest panda density that could be supported by the bamboo crops without pandas overgrazing the bamboo, and thereby decreasing future bamboo productivity. The overall rate of consumption by pandas (cP) is assumed to be directly related to panda density; the per capita rate (c) is assumed to be constant. Other assumed constants are (1) the percentage of the bamboo crop lost to consumption by other predators (mainly rats and insects = i; so total loss is iB) and (2) the annual number of tons harvested by humans (h).

RESULTS AND DISCUSSION

BAMBOO

To estimate vital rates for bamboo, 1-m^2 plots were established in stands of *Q. macrophylla* ($N = 100$) and *I. longiauritus* ($N = 80$), distributed randomly with regard to the elevation, slope gradient, aspect, bamboo density, and other habitat variables. Rates of shoot production, density-independent mortality, and bamboo consumption by pandas and other predators were recorded for each sample plot (table 13.2).

RATE OF INCREASE

By definition, the mean geometric rate of change in population size over time interval t is calculated as:

$$r_{0-t} = [\ln(B_t/B_0)]/t = [\ln(B_t) - \ln(B_0)]/t,$$

where B_o and B_t are bamboo densities at times o and t, and t is the survey duration (1 year). Using the data in table 13.2:

Q. macrophylla: $r_Q = \ln(22.72) - \ln(18.35)$
$= 0.214 = 21.4\%/\text{year}$;

I. longiauritus: $r_I = \ln(20.57) - \ln(16.55)$
$= 0.218 = 21.8\%/\text{year}$.

MAXIMUM BAMBOO DENSITY

In our study area, sympatry by these bamboo species was low enough for interspecific density interactions to be ignored. Within each species, however, shooting rate was negatively related to culm density (C culms/km²), as expressed in the following regression equations. C_{max} was approximated by extrapolating the value of C at which S (shooting rate in culms/km²/yr) = annual culm production = 0:

Q. macrophylla ($N = 100$ plots):
$S_Q = 0.376 - 0.0108 C_Q$ ($R_Q = -0.60$);
$C_{Qmax} = 3.48 \times 10^7$ culms/km²;

I. longiauritus ($N = 80$ plots):
$S_I = 0.495 - 0.0153 C_I$ ($R_I = -0.75$);
$C_{Imax} = 3.24 \times 10^7$ culms/km²,

where R is the correlation coefficient.

The averages of the culm sizes was computed to be:

	WEIGHT (g)	HEIGHT (cm)	BASAL DIAMETER (mm)
Q. macrophylla ($N = 384$)	277.31	281.21	12.18
I. longiauritus ($N = 277$)	83.48	166.51	10.81

Two regression equations were derived between culm weight (W) versus height (H) and base diameter (D):

Q. macrophylla ($N = 127$):
$W_Q = -417.19 + 1.2970 H_Q + 27.07 D_Q$;

I. longiauritus ($N = 73$):
$W_I = -306.40 + 1.8781 H_I + 7.72 D_I$.

Converting culm numerical density (culms/km²) to biomass density (tons/km²): maximum biomass densities were calculated as:

C_{max} (culms/km²) · W (g/culm) × 10^6
$= B_{max}$ (tons/km²);

Q. macrophylla:
$B_{Qmax} = 3.48 \times 10^7 \cdot 277.31 = 9650.39$;

I. longiauritus:
$B_{Imax} = 3.24 \times 10^7 \cdot 83.84 = 2716.42$.

Using equation 13.4, tons of culms/km² are calculated:

Q. macrophylla: $r_B/B_{Qmax} = 0.214/9650.39$
$= 2.22 \times 10^{-5}$ tons/km²;

I. longiauritus: $r_B/B_{Imax} = 0.218/2716.42$
$= 8.03 \times 10^{-5}$ tons/km².

RATES OF BAMBOO CONSUMPTION

Mabian Natural Reserve is home to eight pandas. During our year of observations, they spent approximately 213 days in Q. macrophylla and 137 days in I. longiauritus (Yang 1993). Number of culms eaten/day/panda was estimated as 707 ($N = 9$ scats) for Q. macrophylla and 722 ($N = 3$) for I. longiauritus (Yang 1993), based on the number of culms utilized and scat deposited by pandas on the tracking line (calculated culms/scat) and on the 120-scat/day rate observed by Hu et al. (1990).

Annual numbers of bamboo culms consumed are estimated as:

N = (days/year eating culms) · (culms/day/panda)
$= $ culms/year/panda;

Q. macrophylla:
$N_Q = 213 \cdot 707 = 15.1 \times 10^4$;

I. longiauritus:
$N_I = 137 \cdot 722 = 9.9 \times 10^4$.

Then, mean rates of bamboo consumption were estimated as:

(g/culm) · (culms/year/panda)
$=$ tons/year/panda;

Q. macrophylla:
$c_Q = 277.21 \cdot 15.1 \times 10^4 = 41.76;$

I. longiauritus:
$c_I = 83.48 \cdot 9.9 \times 10^4 = 8.29.$

RATES OF CONSUMPTION OF BAMBOO BY INSECTS AND RATS

These losses were estimated (table 13.2) at $i_Q = 10.1\%$ and $i_I = 10.8\%$.

RATE OF LOSS TO HARVEST BY HUMANS

This rate (h) was estimated as a multiple of the weight of shoots sold in the local market. Over two successive years, we estimated that 20,000 kg and 25,000 kg of shoots were harvested from approximately 7.52 km² of Q. macrophylla around our field station. The mean of 22,500 kg was used to estimate $h_Q = 2.99$ tons/km². Shoots of I. longiauritus are so slender and distributed at such a high altitude (>2600 m) that they were not harvested; $h_I = 0$.

BAMBOO GROWTH EQUATIONS

Using equation 13.4, we calculate that:

Q. macrophylla:
$$d_{BQ}/dt = [(0.214/9650.39) \cdot (9650.39 - B)B]$$
$$- [0.101B_Q + 41.76P + h]$$
$$= [0.113B_Q - 2.22 \times 10^{-5} B_Q^2] - [41.76 \cdot 3.03 + h]$$
$$= [0.113B_Q - 2.22 \times 10^{-5} B_Q^2] - 127 - h; \quad (13.5)$$

I. longiauritus: $dB_I /dt = [(0.218/2716.42)$
$$\cdot (2716.42 - B)B] - [0.108B_I + 8.29P + h]$$
$$= 0.110B_I - 8.03 \times 10^{-5} B_I^2 - 25.1 - h. \quad (13.6)$$

PANDAS

RATE OF INCREASE

From equation 13.2, it can be seen that this logistic-like model is just a rearrangement of that applied to panda population dynamics by Yuan et al. (1990). Based on the age structure of the Mabian panda population (Yang 1993; Yang et al. 1994) and age-specific birth rates of pandas elsewhere (Wei et al. 1989), the birth rate (b) for our population was estimated as 0.193 cubs/panda/year (i.e., 19.3%) for the current age-sex ratio. The mortality rate (d) observed by Wei et al. (1989) was 17.4%/year. For the purposes of this preliminary modeling, these rates and ratios are assumed to remain constant, on average, over the foreseeable future. (Ideally, later iterations of the model will address changes in vital rates and in age-sex ratios.)

We used the observed mean rate of increase:

$$r'_p = b - d = 0.1931 - 0.174 = 0.019$$

to approximate r_p, the intrinsic (density-independent) rate of increase, in the absence of any other basis for predicting how much higher r_p would be. Although the logistic model assumes intraspecific density effects even at $P = 2$, in reality, such effects may not begin for Ursidae until their density approaches K (Taylor 1994).

This value of r_p would be required for calculating the relationships between the dynamics of panda, bamboo, insect, and rat populations. However, at this early stage of modeling, we ignore dynamics in panda population density and analyze sustainable bamboo harvest by people only at equilibrium density for the panda population.

PREDATOR-PREY EQUILIBRIA

PANDAS

Studies at Wolong Nature Reserve showed that the mean size of a giant panda's core area was 33 ha. Most of the panda's nutrition came from its core area (Schaller et al. 1985). Overlap between core areas was negligible. Hence, mean density of these pandas was estimated as:

$$1/33 \text{ ha} = 1/0.33 \text{ km}^2 = 3.03/\text{km}^2.$$

This is assumed to approximate equilibrium density at Wolong, and serves as a first approximation for equilibrium density in our study area. Note that this is more than six times the present density of 0.48/km² (Yang et al. 1994).

BAMBOO

The assumed density-dependent relationship between bamboo abundance and the addition of

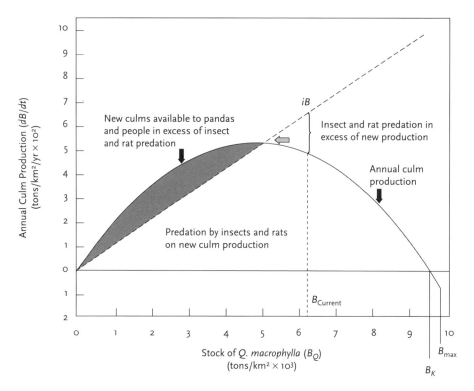

FIGURE 13.1. Bamboo culm biomass available for human harvest at various densities of bamboo, rats, and insects. B_K = equilibrium bamboo density (reproductive rate exactly balanced by mortality rate); B_{max} = the density above which bamboo reproduction ceases, despite continuing mortality. Consumption by insects and rats is represented by iB, rising as a constant proportion of stock density. The relationship shown here for Q. macrophylla is similar for I. longiauritus.

new culms (reproduction) is parabolic—rising at a decreasing rate as the bamboo stock increases from low to moderate densities, then decreasing as stock density continues rising toward carrying capacity B_K (figure 13.1). Culm consumption by insects and bamboo rats is represented by a diagonal line iB with a y-intercept at the origin, rising as a constant proportion of stock density. At the current stock density $B_{Current}$ of 6293 tons/km² for Q. macrophylla, insects and rats are consuming more of this bamboo than new culm production can replace—depicted at the far right of figure 13.1 by the area below the diagonal line but above the parabola. This excessive predation by insects and rats, coupled with bamboo consumption by pandas and human harvesting, tends to decrease overall bamboo stock density, as indicated by the gray arrow at the left. Once bamboo density falls well below about 50% of B_K (i.e., in the darkly shaded area under the parabola but above the diagonal line), culm production would be high enough, and predation by insects and rats low enough, to leave sufficient "surplus" to meet the needs of giant pandas and people without continuing to reduce bamboo density. A new bamboo equilibrium could be created at any bamboo stock density within the shaded range.

In figure 13.2, the shaded area in figure 13.1 is plotted separately, whereupon the diagonal line in figure 13.1 becomes horizontal, running along the x-axis in figure 13.2. As can be seen most clearly in figure 13.2, the amount of bamboo available for pandas and humans would be maximized near 25% of B_{max}, in contrast to total culm production, which is maximized near 50% of B_{max}. The amount of bamboo (cP) needed by an equilibrium density of pandas (P_K) (i.e., cP_K)

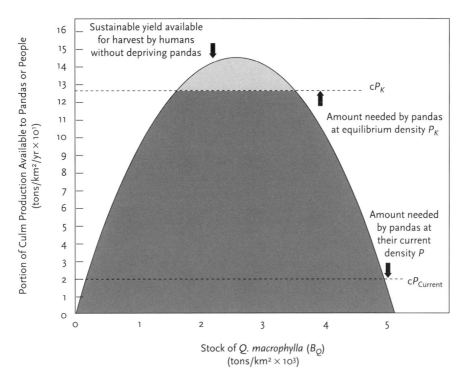

FIGURE 13.2. Sustainable culm harvest by people and pandas, in excess of predation by insects and rats. The entire parabola is equivalent to the shaded area in figure 13.1, where the diagonal iB in that figure runs horizontally along the origin in figure 13.2. The lightly shaded area here is the amount of bamboo culms available for harvest by humans at an equilibrium panda population density. The relationship shown here for *Q. macrophylla* is similar for *I. longiauritus*.

is represented by a horizontal line bisecting the parabola in figure 13.2. Once panda density reaches equilibrium, there would be only a narrow range of bamboo densities in which people would be able to harvest even moderate amounts of bamboo culms without causing the stock density to decline, which would be a problem only at stock densities below 25% of B_{max}.

From equations 13.5 and 13.6, the maximum allowable human harvest of culms is calculated in two steps. First, the maximum value of dB/dt is calculated for each equation by differentiating it with respect to t and setting the resultant equation equal to 0. This yields the bamboo stock density at which dB/dt is maximized. That maximum is then calculated. Second, from that stock density we subtract total bamboo consumption by rats, insects, and an equilibrium population of pandas. Total available bamboo minus the amount consumed by animals is h_{max}, the maximum sustainable harvest by humans:

	B AT MAX dB/dt (tons/ km²)	MAX dB/dt (tons/ km²/ year)	MAXIMUM SUSTAINABLE HARVEST (tons/km²/ year)
Q. macrophylla	2250	144	17.4
I. longiauritus	685	37	12.0

ECOLOGICAL IMPLICATIONS AND CONSERVATION STRATEGIES

MEETING THE NEEDS OF PANDAS

The equilibrium density/km² of pandas (3.03) is more than six times the current density (0.48) at Mabian (Yang et al. 1994), whereas the corresponding equilibrium densities/km² of bamboo are about 60–70% of current levels: 5088.7 tons of *Q. macrophylla* and 1411.95 tons of *I. longiauritus* (Yang 1993). Increasing panda density/km² to 3.03 would tend to reduce bamboo supply to these new equilibrium densities. Bamboo supply thus appears more than adequate to support a several-fold increase in the panda population, so long as losses to rats, insects, and human harvest do not increase dramatically.

MEETING THE NEEDS OF HUMANS

The maximum shoot harvesting that will not disrupt the dynamic equilibrium between the panda and the bamboo is h_{max}. Once h_{max} is exceeded, the combined predation by insects, rats, and pandas would tend to depress bamboo populations to the point that the pandas would eventually starve.

Earlier in the chapter, for each species of bamboo we estimated potential rate of increase, density dependence, and predation loss to insects, rats, and pandas. For pandas, we estimated low-density rates of birth, death, and population increase corresponding to the assumed equilibrium size. Based on those figures, we calculated maximum sustainable human harvest levels of 17.4 tons/km² for *Qiongzhuea* and 12.0 tons/km² for *Indocalamus*. Both are higher than the current harvest of 2.99 tons/km² for *Qiongzhuea* alone. By definition, so long as the harvest level remains below h_{max} for each bamboo species, harvesting should not prevent an increase in the panda population to equilibrium size, much less jeopardize permanent survival of the panda or either species of bamboo.

Years of observation may be required to verify that these model outputs are basically correct, much less to refine them. One approach to refinement is to devise more realistic models—for instance, models that take into account variations over time in (1) vital rates for pandas and bamboo (e.g., as influenced by weather), and (2) in predation by insects and rats. We also need models which more closely mimic population dynamics of rats, insects, and ursids—especially models of response to changes in food supply and competition, such as those derived by Stringham (1983, 1985) for the grizzly bear (*Ursus arctos*). A second way of refining these results would be through cautious adaptive management; for instance, by setting substantially lower quotas and noting responses by pandas and each species of bamboo. This would require continued field study of not only pandas and bamboo, but also the majority of factors that control their vital rates, such as weather and predation on bamboo by insects and rats. Finally, it is essential to keep in mind that these figures for allowable harvest levels of bamboo address only the impacts of lowering the bamboo supply. They do not address the impacts of harvesting itself—for instance, disturbance of the pandas or degradation of their habitat.

ACKNOWLEDGMENTS

This project was supported by the Key Project of the National Science Foundation of China (30230080) and the National Science Fund for Distinguished Young Scholars (30125006). We thank Wenzheng Mei, Jian Qing of the Mabian Forestry Bureau, Ayu Muzi and Qubie Abo of the management department of Mabian Dafengding Nature Reserve, and Wei Wang of Sichuan Normal College for their kindly assistance in the fieldwork. Stringham's contributions were supported by the University of Alaska.

REFERENCES

Hu, J. 2001. *Research on the giant panda*. (In Chinese.) Shanghai: Shanghai Publishing House of Science and Technology.

Hu, J., F. Wei, C. Yuan, and Y. Wu. 1990. *Research and progress in the biology of the giant panda*. (In Chinese.) Chengdu: Sichuan Publishing House of Science and Technology.

Johnson, K. G., G. B. Schaller, and J. Hu. 1988. Comparative behaviour of red and giant pandas in the Wolong Reserve, China. *J Mammal* 69:552–64.

Pan, W., Z. Gao, and Z. Lü. 1988. *The giant panda's natural refuge in Qinling Mountains*. (In Chinese.) Beijing: Peking University Press.

Reid, D. G., and J. Hu. 1991. Giant panda selection between *Bashania fangiana* bamboo habitats in Wolong Reserve, Sichuan, China. *J Appl Ecol* 28:28–43.

Schaller, G. B., J. Hu, W. Pan, and J. Zhu. 1985. *The giant panda of Wolong*. Chicago: University of Chicago Press.

Schaller, G. B., Q. Deng, K. Johnson, X. Wang, H. Shen, and J. Hu. 1989. Feeding ecology of giant pandas and Asiatic black bears in the Tangjiahe Reserve, China. In *Carnivore behavior, ecology and evolution*, edited by J. L. Gittleman, pp. 212–41. Ithaca, N.Y.: Cornell University Press.

Smith, J. M. 1979. *Ecological model*. Beijing: Science Press.

Stringham, S. F. 1983. Roles of adult males in grizzly bear population biology. *Inf Conf Bear Res Man* 5:140–51.

Stringham, S. F. 1985. Responses by grizzly bear population dynamics to certain environmental and biosocial factors. Ph.D. dissertation, University of Tennessee, Knoxville.

Taylor, M. 1994. Density-dependent population regulation of black, brown, and polar bears. In *9th International Conference on Bear Research and Management, Monograph Series No. 3*, edited by M. Taylor, pp. 3–14. Portland, Oregon: International Association for Bear Research and Management.

Wei, F., and J. Hu. 1994. Studies on the reproduction of wild giant pandas in Wolong Nature Reserve. (In Chinese.) *Acta Theriol Sin* 14:243–48.

Wei, F., J. Hu, G. Xu, S. Zhong, M. Jiang, and Q. Deng. 1989. A study on the life table of wild giant pandas. (In Chinese.) *Acta Theriol Sin* 9(2):81–86.

Wei, F., C. Zhou, J. Hu, G. Yang, and W. Wang. 1996a. Bamboo resources and food selection of giant pandas in Mabian Dafengding Natural Reserve. (In Chinese.) *Acta Theriol Sin* 16:171–75.

Wei, F., C. Zhou, J. Hu, G. Yang, and W. Wang. 1996b. Habitat selection by giant pandas in Mabian Dafengding Reserve. (In Chinese.) *Acta Theriol Sin* 16:241–45.

Wei, F., Z. Feng, and J. Hu. 1997a. Population viability analysis computer model of the giant panda population in Wuyipeng, Wolong Natural Reserve, China. *Int Conf Bear Res Man* 9:19–23.

Wei, F., J. Hu, W. Wang, and G. Yang. 1997b. Estimation of daily energy intake of giant pandas and energy supply of bamboo resources in Mabian Dafengding Reserve. (In Chinese.) *Acta Theriol Sin* 17:8–12.

Wei, F., Z. Feng, and Z. Wang. 1999a. Feeding strategy and resource partitioning between giant and red pandas in Xiangling Mountains. *Mammalia* 63:417–30.

Wei, F., Z. Feng, Z. Wang, and J. Liu. 1999b. Association between environmental factors and growth of bamboo species *Bashania spanostachya*, the food of giant and red pandas. (In Chinese.) *Acta Ecol Sin* 19:710–14.

Wei, F., Z. Feng, and Z. Wang. 1999c. Habitat selection by giant and red pandas in Xiangling Mountains. (In Chinese.) *Acta Zool Sin* 45:57–63.

Wei, F., Z. Feng, Z. Wang, and J. Hu. 2000a. Habitat use and separation between the giant panda and the red panda. *J Mammal* 81:448–55.

Wei, F., Z. Wang, and Z. Feng. 2000b. Energy flow through populations of giant pandas and red pandas in Yele Natural Reserve. (In Chinese.) *Acta Zool Sin* 46:287–94.

Yang, G. 1993. Population ecology of giant pandas in Mabian Nature Reserve, Liangshan Mountains. (In Chinese.) M.Sc. dissertation. Sichuan Normal College, Sichuan.

Yang, G., J. Hu, F. Wei, and W. Wang. 1994. The size and activity of the giant panda population in Defengding Nature Reserve. (In Chinese.) *J Sichuan Normal College* 15:114–18.

Yuan, C., J. Hu, H. Zhang, and S. Dong. 1990. The mathematical models of dynamics of the giant panda and bamboo populations and their application to the two populations in Wuyipeng. In *Research and progress in the biology of the giant panda*, edited by J. Hu, F. Wei, C. Yuan, and Y. Wu, pp. 205–22. (In Chinese.) Chengdu: Sichuan Publishing House of Science and Technology.

PANEL REPORT 13.1

Management of Giant Panda Reserves in China

Changqing Yu and Xiangsui Deng

THE ESTABLISHMENT and management of protected areas in range states are critical to the in situ conservation of endangered species. In 1963, the government of China set aside four reserves for the protection of the giant panda *(Ailuropoda melanoleuca)* and associated wildlife at Wanglang, Wolong, Baihe, and Labahe. All four reserves are in Sichuan Province, in the central and southwestern parts of the giant panda's current distribution. During the 1970s and 1980s, additional reserves were established and are known as the Tangjiahe, Jiuzhaigou, Xiaozhaizigou, Huanglongsi, Dafengding, Fengtongzhai, Baishuijiang, and Foping Nature Reserves. In subsequent years, additional reserves were created, bringing the total to over thirty and including a little more than half of the present-day range of the species.

Although these designations of protected areas are commendable steps, they do not by themselves ensure the survival of giant pandas in their remaining natural habitat. The key to protection is effective management of the ecosystems of which the species is a part. Central to effective management is conservation of the biodiversity, integrity, and ecological functioning of the ecosystems in each reserve. Without effective management, the existing causes of endangerment—primarily human overexploitation of essential resources in and around reserves and the resultant fragmentation of the population and its habitat—will be exacerbated. Three important aspects of management that require urgent attention can be identified: reserve design (i.e., creation of an infrastructure that enables staff to perform), training of personnel and long-range planning, and implementation of effective protection programs.

PRESENTATIONS AND DISCUSSION

A panel of experts in management and design convened at *Panda 2000* to discuss issues in reserve planning. A number of management procedures are being implemented in several reserves, as exemplified by Foping and Wanglang Reserves. According to Yange Yong, director at Foping, and Youping Chen, director at Wanglang, these revised procedures include reserve management planning, community-based conservation (participatory management or co-management), and monitoring/patrolling of areas where pandas are known to exist. Prof. Yanling Song of the Chinese Academy of Sciences stressed the importance of better planning and incentives for managers to excel, as well as the need for technical input by management experts.

Changqing Yu of World Wildlife Fund (WWF)–China reviewed the international programs offering support for the preservation of the panda in the wild. Such programs include the Reserve Management Project (funded by the Global Environmental Facility, a joint program of the World Bank, the United Nations Environment Program, and the United Nations Development Program), and monitoring and patrolling projects, training programs, and prioritization of reserve-based activities, all supported by WWF.

The reserve management plan for Sichuan Province, introduced by Xiangsui Deng, director of the Conservation Division in the Sichuan Forestry Department, emphasizes the strict protection of natural resources; specifically, forest protection against logging, mining, livestock grazing and other pressures, fire prevention, and antipoaching measures. Another major component of the plan is the patrolling and long-term monitoring of panda habitat, currently supported by WWF in fifteen reserves. Capacity building, community participation in conservation, research, education, and improved communication were identified as additional components of the provincial plan.

FINDINGS AND RECOMMENDATIONS

In April 2000, the Sichuan Forest Department and WWF jointly conducted a workshop on monitoring and patrolling programs and ways to achieve more effective conservation in the reserves. Participants identified the primary threats facing the reserves, as well as factors that limit the capacity of reserve personnel and the training needed to make them more effective. In addition, the participants devised a list of problems and priority action items for the reserves, grouped into four basic categories:

Primary threats:
- Wildlife poaching
- Gathering of medicinal plants, bamboo shoots, and firewood
- Illegal timber harvesting
- Pasturing of domestic livestock
- Tourism
- Mining

Limiting factors in reserve management:
- Inadequate facilities and equipment
- Inadequate financial resources
- Limited capacity to undertake conservation activities
- Reliance of impoverished locals on forest resources for their livelihood
- Inadequate training opportunities

Priorities for further training:
- Improving skills in species identification, data collection, and use of essential equipment
- Development of more effective monitoring and patrolling techniques
- Education in the principles of conservation biology
- Training in utilization of computers
- Training in data analysis and reporting of results

Priority actions for panda reserves:
- Capacity building
- Increased forest protection and fire prevention
- Investing in infrastructure (facilities, equipment)
- More effective monitoring and patrolling
- Developing ecotourism as a revenue source

There is general agreement that three underlying issues are critical to achieving more effective conservation activities by reserves. These are reserve management planning, community-based conservation, and development of ecotourism as a source of additional funding.

Reserve management planning is contrasted with reserve general programming that is embraced by most reserves today. It is often difficult to distinguish between general programming and management planning. General programming is usually construction oriented—focusing largely on infrastructure development—and is implemented by Chinese forestry survey and planning institutes. Reserve management planning is activity oriented—focusing on implementing conservation—and is facilitated by international project consultations but implemented by the local reserve staff. Reserve management planning involves analyzing the specific set of problems each reserve faces and deciding how to address them in specific situations, setting concrete conservation goals and determining how to achieve them, and designing ways to measure progress toward these goals and adapt approaches to changing conditions. Clearly, stronger emphasis should be placed on the adoption of reserve management planning.

COMMUNITY-BASED CONSERVATION

A reserve is not an island, but a designated area that blends into the larger landscape around its borders. Effective reserve management requires the development of partnerships, especially with the local community. Establishing co-management zones or buffer zones in the areas occupied by the local people is essential to the development of conservation programs (see, e.g., Western et al. 1994).

MONITORING AND PATROLLING

Monitoring, which is management oriented, is designed to continuously assess the well-being of both wildlife and its habitats. Regular patrolling provides the main source of monitoring data and can be integrated with monitoring as a primary activity. Taken together, these activities yield certain essential data, but also protect against poaching, illegal logging, and forest fires. Monitoring is also useful in assessing the effectiveness of conservation approaches. In partnership with local communities and outside specialists, and by communication with other reserves and conservation stations, monitoring and patrolling provide a basis for networking among reserves.

ECOTOURISM

Ecotourism is envisioned as a source of revenue for more effective reserve management. But such programs must be carefully managed to limit damage to wild populations and their habitats. If well managed, ecotourism can contribute to the financial well-being of local communities and provide a benefit to reserves having insufficient sources of revenue. It may also assist in efforts to achieve sustainable development and may raise conservation awareness among ecotourists. An opportunity for ecotourism related to pandas exists, but good examples of this approach cannot yet be found in China.

ACKNOWLEDGMENTS

We thank Yanling Song, Yange Yong, and Youping Chen for their contributions to this report.

REFERENCE

Western, D., R. M. Wright, and S. C. Strum. 1994. *Natural connections: Perspective in community-based conservation.* Washington, D.C.: Island.

PART FOUR

Giant Panda Conservation

Few will doubt that conserving nature requires the support of those who make the laws, shape the policies, and dole out the resources. The understanding that arises from an increased knowledge of nature and natural processes is critical to bending political wills in favor of any given conservation effort. For example, knowing the risk of extinction, given the life history of a taxon, could affect the level of public concern. An update on population numbers is a more obvious datum that is basic to action. As these two examples indicate, much of this volume is aimed at using science to get our facts in order. But even after we have done so, in the end, we must recognize that conservation is about people—their attitudes and their decisions.

Educational efforts constitute a special form of indoctrination that offers the prospect of shaping decisions by present and future leaders, whereas nongovernmental organizations and other entities attempt through government contacts, program development, financial inducements, and such international conferences as *Panda 2000* to reach the existing power structure. In part 4, we group together chapters that contribute to a factual foundation for engendering support for panda conservation, whether at the level of the citizenry at large or of its leaders. We also include examples of international collaboration that exemplify the place of people in advancing the prospect that species and ecosystems in one of our planet's most diverse regions will be sustained.

Human activities and behaviors, so tightly intertwined with the survival of the giant panda *(Ailuropoda melanoleuca)*, are too often overlooked in panda research, according to Liu, Ouyang, Zhang Linderman, An, Bearer, and He (chapter 14). Effective conservation requires an understanding of how human activities impact panda habitat at multiple spatial and temporal scales. In chapter 14, they propose a general framework that integrates panda ecology with human demography, socioeconomics, and human behavior, and apply the framework to Wolong Nature Reserve. Their report examines how panda habitat changed before and after Wolong was established as a nature reserve; how ecological, socioeconomic, and demographic factors affect panda habitats; and how various factors can be modified to reduce human impacts on panda habitats. It uncovers the counterintuitive fact that panda habitat deteriorated after Wolong became a reserve and details the human pressures that caused this to happen. The authors illustrate how integrating ecological, social, economic, behavioral, and demographic information—and cooperation among scientists, government agencies, and local communities—can lead to balanced and informed decisions about how to minimize the impact of humans on panda habitat.

In panel report 14.1, Lü and Liu provide an update on China's national plan for conservation of the giant panda. This plan was first published in 1989 (Chinese Ministry of Forestry and World Wildlife Fund 1989), and has provided the basis for policy decisions by the government of China. However, with the passage of time, new knowledge (e.g., a recent population recensus) and new priorities (e.g., a national ban on logging) have dictated the need for revision, and these authors put forth recommendations for an up-to-date planning document that will guide future decisions.

For giant pandas to survive over the long term, Dinerstein, Loucks, and Lü (chapter 15) develop the thesis that populations of pandas must be managed at a landscape scale that includes core areas of protection, buffer zones, dispersal corridors, and other forested tracts. Together, these elements of the landscape comprise what they have termed panda conservation units (PCU). Ground-truthing the boundaries of PCUs, an activity that has taken place in several parts of the giant panda's range, is an important step toward improving conservation efforts. However, an annual evaluation of the status of efforts at the landscape scale is also a necessity. Their chapter offers a straightforward scorecard for keeping track of progress, or lack thereof, in addressing conservation issues. Scorecard reports identify areas in which progress is lagging and also provide donors with a credible way of judging the effects of their investments in conservation. Above all, the objectivity provided by scorecards helps to ensure that the most important elements of giant panda conservation receive highest priority.

A second activity that was under way during *Panda 2000* is the Chinese government's resurvey of the giant panda and its habitat. In panel report 15.1, Yu and Liu review briefly the history of efforts that led to the estimate of approximately one thousand giant pandas surviving in the wild today. However, much habitat has disappeared since this mid-1980s datum was issued, leading to the conclusion that a new assessment is sorely needed in planning future conservation efforts. After substantial training of personnel and field testing, this report describes the indirect census method of "bite size" bamboo fibers found in panda droppings that was used to distinguish between individuals in an area. With an improved database, it becomes possible to conduct analyses of population trends, metapopulation and habitat fragmentation, human disturbances, and gaps in the conservation effort.

We believe persuasiveness in favor of conserving giant pandas gains support from theoretical treatments of their biology, as offered in chapter 16 by Gittleman and Webster. It is their thesis that the giant panda and related carnivores are important test cases for learning about the processes of extinction and, specifically, for determining whether some species, having certain constellations of traits, are predisposed to die out. Using a complete phylogeny (a supertree) of the carnivores, they show that high trophic level, low population density, slow life history, and, especially, small geographic range size are all significantly and independently associated with high extinction risk. They then focus on life history patterns to show which traits of giant pandas (e.g., birth weight, growth rate) are significantly different from reproductive rates in related terrestrial carnivores. Comparison of these patterns suggests that conservation action must solve the problem of how to manage and better understand the giant panda's unusual life history evolution. In turn, such comparisons are potentially useful for preventing extinction in other species.

The priority that should be given to reintroduction schemes in conserving a species is a recurring issue, no less so in the case of the giant panda. In panel report 16.1, Mainka, Pan, Kleiman, and Lü update criteria used in the consideration of previous reintroduction proposals. Previous evaluations of this strategy indicated that before reintroductions are attempted, more information is needed on the status of the wild population and other key elements, such as rearing captive-born individuals under regimes that endow them with the necessary survival skills. They conclude that although captive breeding has generated a surplus of giant pandas that could be so used, the current consensus is that in situ conservation and research remain as the more viable alternatives at this time.

An outstanding example of international cooperation is seen in recent work by the IUCN's Conservation Breeding Specialist Group (CBSG). In chapter 17, Ellis, Zhang, Zhang, Zhang, Zhang, Lam, Edwards, Howard, Janssen, Miller, and Wildt summarize major findings from a 3-year biomedical survey of China's captive population. In viewing this population of more than one hundred individuals as the ultimate hedge against extinction and perhaps even as a pool from which candidates for return to wild habitat could be produced, there was general agreement that the population was behaviorally and physically compromised and had little prospect of improvement in either quality of life or in numbers under traditional management regimes. Under the aegis of the Chinese Association of Zoological Gardens, the CBSG team recruited specialists in the medical and behavioral fields that evaluated the health and breeding potential of sixty-one individuals in four of China's major holding facilities. Variables found to be impediments to reproductive fitness ranged from genetic overrepresentation of certain individuals to suboptimal nutrition. A number of the findings from this collaboration inspire hopes of developing a healthy, sustainable captive population in China.

China's diverse and, in many ways, unique biomes have suffered great degradation, bringing many species to the brink of extinction. Insofar as the decisionmakers of today must rely on the will of the public, the education of school children and, through them, parents and extended families, looms large as a conservation strategy. In brief report 17.1, Bexell, Luo, Hu, Maple, McManamon, Zhang, Zhang, Fei, and Tian describe avenues developed by a team from Zoo Atlanta through which to deliver just such a conservation message. In collaboration with the strategically located Chengdu Zoo and Chengdu Research Base of Giant Panda Breeding, their report details the positive reception given to wildlife curriculum development, training classes, and other programs designed to educate urban dwellers on the plight of Asian wildlife. A healthy respect for cultural differences and for the importance of trust between representatives of East and West earmark these efforts.

Another tool in the conservation effort—one that entails an unusual level of collaboration between governments—is the loan of giant pandas to institutions outside of China. Whereas, in the 1970s, a new political regime in China was constrained to gain international favor through gifts of its greatest animal treasure; today, loans to the West have the objective of generating funds for use in in situ conservation. In workshop report 17.1, Potter and Stansell probe the political ramifications of international loans, particularly between China and the United States, and how interpretations of statutes becomes a critical aspect of these efforts. They stress that loans of Appendix I animals must conform to the governing principles established through the Convention on International Trade in Endangered Species of Wild Fauna and Flora (CITES), and conformity entails long-term commitments and close coordination between parties.

REFERENCE

Chinese Ministry of Forestry and World Wildlife Fund. 1989. *The national conservation management plan of the giant panda and its habitat.* Beijing: Chinese Ministry of Forestry and World Wildlife Fund.

14

A New Paradigm for Panda Research and Conservation

INTEGRATING ECOLOGY WITH HUMAN DEMOGRAPHICS, BEHAVIOR, AND SOCIOECONOMICS

Jianguo Liu, Zhiyun Ouyang, Hemin Zhang, Marc Linderman, Li An, Scott Bearer, and Guangming He

SINCE THE 1970S, many biological studies on giant pandas (*Ailuropoda melanoleuca*) have been conducted by scientists in China (e.g., Giant Panda Expedition 1974; Hu et al. 1980; Pan et al. 1988; Zhang et al. 1997) and from abroad (e.g., Schaller et al. 1985; Johnson et al. 1988; Reid et al. 1989; Reid and Hu 1991; Schaller 1993). These studies focused primarily on the biology of giant pandas in the field (including population dynamics, movement patterns, reproductive biology, and food habits) and in captivity (e.g., nutrition, reproduction, nursery for newborn pandas). For example, detailed individual information is available from those pandas that were captured, fitted with radio transmitters, and released (Schaller et al. 1985). However, as the main threat to pandas in the wild is the loss and fragmentation of their habitat (Chinese Ministry of Forestry and World Wildlife Fund 1989), biological studies alone are not sufficient for effective panda conservation. Future efforts to conserve wild pandas must focus on the underlying mechanisms influencing habitat loss and fragmentation, if effective policies are to be developed to maintain suitable habitat for viable populations.

Although some researchers have intuitively recognized that human activities are important factors causing the loss and fragmentation of panda habitat (Hu et al. 1980; Pan et al. 1988; Schaller 1993), no quantitative and systematic research had been undertaken to link panda habitats explicitly with human population and activities until a few years ago (e.g., Ouyang et al. 1995, 2000; Liu et al. 1999a,b; An et al. 2001; Liu et al. 2001a). Integrated analyses would have made many conservation policies and efforts more effective. For example, the Chinese government and some international organizations have tried to relocate residents living in Wolong Nature Reserve in Sichuan Province. Although some residents were relocated, many of them returned (Li et al. 1992; Liu et al. 2001a). Even relocation within the reserve was difficult to achieve. In the early 1980s, the Chinese government and the World Food Program built a large apartment complex in an area unsuitable for giant pandas within the reserve. The hope was to have local residents move voluntarily from the core habitat areas of Wolong to the apartment complex. However, not a single household did so. Liu et al. (1999a) found that in this instance, the elderly were accustomed to their lifestyle and did not want to relocate. Furthermore, there was insufficient land near the apartment complex for would-be migrants to farm. Because most of the local residents were farmers, they could not survive without land. These examples illustrate the need to understand the attitudes and needs of local residents before a conservation project is undertaken.

Integrated human–panda research is urgently needed to design and implement feasible policies. Studying the linkages between human dimensions and panda habitats requires a new paradigm that is different from traditional panda studies, encompassing not only ecological components but also major human dimensions (e.g., human demographics, behavior, socioeconomics) (Liu 2001). In this chapter, we provide an overview of this approach. We first introduce a conceptual framework, summarize some methods and results, and then describe some of our ongoing efforts. Finally, we discuss the perspectives of this new paradigm.

CONCEPTUAL FRAMEWORK OF THE NEW PARADIGM

Like many other wildlife species, the giant panda depends on forests for its habitat. Significant areas of forests in panda ranges have been altered due to human factors (Liu et al. 2001a,b). Human factors include demographic (e.g., population size and structure, household distribution), economic (e.g., income, expenses, production, consumption), social (e.g., needs and wants, perceptions and attitudes toward wildlife conservation), and behavioral (e.g., timber harvesting, fuelwood collection for cooking and heating). These factors are the primary forces influencing the rate and location of human impacts on forest systems and, consequently, affecting the spatial and temporal configuration of panda habitat (figure 14.1).

All three components (human, forest, and habitat systems) in the system framework can be directly or indirectly influenced by one another, and by external government policies and natural factors. For example, human activities are affected by other human factors, such as human needs. Human activities influence forests directly and panda habitat indirectly. Changes in forest structure, function, integrity, and dynamics can all have an impact on panda habitat quantity, quality, timing (when the habitat is available), and location. Government policies can significantly affect various aspects of the human system. However, the policymaking process and its effectiveness can also be shaped by the human system and panda habitat conditions. When panda habitat conditions are seriously deficient and human attitudes toward panda conservation are positive, government policies may be more favorable to panda conservation. Furthermore, the human system may be constrained by feedback from the forest system. For example, after all trees in a forest are harvested, local residents must adopt a different lifestyle without the use of timber and fuelwood. In addition, natural features of the physical environment (e.g., elevation, landslides) impact humans, forests, and panda habitats.

This conceptual framework guided our assessment of the impact of human factors on panda habitats in Wolong Nature Reserve. All components in the framework were incorporated into geographic information systems (GIS) as data layers and integrated into systems models as driving variables, state variables, or parameters. The integration, analysis, and modeling of these data helped us to answer many questions, such as:

1. How did panda habitats change before and after Wolong was established as a nature reserve?
2. How did ecological, socioeconomic, and demographic factors affect panda habitats?

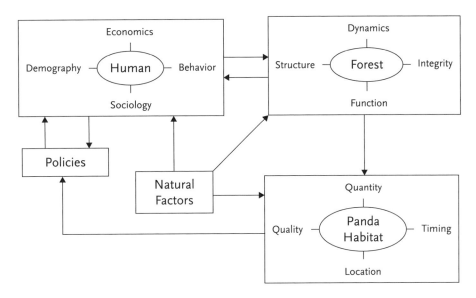

FIGURE 14.1. Framework of the new paradigm that integrates ecology with human demographics, behavior, and socioeconomics for panda research and conservation. Modified from Liu et al. (1999a).

3. How can various factors be modified to reduce human impacts on panda habitats?

METHODS FOR THE NEW PARADIGM

We chose Wolong Nature Reserve as our study site for four major reasons. First, it is one of the largest nature reserves (~200,000 ha) designated for conserving giant pandas and contains approximately 10% of the wild panda population (Zhang et al. 1997). Second, like many other nature reserves, there are local people residing in Wolong (more than 4,000 local residents in over 900 households in 1998). Third, Wolong is a "flagship" nature reserve and has received exceptional financial and technical support from the Chinese government and many international organizations since its creation in 1975. It is also part of the international "Man and Biosphere" reserve network (He et al. 1996). Fourth, many biological studies on giant pandas have been conducted in the reserve, and there is a good record of economic and demographic statistics. These previous studies and the data they produced provided a good foundation for our study.

Our new paradigm takes four general approaches: a systems approach, multiscale approach, interdisciplinary approach, and an integrated approach. The systems approach considers not only ecological factors, but also human demographic, behavioral, and socioeconomic factors; not only factors inside the reserve, but also those outside the reserve; not only what happened in the past and what is happening at present, but also what will happen in the future. The multiscale approach deals with issues at multiple spatial scales (e.g., patch, landscape), temporal scales (e.g., days, seasons, years, decades), and socioeconomic scales (e.g., individual, household, group [age, sex, educational levels], village, township, reserve). Our research team is interdisciplinary, consisting of researchers from such diverse fields as ecology, forestry, economics, sociology, geography, remote sensing, GIS, systems modeling and simulation, and reserve planning and management.

We used an integrated approach for data collection, data management, data analysis, data integration, and information dissemination. In terms of data collection, we combined field studies, interviews, government statistics and documents, information from the literature, and data from remote sensing (satellite and aerial photographs) and global positioning systems (GPS). The remotely sensed data included Corona photographs from 1965, LANDSAT Multispectral Scanner (MSS) data from 1974 (the year before the reserve was established), and LANDSAT Thematic Mapper data from 1997 (Liu et al. 2001a). The Corona data are stereo-pair photographs acquired on January 20, 1965, as part of the Corona photoreconnaissance satellite project (USGS Eros Data Center, Sioux Falls, South Dakota [Liu et al. 2001a]). Both LANDSAT MSS (January 3, 1974) and LANDSAT TM data (September 27, 1997) were obtained from China's Satellite Ground Station (Beijing, China). Four images of IKONOS data (different times between August 31 and November 16, 2000) with 1-m resolution were acquired through the National Aeronautics and Space Administration (NASA). The remote sensing data (Liu et al. 2001a) were georeferenced using highly accurate GPS receivers (Trimble Pathfinder). Field data were collected and georeferenced to the remotely sensed data using GPS locations to allow classifications of land covers, relate the spectral characteristics of the remotely sensed data to the spatial distribution of understory bamboo, and measure human influences and landscape structures. A Digital Elevation Model (DEM) was also developed from topographic maps and referenced to ground control points. GPS and satellite data were used to assist in gathering some socioeconomic data, such as household locations.

Ecological and geographical data included slope, aspect, elevation, vegetation community structure, plant composition, forest vertical structure, plant species richness, plant species diversity, bamboo biomass, bamboo density and species, forest canopy cover, mid-story cover, and understory cover (Ouyang et al. 2000; Linderman et al., in press). Data on the use of habitat by giant pandas (e.g., spatial distribution of panda feces) are also being collected (Bearer et al. 2000).

To understand the underlying mechanisms of habitat change, we have collected a large volume of social, economic, human demographic, and human behavioral data. Demographic data first became available in the 1960s, but in that decade and the next, the only available information concerned human population size. Three censuses (population censuses in 1982 and 2000 and an agriculture census in 1996) provided more detailed information (e.g., population structure, economic data). In addition, other demographic data (deaths, births, marriages, migrations) were available on an annual basis in most of the years in the 1980s and 1990s. Collection of social and behavioral data (e.g., perceptions, attitudes, activities) began in 1996. Economic data (e.g., income, expenses, production, consumption) have been available since the 1970s. We also surveyed local residents to identify economic activities of different age groups (Liu et al., unpubl. data) and investigate the amount of fuelwood consumption at the household level (An et al. 2001).

After the data were collected, they were entered into and managed using a relational database program (ACCESS) and GIS. Data were analyzed using spatial statistics, GIS, and statistical packages, such as SAS, S-Plus, and SPSS. Classification of remotely sensed data was initially accomplished through visual interpretation of a limited set of spectral data (to allow consistent analyses of different remote sensing images for a multitemporal study), unsupervised classification (for comparison analyses in a multitemporal study), and supervised classification using recent ground-truthing (for current landscape analyses).

Results from data analysis were then integrated into systems models using systems modeling and simulation techniques (Liu et al. 1999a), as well as decision support systems (DSS) (He et al. 2000). The model results are useful for decisionmakers in choosing desirable management alternatives to achieve both ecological and economic goals. We have developed systems models to link panda habitat data to socioeconomic and demographic data. In addition, we used computer simulations to project possible human demographic and ecological consequences of different policy scenarios (Liu et al. 1999a). For example, under different birth rates, emigration rates, emigrant composition (e.g., young versus old people), and levels of fuelwood consumption, we were able to project the future dynamics of human population sizes and structures, as well as panda habitats, over a period of fifty years (Liu et al. 1999a).

SOME RESULTS

Many results have been generated from our research (see, e.g., Liu et al. 1999a,b, 2001a,b; Ouyang et al. 2000; An et al. 2001; Linderman et al., in press). Here we present some of these results, including overall habitat changes within Wolong Nature Reserve, the influence of the human population and behaviors on the habitat, and how policy could potentially alter those behaviors for conservation purposes.

Both the quantity and quality of the panda habitat in the reserve continued to decline after the reserve was established (figure 14.2). More unexpectedly, rates of loss and fragmentation of high-quality habitat were higher after the reserve's establishment. The loss and fragmentation of panda habitats in Wolong were directly due to forest loss and fragmentation, as large tracts of forests were divided into smaller tracts and forest fragments were reduced in size.

Human population and behavior were the ultimate reasons for the much higher rates of loss and fragmentation of high-quality habitat after Wolong was designated as a nature reserve. As figure 14.3 shows, the number of local residents increased approximately 70% between 1975 and 1996, and the number of households more than doubled. The population structure has also changed dramatically (Liu et al. 1999b). For example, the number of children (0–19 years old) declined by almost 17% between 1982 and 1996, whereas the number of young adults (20–34 years old) almost doubled (~98%) during the same period. The main labor force (20–59 years old) increased about 60% from 1982 to 1996. Finally, the number of seniors (≥60 years) increased by almost 25%. The sex ratio (male:

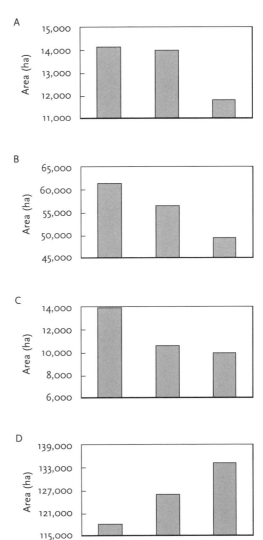

FIGURE 14.2. Change in the amount of panda habitat in Wolong Nature Reserve before and after the reserve was established in March 1975. (A) Highly suitable habitat; (B) suitable habitat; (C) marginally suitable habitat; (D) unsuitable habitat. From Liu et al. (2001a), with permission.

female) of young children (0–4 years of age) was 0.98:1 in 1982 but changed to 1.20:1 in 1996. In the age class of the oldest residents (>70 years), the sex ratio was 0.57:1 in 1982, but changed to 0.98:1 in 1996. More of the men lived longer in 1996 than in 1982.

The local resident population in Wolong will continue to increase, given the current birth rates, death rates, and migration rates (Liu et al.

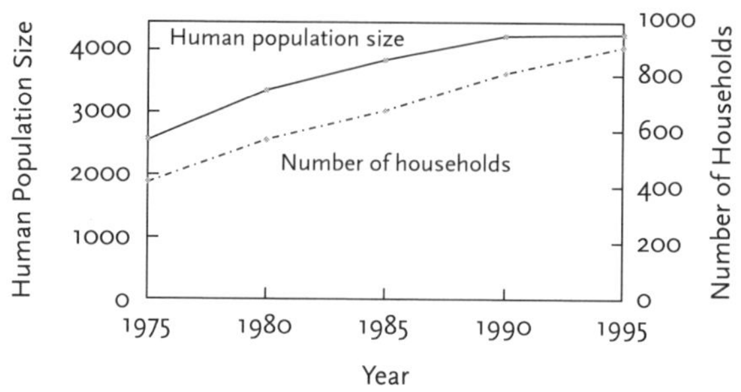

FIGURE 14.3. Human population and household dynamics in Wolong from 1975 to 1995. From Liu et al. (1999a), with permission.

1999a). Based on our projection (Liu et al. 1999a), the Wolong population will increase from 4320 in 1998 to 5960 in the year 2047. If the rate of emigration (all people in a household, both young and old) increases to 3%, the population will be reduced to 1671 by the year 2047. However, emigration of young people would be more effective in reducing human population size, because these emigrants would then not have children on the reserve, thus reducing the total number of potential residents and emigrants in the future. For example, relocating only 22% of the young people (17–25 years of age) would lead to a total population of 762 in the year 2047. Over a period of fifty years, the household emigration approach and the youth emigration approach would lead to a total of 4553 and 2189 local residents, respectively. Also, we found that, unlike the older residents in Wolong, the vast majority of young people were willing to move out of the reserve, especially if they were given the opportunity for higher education and jobs in the city. Furthermore, although senior residents themselves generally do not want to relocate, they are supportive of relocation by their children or grandchildren. In fact, they would feel very proud if their children or grandchildren could go to college. This shows that youth emigration would not only be more effective, but also more socially feasible.

Among various human activities, fuelwood consumption is the main factor affecting panda habitat at Wolong. Fuelwood is the major energy source for heating as well as cooking food for both humans and pigs (the main livestock in the reserve). Fuelwood consumption takes place at the household level and varies with household size, composition, economic status, and social attitudes. For example, a household with a senior resident (>60 years of age) consumed 19% more fuelwood for heating in the winter than a household without a senior (An et al. 2001). Households with a senior resident start the heating season earlier and end the heating season later in the year. An et al. (2001) developed a model to predict fuelwood consumption under different socioeconomic and demographic conditions. The model accurately estimated the actual amounts of fuelwood consumption. The amount of fuelwood consumption for heating had no significant relationship with household size, but the amount of fuelwood used in cooking was positively related to household size.

ONGOING WORK

In this chapter, we have highlighted several aspects of our ongoing work, including mapping the spatial distribution of bamboo, impacts of forest harvesting on giant panda habitat use, habitat restoration, and DSS. We are expanding this work, using the paradigm just described.

Bamboo is the staple food of pandas (see Long et al., chapter 6). The significance of bamboo distribution to panda populations has been well recognized (Reid et al. 1989; Reid and Hu 1991; Taylor and Qin 1997). However, because bamboo is an understory species mostly distributed in areas of complex topography, its distribution is difficult to map. To date, there is only cursory information gathered by ground surveys (but see Yong et al., chapter 10, and Liu et al., chapter 11). Although detailed maps exist for some small areas, large-scale maps, based on information from more general field surveys, are less useful. Remote sensing is a widely used method for large-scale data acquisition, but methods to map such understory species as bamboo, even including the use of aerial photography, have not been successful (Morain 1986; DeWulf et al. 1988; Porwall and Roy 1991). Combining field ground-truthing data with neural networks (an artificial intelligence technique), we were able to map the bamboo distribution for the entire Wolong Nature Reserve with an overall accuracy of 80% (Linderman et al., in press). The information for bamboo distribution is being incorporated into further habitat analysis (Linderman et al., unpubl. data). Furthermore, the method used in bamboo mapping will be very useful for mapping bamboo distribution in other parts of the panda's range and in studies of biodiversity, particularly understory vegetation and species that depend on understory vegetation as food or cover.

As noted in our review of results achieved to date, high-quality habitat for giant pandas has decreased markedly over the past several decades, due to human activities occurring within the original range of the giant panda. Of these activities, timber extraction and fuelwood collection greatly contributed to the reduction and fragmentation of their habitat. Thus, it is critical to ascertain how these harvesting activities have affected the use of habitats by giant pandas. Monitoring of panda activity (e.g., using feces and feeding sites as indicators) is being conducted in harvested and nonharvested areas throughout Wolong Nature Reserve. These data are being collected in a spatial format that will allow for comparisons with bamboo distribution and human demographic models, and will also provide necessary information for a landscape-level model of habitat use to be constructed. Preliminary results suggest that pandas avoid large, recently harvested areas (Bearer et al. 2000).

To effectively restore panda habitat in Wolong and other areas in Wenchuan County, Ouyang et al. (2001) began developing a landscape approach to China's national "return-steep-cropland-to-forest" program and "natural forest conservation" program. As the names indicate, these two programs have great potential for panda habitat restoration. Thus, we are creating integrated methods for evaluating the ecological and economic consequences of these two programs. Also, we are exploring integrated approaches to better plan and implement these programs in the Wolong-Wenchuan region for panda habitat restoration. Our new paradigm provides a good basis for such restoration efforts.

To better manage nature reserves for giant pandas (see Yu and Deng, panel report 13.1), we are developing a DSS (He et al. 2000) that integrates ecological, social, economic, behavioral, and demographic information. It provides a platform for conservation biologists, local residents, government officials, reserve managers, panda lovers, and perhaps other stakeholders to share their knowledge and concerns, evaluate long-term consequences of different policy scenarios, and make balanced and informed decisions about how to minimize the impact of humans on panda habitats. The DSS is web-based and integrated with spatial technologies, such as GIS, and dynamic systems modeling.

DISCUSSION

Our new paradigm lays a good foundation for understanding spatiotemporal patterns and mechanisms of panda habitat changes in Wolong and other reserves. It also provides an integrated framework to guide efforts for the conservation and restoration of panda habitat. Because the paradigm encompasses many disciplines, it addresses fundamental issues that no single discipline can address by itself. Thus, we advocate greater participation and collaboration from other disciplines in future endeavors that would facilitate panda research and conservation.

Our studies indicate that it is critical to explore the primary factors that cause habitat loss and fragmentation. Understanding the basis of these problems is an important first step toward the development and implementation of feasible and effective strategies for solutions. It is essential that scientists work with various stakeholders, especially government agencies and the local people, in their research. On the one hand, scientists have obligations to provide scientific information to government agencies, so that sound management decisions can be made. On the other hand, government agencies should encourage and support more interdisciplinary research that cuts across boundaries of the natural and social sciences.

The needs and perspectives of local people must be better appreciated. It is important to provide them with economic incentives to bring conservation efforts to fruition, because local residents are the main cause of habitat loss and fragmentation. If their needs for food and fuelwood are not adequately addressed, local residents will continue to cut down the forests that are home to the pandas. It is encouraging that Wolong Nature Reserve is implementing a natural forest conservation program that pays a household to take care of a specific forest area. Wolong has also built an "eco-hydropower" station designed to provide electricity to local residents for cooking and heating. Our research indicates that the price of electricity must be affordable and the supply of electricity must be stable for local residents to switch from fuelwood consumption to electric power (An et al. 2002). The natural forest conservation program and the eco-hydropower station program appear promising, but their effectiveness remains to be seen. Such long-term solutions as encouraging and helping more young people to move out of the reserve will reduce human pressures inside and around the reserve. Through a new partnership among scientists, policymakers, reserve managers, and local people, innovative and feasible conservation policies can be successfully developed and implemented, and habitats for endangered species like the giant panda can be effectively saved, efficiently restored, and sustainably managed.

ACKNOWLEDGMENTS

We appreciate the logistical support of the Wolong Nature Reserve, especially the assistance from Jinyan Huang, Yingchun Tan, Jian Yang, and Shiqiang Zhou. This chapter was written while the senior author was on sabbatical at the Center for Conservation Biology (CCB) at Stanford University. The hospitality of the CCB staff (especially Carol Boggs, Gretchen Daily, Anne Ehrlich, and Paul Ehrlich) is greatly appreciated. We also thank Don Lindburg for his persistent efforts and patience, which made this chapter possible, and for his and an anonymous reviewer's helpful comments on an earlier draft. We gratefully acknowledge financial support from the National Science Foundation, the National Institutes of Health (National Institute of Child Health and Human Development, grant R01 HD39789), NASA, the American Association for the Advancement of Science, the John D. and Catherine T. MacArthur Foundation, the St. Louis Zoo, Michigan State University, the National Natural Science Foundation of China, the Ministry of Science and Technology of China (grant G2000046807), and China Bridges International.

REFERENCES

An, L., J. Liu, Z. Ouyang, M. A. Linderman, S. Zhou, and H. Zhang. 2001. Simulating demographic and socioeconomic processes on household level and implications for giant panda habitats. *Ecol Model* 140:31–50.

An, L., F. Lupi, J. Liu, M. Linderman, and J. Huang. 2002. Modeling the choice to switch from fuelwood to electricity. *Ecol Econ* 42:445–57.

Bearer, S., L. Scott, J. Liu, G. He, J. Huang, and Y. Tan. 2000. The effects of timber harvesting and fuelwood collection on giant panda habitat use. Paper presented at *Panda 2000, Conservation Priorities for the New Millennium*, October 18, San Diego, California.

Chinese Ministry of Forestry and World Wildlife Fund. 1989. *The national conservation management plan of the giant panda and its habitat.* Beijing: Chinese Ministry of Forestry and World Wildlife Fund.

De Wulf, R. R., R. E. Goossens, J. R. MacKinnon, and S. C. Wu. 1988. Remote sensing for wildlife management: Giant panda habitat mapping from LANDSAT MSS Images. *Geocarto Int* 1:41–50.

Giant Panda Expedition. 1974. A survey of the giant panda *(Ailuropoda melanoleuca)* in the Wolong Nature Reserve, Northern Szechuan, China. (In Chinese.) *Acta Zool Sin* 20:162–73.

He, N., C. Liang, and X. Yin. 1996. Sustainable community development in Wolong Nature Reserve. *Ecol Econ* 1:15–23.

He, G., J. Liu, Z. Ouyang, M. Linderman, L. An, S. Bearer, H. Zhang, and J. Yang. 2000. A preliminary Decision Support System (DSS) prototype for giant panda reserve management. Paper presented at *Panda 2000, Conservation Priorities for the New Millennium*, October 18, San Diego, California.

Hu, J., Q. Deng, Z. Yu, S. Zhou, and Z. Tian. 1980. Biological studies of the giant panda, golden monkey, and some other rare and prized animals. (In Chinese.) *Nanchong Teacher's College J* 2:1–39.

Johnson, K. G., G. B. Schaller, and J. Hu. 1988. Responses of giant pandas to a bamboo die-off. *Nat Geogr Res* 4:161–77.

Li, C., S. Zhou, D. Xiao, Z. Chen, and Z. Tian. 1992. *The animal and plant resources and protection of Wolong Nature Reserve*, edited by Wolong Nature Reserve and Sichuan Normal College, pp. 311–72. (In Chinese.) Chengdu: Sichuan Publishing House of Science and Technology.

Linderman, M., J. Liu, J. Qi, L. An, Z. Ouyang, J. Yang, and Y. Tan. In press. Using artificial neural networks to map the spatial distribution of understory bamboo from remote sensing data. *Int J Remote Sensing*.

Liu, J. 2001. Integrating ecology with human demography, behavior, and socioeconomics: Needs and approaches. *Ecol Model* 140:1–8.

Liu, J., Z. Ouyang, W. Taylor, R. Groop, Y. Tan, and H. Zhang. 1999a. A framework for evaluating effects of human factors on wildlife habitat: The case of the giant pandas. *Conserv Biol* 13:1360–70.

Liu, J., Z. Ouyang, Y. Tan, J. Yang, and S. Zhou. 1999b. Changes in human population structure and implications for biodiversity conservation. *Pop Environ* 21:45–58.

Liu, J., M. Linderman, Z. Ouyang, L. An, J. Yang, and H. Zhang. 2001a. Ecological degradation in protected areas: The case of Wolong Nature Reserve for giant pandas. *Science* 292:98–101.

Liu, J., M. Linderman, Z. Ouyang, and L. An. 2001b. The pandas' habitat at Wolong Nature Reserve—Response. *Science* 293:603–4.

Morain, S. A. 1986. Surveying China's agricultural resources: Patterns and progress from space. *Geocarto Int* 1:15–24.

Ouyang, Z., H. Zhang, Y. Tan, K. Zhang, H. Li, and S. Zhou. 1995. Application of GIS in evaluating giant panda habitat in Wolong Nature Reserve. (In Chinese.) *China's Biosphere Reserve* 3:13–18.

Ouyang, Z., J. Liu, and H. Zhang. 2000. Community structure of the giant panda habitat in Wolong Nature Reserve. (In Chinese.) *Acta Ecol Sin* 20:458–62.

Ouyang, Z., J. Liu, X. Wang, H. Miao, M. Shen, J. Fu, Y. Tan, Z. Li, and H. Zheng. 2001. *Habitat restoration in Wolong-Wenchuan region: A landscape approach.* Proposal for WWF. Gland, Switzerland: World Wildlife Fund.

Pan, W., Z. Gao, Z. Lü, Z. Xia, M. Zhang, L. Ma, G. Meng, X. Zhe, X. Liu, H. Cui, and F. Chen. 1988. *The giant panda's natural refuge in the Qinling Mountains.* (In Chinese.) Beijing: Peking University Press.

Porwall, M. C., and P. S. Roy. 1991. Attempted understory characterization using aerial photography in Kanha National Park, Madhya Pradesh, India. *Environ Conserv* 18:45–50.

Reid, D. G., and J. Hu. 1991. Giant panda selection between *Bashania fangiana* bamboo habitats in Wolong Reserve, Sichuan, China. *J Appl Ecol* 28:228–43.

Reid, D. G., J. Hu, S. Dong, W. Wang, and Y. Huang. 1989. Giant panda *(Ailuropoda melanoleuca)* behaviour and carrying capacity following a bamboo die-off. *Biol Conserv* 49:85–104.

Schaller, G. B. 1993. *The last panda.* Chicago: University of Chicago Press.

Schaller, G. B., J. Hu, W. Pan, and J. Zhu. 1985. *The giant pandas of Wolong.* Chicago: University of Chicago Press.

Taylor, A. H., and Z. Qin. 1997. The dynamics of temperate bamboo forests and panda conservation in China. In *The bamboos*, edited by G. P. Chapman, pp. 189–211. San Diego: Harcourt Brace.

Zhang, H., D. Li, R. Wei, C. Tang, and J. Tu. 1997. Advances in conservation and studies on reproductivity of giant pandas in Wolong. (In Chinese.) *Sichuan J Zool* 16:31–33.

PANEL REPORT 14.1

China's National Plan for Conservation of the Giant Panda

Zhi Lü and Yongfan Liu

CHINA'S CONSERVATION management plan for the giant panda *(Ailuropoda melanoleuca)* (Chinese Ministry of Forestry and World Wildlife Fund 1989) has been in effect since 1993. Over the years since the publication of this document, and guided by the activities outlined therein, in situ conservation has made a great deal of progress. For example, twenty new reserves have been established or designated, reserve staff have received different types of training, integrated conservation and development projects are yielding positive results (with an emphasis on local participation), a third national survey on panda population and habitats has been completed, and international organizations have become involved in the conservation effort. Moreover, a national ban on logging in natural forests since 1998 has provided new opportunities for habitat protection and restoration.

Consequently, the focus of China's national policy today is environmental protection and ecological restoration. Meanwhile, although pressures for panda conservation have continued unabated, overwhelming economic growth and development raise new challenges. Therefore, although planning is a dynamic process that responds to changing realities, a basic, up-to-date document that offers guidance in light of new policies and the lessons learned from the past is essential to the prioritizing of actions.

An effective National Conservation Plan rests on the following components:

- Involvement of more of the stakeholders at the provincial, county, and reserve level in land-use planning;
- Greater involvement of wildlife experts in designing and updating plans;
- Utilization of data collected from the recently completed national survey of giant pandas and their habitats;
- Evaluation and monitoring that is based on clear and measurable criteria for progress;
- A continuing role for captive breeding;
- Development of clear criteria for prioritizing projects;
- Designation of new reserves, based on the updated knowledge from the last national survey, and following discussions with local communities in each area; and
- Identification of sustainable funding sources, including major support from gov-

ernment entities in China, for support of a national protection plan for forests and for China's "Grain-to-Green" program (regeneration of hillside farmland).

LAND-USE PLANNING IN PANDA HABITAT

Land-use planning throughout their range is essential for the long-term viability of giant pandas. We must identify areas for protection, including currently suitable and intact habitat, degenerated habitat that can be restored, corridors or linkages that connect habitat patches in areas where protection and restoration are possible, and habitat having the potential for future population deployment.

Key areas and corridors identified as locations for the establishment of future reserves require strict protection. Meanwhile, other habitats occupied by pandas may be designated as multiuse protected areas, where sustainable economic development is required. Effective land-use planning is based on a participatory process for stakeholders and local communities. Research on the development of suitable techniques for habitat restoration must be supported.

There are several measures that are essential to the protection and management of the wild population. These include the prevention of poaching, additional research on habitat selection and utilization by wild populations, and the use of radiotelemetry in tracking key individuals. Also, further genetic study, especially on isolated populations, will enable the development of effective techniques for population management.

The same national plan that guides the host country's activities should guide those of international stakeholders as well. Recognition of the national mandate will facilitate open communication between China and the international community. Mechanisms for ensuring the ascendancy of the national plan, such as review by peers or other appropriate entities, and international communication about its implementation must be an ongoing aspect of the plan.

REFERENCE

Chinese Ministry of Forestry and World Wildlife Fund. 1989. *The national conservation management plan of the giant panda and its habitat.* Beijing: Chinese Minisitry of Forestry and World Wildlife Fund.

15

Biological Framework for Evaluating Future Efforts in Giant Panda Conservation

Eric Dinerstein, Colby Loucks, and Zhi Lü

THE CHINESE GOVERNMENT began to take steps to protect wild panda populations in the early 1960s with the establishment of several nature reserves primarily to conserve giant pandas *(Ailuropoda melanoleuca)*. However, it was not until 1989 that China's Ministry of Forestry and World Wide Fund for Nature (WWF) presented a joint national conservation management plan for the giant panda and its habitat to the government of China (Chinese Ministry of Forestry and World Wide Fund 1989). The Chinese government eventually ratified the plan in 1992, and for the past decade, this plan has guided panda conservation activities. During this time, China took great strides in conserving wild pandas by establishing new nature reserves, altering forestry practices, and enhancing its captive breeding program. However, pandas and their habitat are still at risk of extinction. As we begin a new century, we must develop a long-term vision for the future conservation of wild pandas.

The giant panda's historical range once extended throughout most of the subtropical evergreen forests of southeastern and eastern China (see figure 9.1 in Hu and Wei, chapter 9). As the climate changed and human activities intensified, the fertile plains of this region were converted from forests to agricultural fields. Consequently, the giant panda suffered a large contraction of its range. Today, the panda survives at the periphery of its historical range in the temperate deciduous and conifer forests at the edge of the Tibetan Plateau in Sichuan and Gansu Provinces, and in the Qinling Mountains in Shaanxi Province.

Channell and Lomolino (2000) found that the majority of the world's endangered species survive at the periphery of their historical ranges. As a species moves from the core to the periphery of its range, the remaining populations survive in less favorable habitats and at lower densities (Brown 1984; Gaston 1990; Brown et al. 1995). These populations also tend to be more fragmented and less likely to exchange genetic material with each other, increasing the probability of extinction of each isolated population (MacArthur and Wilson 1967; Brown and Kodric-Brown 1977; Pimm et al. 1988; Tracy and George 1992).

Unfortunately, despite several decades of conservation effort and millions of dollars of support, giant panda populations show most of these characteristics. This is mainly caused by increasing demands on resources, due to China's rapidly growing economy and population. However, institutional complexity within China has also contributed to endangering the pandas. Several organizations at the national level oversee panda conservation efforts, such as the State Forestry Administration (formerly the Ministry of Forestry) and the Ministry of Construction, and each organization has its own priorities, planning structure, and mandates for success. As a result, lack of funding for many reserves, limited direction for implementation of management plans, and the need for local or county governments to develop their own revenue remain important problems. To establish a successful conservation plan means negotiating with each of the various political institutions and planning levels (i.e., national, provincial, and county).

To make panda conservation more effective, agreement on a long-term biologically based vision for panda conservation will be necessary. In broad terms, such a vision would entail a network of well-protected core areas, with extensive forest cover and a broad range of bamboo species in the understory; connectivity between reserves; elimination of factors contributing to habitat loss and fragmentation; and powerful economic or regulatory incentives to maintain healthy wild panda populations. To achieve these goals, we propose the establishment of a timetable of conservation goals across a variety of spatial scales and development of a framework to monitor success in meeting these conservation targets. Through a participatory process, an agreed agenda will hopefully be achieved. This chapter, however, focuses only on addressing the panda's biological requirements that should be made part of these goals.

TIMELINE OF CONSERVATION GOALS

We present an ambitious timeline of conservation targets which, when taken together, will help ensure that pandas will be found in the wild for generations to come. These actions go beyond protecting existing habitat in that they would promote restoration of lost habitat, increase connectivity between nature reserves, and expand the panda's current range. To evaluate and monitor success in meeting these conservation goals, we have developed a multiscale framework for monitoring conservation activities:

Short-term (1–5 years):

1. Complete third national survey and disseminate results to all interested parties
2. Revise national management plan based on national survey results
3. Expand protected areas system
4. Continue to implement "Grain-to-Green" and logging ban policies
5. Restore degraded habitat

Longer-term (5–20+ years):

1. Develop comprehensive spatial database on panda distribution and habitat quality
2. Add to protected areas system all remaining panda habitat
3. Adapt national management plan to provide sustainable funding to all reserves
4. Expand pandas' range through habitat restoration

FROM ECOREGIONS TO SITES: CONSERVATION AT MULTIPLE SCALES

We believe that there are three important scales at which to focus future conservation activities: ecoregion, landscape, and site. An ecoregion is defined as (Dinerstein et al. 2000: 15):

> a large area of land or water that contains a geographically distinct assemblage of natural communities that (i) share a large majority of their species and ecological dynamics, (ii) share similar environmental conditions, and (iii) interact ecologically in ways that are critical for their long-term persistence.

The landscape scale is defined as a group of habitat blocks that have the potential to contain interacting panda populations (sensu Dinerstein et al. 1997). The site scale typically involves a single nature reserve or habitat block.

Because pandas live in the most biologically rich temperate forests in the world, their protection and ultimate survival will also conserve a myriad of other unique species and ecological processes that occur there (Olson and Dinerstein 1998). To develop a comprehensive conservation strategy for the panda, and by extension, the unique biodiversity in these forests, the current system of protected areas, as well as the remaining habitat blocks, must be evaluated at multiple spatial scales. In general, conservation activities that occur at the site scale, such as the management of a nature reserve, are focused on preserving the biodiversity within the boundaries of the site. Nature reserve managers rarely view how their reserves contribute to an overarching conservation plan. However, large-scale prioritization activities (ecoregion and landscape) can be spatially and temporally integrated to achieve an overarching long-term conservation goal. This is not to say that site-based conservation activities are not important—they are essential. However, site-based conservation activities should be grounded within the requirements of larger spatial scales.

Although a large-scale conservation vision has a number of advantages over site-based conservation efforts, a comprehensive conservation plan must coordinate activities across multiple spatial scales. For example, a suite of conservation activities at several sites in an ecoregion may still fail to meet the habitat connectivity requirements of pandas. Connectivity through linkage zones or corridors is achieved at an intermediate landscape scale. For linkage zones to function, there must be viable habitat and healthy panda populations to be connected. The maintenance of healthy panda populations within a specific area is a site-based process. Thus, for the overall conservation of pandas to succeed, the actions at each of these scales must function in a unified manner.

DEVELOPING A CONSERVATION FRAMEWORK AT MULTIPLE SCALES

ECOREGION SCALE

Developing a suite of conservation activities at the ecoregion scale will generally involve actions mandated through provincial or national legislation (table 15.1). At the ecoregion scale, the maintenance and protection of the remaining forests, reduction of further fragmentation of mountain habitats, restoration of connectivity, and development of an overarching biologically based vision are the key elements to successful conservation of panda populations. For example, the current logging ban in natural forests and restoration of forest habitat from agricultural lands in the "Grain-to-Green" program are national policies that protect and potentially restore panda habitat. This legislation was enacted in 1998, and provides watershed protection for the provision of clean drinking water, and may mitigate the effects of soil erosion and flooding in the fertile Yangtze River valley. Therefore, this legislation provides both conservation of panda habitat and positive economic incentives (reduced crop loss from flooding). Conversely, the national policy to promote migration and development of western China will serve to further fragment and destroy remaining panda habitat and reduce its connectivity.

One method of developing a long-term conservation plan for pandas is to conduct an ecoregion-wide priority-setting exercise (Dinerstein et al. 2000). To best influence national policies, it is imperative to present recommendations in the context of a long-term vision. A comprehensive management plan can map and identify important linkage zones, habitat blocks, or highly sensitive areas. Such a plan allows the identification of locations where threats (e.g., infrastructure development) will cause the most damage, and the proposal of alternative options.

LANDSCAPE SCALE

Conservation at the landscape scale involves land-use planning of core areas (protected areas),

TABLE 15.1
Framework for Panda Conservation Activities at Three Different Scales

KEY VARIABLES	EVALUATION CRITERIA AND ACTIONS
Ecoregion scale	Maintain logging ban and strengthen national policies that keep the region under natural forest cover
1. Maintain forest cover	Stop road-building, hydro-development, and migration of people into core conservation areas
2. Develop habitat connections	Support further extension of the "Grain-to-Green" program and restoration of natural habitat
3. Expand panda range or habitat	Develop a vision of a giant panda wilderness containing the entire habitat and all populations
4. Integrate panda conservation into a biodiversity vision for key ecoregions	Periodic analysis and evaluation of panda monitoring from nature reserves to track trajectory of wild panda populations
Landscape scale	
1. Maintain forest cover and presence of bamboo	Use satellite imagery to identify remaining forest cover Develop maps of bamboo species distribution
2. Identify distribution of pandas (National Survey)	Use results of national survey and habitat assessments to identify potential sites for future nature reserves
3. Develop dispersal corridors and habitat connections	Restrict access to core habitats Limit logging in key landscapes Improve connectivity of nature reserves by conserving additional unprotected habitat
Site scale	
1. Maintain forest cover and presence of bamboo	Develop a GIS system to input data on distribution of bamboo, pandas (or signs of pandas), and habitat
2. Monitor the presence of giant pandas and habitat use	Develop management plan for reserves, especially with respect to tourism and its impacts
3. Monitor population size, fecundity, and reproductive success	Establish a monitoring and evaluation protocol to assess success of management plan on panda populations, habitat, and food source; implement national and provincial policies

buffer zones, linkage zones, and multiple-use areas within a mosaic of land uses that maintain or restore connectivity of habitats. As with the ecoregion scale, the important variables governing the persistence of panda populations are the presence of forest cover and bamboo and the ability of pandas to disperse across the landscape (see table 15.1). It is at the landscape scale that many of the national and provincial policies are implemented, and therefore development of conservation landscapes involves working within the bounds of provincial and county mandates.

At the landscape scale, the use of satellite imagery and remote sensing techniques within a geographic information system (GIS) will have broad applications to land-use planning. It is possible to identify remaining habitat, potential linkage areas between panda populations, existing protected areas, and target the type and severity of activities that may threaten the habitat through fragmentation or destruction (Loucks et al. 2003). Using this information it will be possible to make specific recommendations for the expansion of the protected areas system, monitor panda movement and use of buffer and corridor areas, and propose areas for restoration.

SITE SCALE

At the site scale (e.g., within a nature reserve), the key variables and actions focus on monitoring and evaluation of the health of the panda population and the persistence of bamboo habitat (see table 15.1). Healthy and reproducing panda populations with the potential for migration should be the objective of local activities.

Managers of protected areas should establish an effective and continuous monitoring system of their giant panda populations, as well as other important wildlife within the reserve's boundaries. These data should be incorporated into a GIS database, so that over time, it will be possible to evaluate the number, movement, and reproductive success of the nature reserve's panda population. A GIS system will allow sophisticated analyses of such spatially explicit data as the location of feces, bamboo habitat, den sites, habitat types, and resource utilization by the resident panda population.

CONCLUSION

The conservation of giant pandas in the wild has made significant progress since the ratification of China's management plan in 1992. However, much work remains to be done across multiple scales to ensure that pandas will be roaming the mountains of Sichuan, Gansu, and Shaanxi Provinces at the beginning of the next century. We propose taking a multiscale approach to panda conservation activities. A large-scale ecoregion plan was recently completed for a region that includes all of the panda's remaining wild habitat. In March 2002, more than 120 experts from the Sichuan Western Development Office, the World Wildlife Fund, Conservation International, The Nature Conservancy, the Chinese Academy of Forestry, the Chengdu Institute of Biology, and the Sichuan Forestry Department came together to identify the top priorities for conservation across this region. The goal of this meeting was to establish a context, both spatially and policy-wise, for smaller-scale (landscape and site) activities, such as the ones presented in table 15.1.

ACKNOWLEDGMENTS

We thank the Giant Panda Conservation and Research Center, Peking University, for guiding us in our fieldwork; WWF—China for assistance in the development of the manuscript; and the WWF—U.S. Conservation Science Program staff for reviewing and improving the manuscript.

REFERENCES

Brown, J. H. 1984. On the relationship between the abundance and distribution of species. *Am Nat* 124:255–79.

Brown, J. H., and A. Kodric-Brown. 1977. Turnover rates in insular biogeography: Effects of immigration on extinction. *Ecology* 58:445–49.

Brown, J. H., D. W. Mehlman, and G. C. Stevens. 1995. Spatial variation in abundance. *Ecology* 76:2028–43.

Channell, R., and M. V. Lomolino. 2000. Dynamic biogeography and conservation of endangered species. *Nature* 403:84–86.

Chinese Minisitry of Forestry and World Wildlife Fund. 1989. *The national conservation management plan of the giant panda and its habitat.* Beijing: Chinese Minisitry of Forestry and World Wildlife Fund.

Dinerstein, E., E. Wikramanayake, J. Robinson, U. Karanth, A. Rabinowitz, D. Olson, T. Mathew, P. Hedao, and M. Connor. 1997. *A framework for identifying high priority areas and actions for the conservation of tigers in the wild.* Washington, D.C.: World Wildlife Fund.

Dinerstein, E., G. Powell, D. Olson, E. Wikramanayake, R. Abell, C. Loucks, E. Underwood, T. Allnutt, W. Wettengel, T. Ricketts, H. Strand, S. O'Connor, and N. Burgess. 2000. *A workbook for conducting biological assessments and developing biodiversity visions for ecological conservation.* Washington, D.C.: World Wildlife Fund.

Gaston, K. J. 1990. Patterns in the geographical ranges of species. *Biol Rev* 65:105–29.

Loucks, C., L. Zhi, E. Dinerstein, Wang D., Fu D., and Wang H. 2003. The giant pandas of the Qinling

Mountains, China: A case study in designing conservation landscapes for elevational migrants. *Conserv Biol* 17:558–65.

MacArthur, R. H., and E. O. Wilson. 1967. The theory of island biogeography. *Monogr Pop Biol* 1:1–203.

Olson, D. M., and E. Dinerstein. 1998. The Global 200: A representation approach to conserving the earth's most biologically valuable ecoregions. *Conserv Biol* 12:502–15.

Pimm, S. L., H. L. Jones, and J. Diamond. 1988. On the risk of extinction. *Am Nat* 132:757–85.

Tracy, C. R., and T. L. George. 1992. On the determinants of extinction. *Am Nat* 139:102–22.

PANEL REPORT 15.1

National Survey of the Giant Panda

Changqing Yu and Shaoying Liu

THE FIRST SURVEY of giant pandas *(Ailuropoda melanoleuca)* took place from roughly 1974 to 1977. According to a Chinese government report entitled *Conservation and Survey Meeting for Rare and Treasured Species in Important Provinces, Cities and Autonomous Districts,* other rare species, such as golden monkeys *(Rhinopithecus* spp.) and musk deer *(Moschus chrysogaster),* were also censused. The distribution of giant pandas from county to county and their relative density was eventually reported and used in making assessments of the giant panda's status in the wild.

A second survey of giant pandas was carried out from 1985 to 1988. This was a cooperative effort between China's Ministry of Forestry (MOF) and the World Wildlife Fund (WWF). The results of this survey were used to develop a national conservation management plan for the giant panda (Chinese Ministry of Forestry and World Wildlife Fund 1989), which has been used as a guide by the authorities in China for implementation of national and international conservation programs. Nearly a decade later, new questions about the status of the wild population and its habitat were posed during an analysis of the plan's effectiveness in meeting its conservation goals. For example, had major threats to the species (e.g., loss of habitat, poaching) been ameliorated? Would an updating of strategic conservation plans be required as a result of new information?

A memorandum of understanding (MOU) between MOF and WWF was signed in 1997, agreeing to a multiyear survey of giant pandas in selected areas, with funds to be provided by both parties to the agreement. As a first effort, in 1998, WWF and the Sichuan Forestry Department jointly conducted a survey of habitat in Pingwu county, an area covering about 12% of the pandas' present-day range. In the following year, it was agreed that WWF and the State Forestry Administration (SFA, which replaced the MOF) would jointly carry out a national survey on the giant panda population and the assessment of panda habitats. A pilot study on methodology was completed in Qingchuan county in January 2000. Another MOU between SFA and WWF was negotiated shortly thereafter, specifying cooperation in a national survey that would be completed in June 2002. By June 2000, further surveys had been implemented in the Minshan and Qionglai regions.

The purpose of this greatly expanded effort is to evaluate the status of the giant panda population and its habitat and establish a database from survey results that could be used in strategic planning for and implementation of panda conservation efforts.

REVIEW OF SURVEY METHODS

A panel was convened at *Panda 2000* to discuss methods and provide an update on progress of the National Survey. An introduction to methods of survey data collection and analysis and data management and storage was provided by Jianguo Liu of Michigan State University. Shaoyang Liu, Sichuan Forestry Academy, explained the new methodology to be used in conducting the survey (see also Pan et al., chapter 5). Xuehua Liu (International Institute for Aerospace Survey and Earth Sciences, the Netherlands) discussed the importance of timing in conducting surveys, based on her experiences at Foping Natural Reserve. A progress report on efforts in Sichuan Province through mid-2000 was presented by Kaiqing Shao (Wildlife Survey and Management Station, Sichuan Forestry Department), and Changqing Yu of WWF—China reviewed the expected results from the survey and its progress to date.

APPLICATION OF RESULTS

The major applications of the results from the survey are as follows:

- An updated estimate of the wild population, including size and density, population trends, and the means for conducting a meta-population analysis;
- Assessment of available habitat, including area, suitability, fragmentation, and adverse human impacts;
- Analysis of significant problems facing the wild population (i.e, the source and nature of threats to their survival), including poaching and loss of essential resources;
- Determination of the status of other rare species within the range of the giant panda;
- Identification of the need for capacity building, including training of forestry staff in data collection and processing for panda conservation, improving the efficacy of current management practices, and providing appropriate field equipment for those involved in habitat protection;
- Establishing a basis for evaluating existing conservation efforts, including the adequacy of the panda reserve system, local community involvement in conservation, infrastructure needs, and implementation of the "Grain-to-Green" national initiative that arose out of the recent prohibition of logging in the pandas' range.

ACKNOWLEDGMENTS

We thank Jianguo Liu, Xuehua Liu, and Kaiqing Shao for their contributions to this chapter.

REFERENCE

Chinese Minisitry of Forestry and World Wildlife Fund. 1989. *The national conservation management plan of the giant panda and its habitat*. Beijing: Chinese Minisitry of Forestry and World Wildlife Fund.

16

The Legacy of Extinction Risk

LESSONS FROM GIANT PANDAS AND OTHER THREATENED CARNIVORES

John L. Gittleman and Andrea J. Webster

ALL SPECIES ultimately go extinct. Currently, over 1100 mammal species (~25%) are threatened with the likelihood of extinction (Hilton-Taylor 2000). As extinction is a fundamental evolutionary process, not all of these species will survive. Today, however, the pattern and process of extinction is clearly different than in the geological past, as anthropogenic factors have become increasingly and distressingly influential. The risk of extinction now is a dual process of species' biological characteristics adapting to extreme human effects of habitat loss, overexploitation, invasives, and other secondary "chains of extinction" (Diamond 1989). A primary task for conservation biology is to better understand this dual connection, such that there develops a more proactive science, not merely reacting to forces of species decline but establishing preventive measures to stem the likelihood of greater numbers of extinctions. To do this, we must utilize as many approaches as possible, including some that seem tangential to the goals and action of conservation.

In this chapter, we consider how phylogenies or evolutionary trees can lend insight into what biological characteristics contribute to extinction risk. Specifically, we use the recent advent of "supertrees" (see Bininda-Emonds, chapter 1), complete phylogenies based on all molecular and morphological data available for a taxon, to present some evolutionary patterns of what traits extinction-prone species in the order Carnivora have in common. We then examine whether the traits that are correlated with endangered status in general are also found in the giant panda *(Ailuropoda melanoleuca)*. Our view is that these types of comparative results can be useful in two prescriptive ways. One is that by taking a broad, comparative approach, closely related (sister) taxa may help to inform us about poorly known species. The second is that, for such species as the giant panda, which we know (unfortunately) to be threatened, we can more carefully study, monitor, and preserve the precise biological characteristics of species that are declining.

SUPERTREES: APPLICATIONS TO CONSERVATION

From historical uses, such as Darwin's presentation of only one figure illustration (a phylogeny) in the *Origin*, or Haeckel's demonstration that ontogeny recapitulates phylogeny, to modern uses regarding the history of life or the origin of acquired immune deficiency syndrome (AIDS), the promise of phylogenies in biology is tremendous (Harvey et al. 1996; Hillis 1997; O'Brien et al. 2001). Today, in conservation, phylogenies are increasingly being used for a variety of problems ranging from the identification of taxa that have high speciation and extinction rates (Barraclough et al. 1999) to the preservation of evolutionary history residing in biodiversity hotspots (Sechrest et al. 2002) to selection criteria for measuring losses of biodiversity (Vázques and Gittleman 1998; Purvis and Hector 2000). The frustration has always been that the number of available phylogenies for many taxa is low and agreement among them contentious. To some degree, these problems have been assuaged, at least to the point that applications of phylogenies outweigh the negatives. Now, more phylogenies are available than ever, with over 16,000 papers published in the past thirty years on mammals alone (Liu et al. 2001). Congruence among phylogenies is, and will always be, more problematic; disagreements among molecular and morphological evidence, which taxa should be included due to questions of monophyly, or algorithms for discriminating among alternative phylogenies are important issues to resolve in systematics.

A recent development has permitted a way around these problems, so that phylogenetic information, although not perfect, can be used. These are "supertrees" (for a review, see Bininda-Emonds et al. [2002]). Any complete phylogeny that contains all species from a combination of phylogenies is a supertree. The basic idea is that the supertree provides a thorough collection of phylogenies for a given (monophyletic) group. This collection is then placed in a large matrix constructed from a simple set of rules, whereby each taxon in a clade is related and those not in the clade are unrelated (Sanderson et al. 1998). Parsimony is applied to the matrix to form a complete tree. Obviously, as in any phylogenetic reconstruction, there are various ways of assembling the details of a supertree. For example, some are dependent on preferred character information (e.g., molecular rather than morphological), particular algorithms for combining independent trees, or the use of phylogenies that are in agreement (Baum 1992; Ragan 1992; Sanderson et al. 1998; Springer and de Jong 2001). Significantly, recent work shows that, for some groups (e.g., carnivores), there is actually remarkably good agreement among different types of characters (Bininda-Emonds 2000) and variations in the algorithms used to reconstruct supertrees seem to have only minor effects, given that there is relative congruence among the original source trees (Purvis and Webster 1999; Bininda-Emonds and Sanderson 2001).

At present, supertrees have been assembled for primates (Purvis 1995), carnivores (Bininda-Emonds et al. 1999), bats (Jones et al. 2002) and, at the family level, all eutherian orders (Liu et al. 2001). The one employed here is for the carnivores, which is a complete tree for all 271 species of Carnivora (figure 16.1; see also Bininda-Emonds, chapter 1). It is derived from 166 phylogenies, with the overall supertree being fairly well resolved. Indeed, with respect to the giant panda and other ursids, the percentage resolution is 85.7% among twenty-eight source trees and the number of elements per tree is the third highest among the ten major carnivore families.

The critical feature of using the carnivore supertree in the present context is that a complete species-level phylogeny permits large-scale evolutionary studies about extinction risk that otherwise could not be performed (Purvis 1996; Gittleman 2001). Extinction occurs over time periods that are beyond population-level analyses. Furthermore, at least twenty-four biological traits have been identified with processes of extinction (McKinney 1997), which means that to have any meaningful analyses of the factors related to extinction in a given taxon, it is necessary to carry out tests across multiple taxa that have evolved independently. In addition, once we have identified correlates of extinction risk, it then would be helpful if we could better understand how these traits are evolving. For example, if we detect that small litter sizes and long

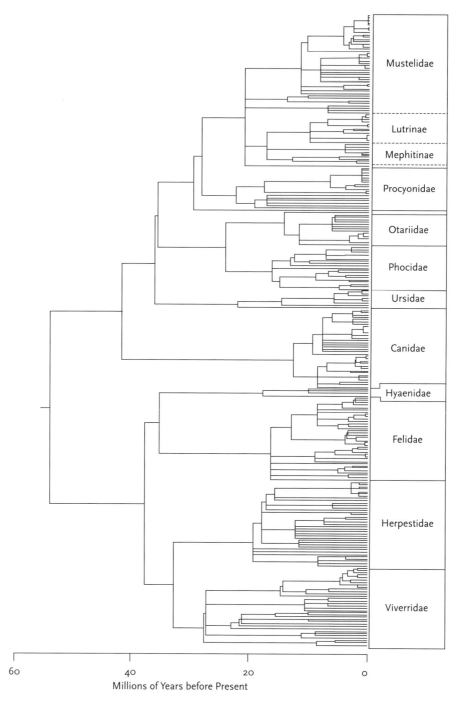

FIGURE 16.1. Complete supertree of the Carnivora. From Bininda-Emonds et al. (1999), with permission.

TABLE 16.1
Multiple Regression Model across Primates and Carnivores Predicting Extinction Risk in Declining Species

PREDICTOR	COEFFICIENT	t
Geographic range	−0.291	−7.65***
Trophic level	0.402	4.00***
Population density	−0.113	−2.06*
Gestation length	1.590	2.77**
Body mass	−0.002	−0.02
Order[1]	−0.084	−1.53
Body mass × order[1]	0.704	3.33***
Gestation length × order[1]	−2.790	−2.80**

SOURCE: Data from Purvis et al. (2000).
NOTES: Sample size is 120 species, 112 contrasts. The model accounts for 47.6% of the total variance. All tests are two-tailed: *, $p = 0.05$; **, $p = 0.01$; ***, $p = 0.001$.
[1] Carnivore = 0, primate = 1.

gestation lengths are indicative of species with slow reproductive rates and, in turn, increased extinction risk, then it would also be useful to know whether the traits of these species are evolving at slower rates. In this way, supertrees help us to unify current factors related to extinction with longer evolutionary patterns, both of which are important for understanding how and why species become threatened (Cracraft 1992; Eldredge 1999; Purvis et al. 2001).

CORRELATES OF EXTINCTION RISK

Some species have biological traits that increase their risk of extinction. There are many reviews of these characteristics and their underlying causal explanations (Lawton 1995; McKinney 1997; Simberloff 1998; Owens and Bennett 2000). Surprisingly, until recently, no statistical analysis revealed precisely which traits are actually more important than others, nor how these traits are differentially involved among taxa. Aside from lacking sufficiently large databases to carry out such studies, the rate-limiting step in most cases was the lack of a complete phylogeny. Consequently, supertrees have greatly aided in direct tests of the biological correlates of extinction risk in bats, carnivores, and primates (Purvis et al. 2000, 2001; Jones et al. 2003), with other taxa currently being assessed. The following is a brief summary of this work and its relevance to giant pandas and carnivores.

Purvis et al. (2000) used the IUCN Red List as a proxy for extinction risk (see Hilton-Taylor 2000). Essentially, each of the threat levels given in the Red List was coded as a continuous character ranging from "lower risk" and "least concerned" to "extinct in the wild." A database was collated from the published literature on eight biological traits for each primate and carnivore species. Using the independent contrasts statistical analysis for each order (Purvis and Rambaut 1995), multiple regression models were developed to find the best predictive variables for extinction risk. Among the many hypothesized traits, across carnivores and primates, four in particular explained up to 50% of the total between-species variation: high trophic level, low population density, slow life history, and especially a small geographic range size (table 16.1). There are also significant differences between the orders. Larger body mass is associated with extinction risk in

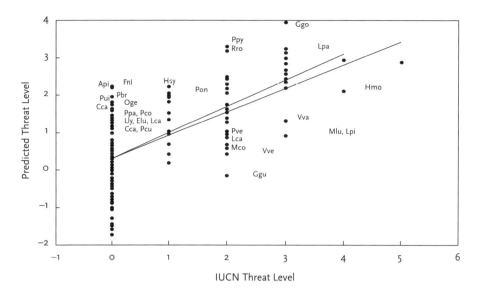

FIGURE 16.2. Predictions from multiple regression models for carnivores (upper line) and primates (lower line), plotted against IUCN threat categories. IUCN threat codes are scored as: 0, Lower Risk, Least Concern; 1, Lower Risk, Near Threatened; 2, Lower Risk, Conservation Dependent and Vulnerable; 3, Endangered; 4, Critically Endangered; 5, Extinct in the Wild and Extinct. Lines are least-squares regressions for each order. Species with residuals greater than 1.0 are coded on the graph as follows: Primates: *Alouatta pigra* (Api), *Cebus capucinus* (Cca), *Gorilla gorilla* (Ggo), *Hylobates moloch* (Hmo), *Hylobates syndactylus* (Hsy), *Lemur catta* (Lca), *Mirza coquereli* (Mco), *Papio ursinus* (Pui), *Pongo pygmaeus* (Ppy), *Propithecus verreauxi* (Pve), *Rhinopithecus roxellana* (Rro), *Varecia variegata* (Vva). Carnivores: *Caracal caracal* (Cca), *Enhydra lutris* (Elu), *Felis nigripes* (Fni), *Gulo gulo* (Ggu), *Lycaon pictus* (Lpi), *Lynx canadensis* (Lca), *Lynx lynx* (Lly), *Lynx pardinus* (Lpa), *Mustela lutreola* (Mlu), *Oncifelis geoffroyi* (Oge), *Panthera onca* (Pon), *Parahyaena brunnea* (Pbr), *Panthera pardus* (Ppa), *Pseudalopex culpaeus* (Pcu), *Puma concolor* (Pco), *Vulpes velox* (Vvu). Species classified by IUCN as Least Concern, with negative residuals, are not annotated, because there is no subdivision in the IUCN system among nonthreatened species, and these residuals therefore have little meaning.

primates, whereas gestation length is more important in carnivores. Taken together, these results show that some factors are more critically involved with processes of extinction than others. Specifically, characteristics that are indicative of population decline and small geographic ranges (rarity) are most salient, as indicated in other studies (Manne et al. 1999). However, some characteristics may only be important for particular taxa, such that biological correlates of extinction risk are, to a degree, taxon specific.

Given that multiple biological characteristics are related to the IUCN Red List, to what degree can our model predict which species are riskprone? Figure 16.2 shows the predicted model plotted against the IUCN Red List. Overall, the fit is remarkably good. The interesting patterns are revealed when we isolate species that fit above or below the regression line. For example, many of the felids, along with the giant panda, are predicted to have a higher extinction risk than currently listed. It is precisely these species that we would suggest should have greater protection and perhaps continual reassessment of their level of threat based on biological correlates. A good example is the sea otter (*Enhydra lutris*; denoted as Elu in figure 16.2), which was reduced in level of threat, but due to recent population declines is again being upgraded to threatened status (Estes et al. 1998). Conversely, there are several species that are classified as more threatened than their biology would suggest; these

TABLE 16.2
Characteristics of Giant Pandas Relative to Observed Correlates of Extinction in Bats, Primates, Carnivores, and Birds

TAXON	LARGE SIZE	SMALL RANGE	SLOW LIFE HISTORY	HIGH TROPHIC LEVEL	LOW DENSITY
Giant pandas	Yes	Yes	Yes	No	Yes
Bats	No	Yes	Yes	—	—
Primates	Yes	Yes	No	Yes	Yes
Carnivores	No	Yes	Yes	Yes	Yes
Birds	Yes	—	Yes	—	—

SOURCES: Data for bats from Jones et al. (2002), primates and carnivores from Purvis et al. (2000), and birds from Owens and Bennett (2000).

NOTE: —, Not included in the analysis.

include the wolverine *(Gulo gulo)* and wild dog *(Lycaon pictus)*, which are currently experiencing massive habitat losses.

In sum, a phylogenetic analysis combined with a database of biological traits associated with extinction risk can be used in a predictive manner for explaining why some species are at risk and some are not. Across the taxa that have been analyzed phylogenetically for correlates of extinction risk (bats, primates, carnivores, birds), large body size, small geographic range distribution, slow life history, high trophic level, and low population density are significant (table 16.2). The giant panda shares all but one of these traits (trophic level). Thus, it is unsurprising that the giant panda is highly endangered. From this comparative analysis, we can now turn to the next important question involving how these traits differ in closely related taxa (ancestral carnivores) and whether they appear to have evolved in a different manner.

TRAITS OF EXTINCTION IN GIANT PANDAS

The giant panda is unique because of its bamboo diet and yet carnivore ancestry, its relatively large radial sesamoid (the so-called "panda's thumb"), and, of course, the adorable black-and-white coloration. These pronounced traits are not typically associated with endangered status, although the conspicuous coloration undoubtedly increases the chances of poaching. The question is what characteristics are unusual for the giant panda that seem to relate to extinction risk and, in particular, are distinctive in comparison with other carnivores. As this volume conveys, a definitive answer cannot be given because we simply do not have all relevant information on behavior, ecology, life histories, and digestive physiology of the giant panda and other carnivores. There are, however, some clear patterns if we simply compare certain functional traits, such as life histories. It is well known that giant pandas have a relatively slow reproductive rate (Hu and Wei 1990; Schaller et al. 1995; Zhu et al. 2001). Reproductive rate in mammals is a composite of many traits, including gestation length, growth rate, age that eyes first open, weaning age, age at sexual maturity, interbirth interval, and longevity. Which traits seem to be different in giant pandas compared with other carnivores? The answer again involves using phylogenies to make comparisons.

An earlier study analyzed ten life history traits across the carnivores to test which ones were significantly different in the giant panda (Gittleman 1994). Surprisingly, most life histories in the giant panda are very similar to those of other bears and carnivores. For example, the age at sexual maturity, weaning age, and even birth weight are not significantly different from expected patterns, once body size (allometry) is taken into account. Stark differences appear with early life histories, such as a slow growth rate (individual and total growth of litter) and a slightly long gestation length. In comparing other species that also have these markedly different life histories, there does not seem to be any consistent association with ecology, mating system, or phylogenetic heritage. The primary explanation seems to rest with age-specific mortality rates: taxa with slow life histories (e.g., most bears, many large cats) have relatively low mortality rates. Comparative trends across many mammal groups thus now show that life history variation is closely related to mortality schedules (Purvis and Harvey 1995). For example, across primates (Ross and Jones 1999) and carnivores (Gittleman 1993, 1994), species with high mortality rates reproduce at an early age and have high birth rates. The problem in the giant panda is that it deviates from this mammalian pattern: from the data available, mortality is relatively high, which would predict that life histories should be comparatively fast. These comparisons indicate that critical factors involved in mortality, either intrinsic (nutritional or reproductive physiology) and/or extrinsic (stress due to habitat loss), are more intense than would be predicted from life history evolution across mammals in general and carnivores in particular.

BIOLOGICAL TRAITS AND EVOLUTIONARY RATES

Large-scale taxonomic differences often result from changes in evolutionary rates. Evolutionary rates may occur fairly quickly, as in hourly generational changes in laboratory strains of fruit flies *(Drosophila)* to lineages of vertebrate fossils that change over millions of years (Gingerich 1993; Carroll 1997). With complete phylogenies, it is possible to not only measure more accurately the tempo and mode of how lineages change but also how specific traits change among individual taxa (Gittleman et al. 1996). For example, evolutionary rates could be accelerated by reducing gestation length, age at sexual maturity, or interbirth interval. In the present context, it is important to ask whether evolutionary rates in the bear clade, including the giant panda, are different from other carnivore clades and which traits are evolving most rapidly. We can answer these questions by comparing sister taxa (see Martins 1994; Purvis et al. in press], or Webster et al. [2003] for further discussion about measuring evolutionary rates). Assuming that characters change by an assumed (Brownian) model of evolution, we can calculate the difference between the amount of change in a trait on one branch and that on another branch. The model of evolution is important: Brownian motion assumes there is no directionality to trait variation over time, thus resulting in the expectation that traits, although showing variation in the short term, are expected to have a net variation of zero in the long term. Sister-taxon comparisons are used because a pair of branches descends from each node to give an independent estimate of rate, so rates can be compared between groups.

Using supertrees, Purvis et al. (in press) showed that carnivores evolve at slow rates: body mass, age at maturity, gestation length, and interbirth interval all evolve at significantly reduced rates in carnivores compared with primates. The one exception is litter size, which is most likely due to most primates having a litter size of one and therefore showing little variance in this trait. What is important about the giant panda, along with related ursids, is that when this clade is compared with other carnivores, gestation length and interbirth interval evolve significantly faster (table 16.3) than Lutrinae/Mephitinae, Canidae, Felidae, and Viverridae. Significantly, this deviation does not seem to involve large body mass: significant differences in evolutionary rates among life histories does not consistently correlate with size per se. Combined with the above comparative results indicating that the giant panda has a number of

TABLE 16.3
Rates of Evolution of Life History Traits in Different Carnivore Clades

CLADE	ASM	BM	GL	IBI	LS
Mustelinae	1.39	24.17	0.81	1.20	29.05
Lutrinae-Mephitinae	1.96	7.29	0.25	0.20	17.56
Phocidae-Otariidae	—	6.09	—	1.08	—
Canidae	0.94	8.27	0.06	0.27	50.15
Felidae	2.12	23.58	0.20	2.83	42.34
Hepestidae	0.87	9.41	0.47	2.20	22.41
Viverridae	1.09	2.93	0.18	0.47	17.97
Ursidae	1.29	14.38	1.04	4.32	12.44
Significance level	ns	ns	*	*	ns

NOTES: ASM, age at sexual maturity; BM, body mass; GL, gestation length; IBI, interbirth interval; LS, litter size, ns, not significant; *, $p = 0.05$; —, no data.

extinction-prone traits, the panda is characterized by extreme variability in evolutionary rates for some fundamental reproductive characteristics (gestation length, interbirth interval). These phylogenetic patterns may indicate that, as in other taxa (McKinney and Gittleman 1995), the giant panda is showing features that indicate responses to environmental stress.

CONCLUSIONS

It is depressingly true that not all species will survive. Giant pandas, due to their charismatic nature, endearing biological qualities, and our willingness to protect them, may not go extinct in our lifetimes. Nevertheless, we can certainly learn from the panda's proclivity to extinction risk and from many other species that are similarly risk-prone. A broad-scale phylogenetic approach permits us to find those particular biological traits that consistently correlate with risk across independent lineages. In essence, phylogenies can provide a short cut to which species are likely to be endangered and perhaps go extinct in the near future. Conservation biology needs as many tools as possible. Using phylogenies to examine the comparative biology of threatened species like the giant panda will not only help this magnificent species but also leave a legacy to protect many other similar species.

ACKNOWLEDGMENTS

We thank Devra Kleiman for organizing the symposium at which the original paper was delivered, and Don Lindburg for an invitation to the *Panda 2000* conference and for his patience in considering this chapter. We also thank Kate Jones for suggesting this collaboration.

REFERENCES

Barraclough, T. G., A. P. Vogler, and P. H. Harvey. 1999. Revealing the factors that promote speciation. In *Evolution of biodiversity*, edited by A. E. Magurran and R. M. May, pp. 202–19. Oxford: Oxford University Press.

Baum, B. R. 1992. Combining trees as a way of combining data sets for phylogenetic inference, and the desirability of combining gene trees. *Taxon* 41:3–10.

Bininda-Emonds, O. R. P. 2000. Factors influencing phylogenetic inference: A case study using the mammalian carnivores. *Mol Phylogenet Evol* 16: 113–26.

Bininda-Emonds, O. R. P., and M. J. Sanderson. 2001. Assessment of the accuracy of matrix representation with parsimony supertree construction. *Syst Biol* 50:575–79.

Bininda-Emonds, O. R. P., J. L. Gittleman, and A. A. Purvis. 1999. Building large trees by combining phylogenetic information: A complete phylogeny of the extant Carnivora (Mammalia). *Biol Rev* 74: 143–75.

Bininda-Emonds, O. R. P., J. L. Gittleman, and M. A. Steel. 2002. The (super)tree of life: Procedures, problems, and prospects. *Annu Rev Ecol Syst* 33: 265–89.

Carroll, R. L. 1997. *Patterns and processes of vertebrate evolution.* New York: Cambridge University Press.

Cracraft, J. 1992. Explaining patterns of biological diversity: Integrating causation at different spatial and temporal scales. In *Systematics, ecology, and the biodiversity crisis,* edited by N. Eldredge, pp. 59–76. New York: Columbia University Press.

Diamond, J. 1989. Overview of recent extinctions. In *Conservation for the twenty-first century,* edited by D. Western and M. Pearl, pp. 37–41. New York: Oxford University Press.

Eldredge, N. 1999. Cretaceous meteor showers, the human ecological "niche," and the sixth extinction. In *Extinctions in near time,* edited by R. D. E. Macphee, pp. 1–14. New York: Kluwer.

Estes, J. A., M. T. Tinker, T. M. Williams, and D. F. Doak. 1998. Killer whale predation on sea otters linking oceanic and near-shore ecosystems. *Science* 282:473–76.

Gingerich, P. D. 1993. Quantification and comparison of evolutionary rates. *Am J Sci* 293A: 453–78.

Gittleman, J. L. 1993. Carnivore life histories: A reanalysis in the light of new models. In *Mammals as predators,* edited by N. Dunstone and M. L. Gorman, pp. 65–86. Oxford: Oxford University Press.

Gittleman, J. L. 1994. Are the pandas successful specialists or evolutionary failures? *BioScience* 44: 456–64.

Gittleman, J. L. 2001. Hanging bears from phylogenetic trees: Investigating patterns of macroevolution. *Ursus* 11:29–40.

Gittleman, J. L., C. G. Anderson, M. Kot, and H.-K. Luh. 1996. Comparative test of evolutionary liability and rates using molecular phylogenies. In *New uses for new phylogenies,* edited by P. H. Harvey, A. J. L. Brown, J. Maynard Smith, and S. Nee, pp. 289–307. Oxford: Oxford University Press.

Harvey, P. H., A. J. L. Brown, J. Maynard Smith, and S. Nee. 1996. *New uses for new phylogenies.* Oxford: Oxford University Press.

Hillis, D. M. 1997. Biology recapitulates phylogeny. *Science* 276:218–19.

Hilton-Taylor, C., ed. 2000. *2000 IUCN red list of threatened species.* Gland, Switzerland, and Cambridge: IUCN—The World Conservation Union.

Hu, J., and F. Wei. 1990. Development and progress of breeding and rearing giant pandas in captivity within China. In *Research and progress in biology of giant pandas,* edited by H. Jinchu, pp. 322–25. Sichuan: Sichuan Publishing House of Science and Technology.

Jones, K. E., A. Purvis, A. Maclarnon, O. R. P. Bininda-Emonds, and N. Simmons. 2002. A phylogenetic supertree of the extant bats (Mammalia: Chiroptera). *Biol Rev* 77:223–59.

Jones, K. E., J. L. Gittleman, and A. Purvis. 2003. Biological correlates of extinction risk in bats. *Am Nat* 161:601–14.

Lawton, J. H. 1995. Population dynamic principles. In *Extinction rates,* edited by J. H. Lawton and R. M. May, pp. 147–63. Oxford: Oxford University Press.

Liu, F.-G. R., M. M. Miyamoto, N. P. Freire, P. W. Ong, M. R. Tennant, T. S. Young, and K. F. Gugel. 2001. Molecular and morphological supertrees for eutherian (placental) mammals. *Science* 291: 1786–89.

Manne, L. L., T. M. Brooks, and S. L. Pimm. 1999. Relative risk of extinction of passerine birds on continents and islands. *Nature* 399:258–61.

Martins, E. 1994. Estimating the rate of phenotypic evolution from comparative data. *Am Nat* 144:193–209.

McKinney, M. L. 1997. Extinction vulnerability and selectivity: Combining ecological and paleontological perspectives. *Annu Rev Ecol Syst* 28:495–516.

McKinney, M. L., and J. L. Gittleman. 1995. Ontogeny and phylogeny: Tinkering with covariation in life history, morphology and behaviour. In *Evolutionary change and heterochrony,* edited by K. J. McNamara, pp. 21–46. New York: John Wiley and Sons.

O'Brien, S. J., E. Eizirik, and W. J. Murphy. 2001. On choosing mammalian genomes for sequencing. *Science* 292:2264–65.

Owens, I. P. F., and P. M. Bennett. 2000. Ecological basis of extinction risk in birds: Habitat loss versus human persecution and introduced predators. *Proc Nat Acad Sci USA* 971:12144–48.

Purvis, A. A. 1995. Composite estimate of primate phylogeny. *Phil Trans Roy Soc Lond B* 348:405–21.

Purvis, A. A. 1996. Using interspecies phylogenies to test macroevolutionary hypotheses. In *New uses for new phylogenies,* edited by P. H. Harvey, A. J. L. Brown, J. Maynard Smith, and S. Nee, pp. 153–68. Oxford: Oxford University Press.

Purvis, A. A., and P. H. Harvey. 1995. Mammal life history: A comparative test of Charnov's model. *J Zool (Lond)* 237:259–83.

Purvis, A. A., and A. Hector. 2000. Getting the measure of biodiversity. *Nature* 405:212–19.

Purvis, A. A., and A. Rambaut. 1995. Comparative analysis by independent contrasts (CAIC): An Apple Macintosh application for analysing comparative data. *Computer Applications Biosci* 11:247–51.

Purvis, A. A., and A. J. Webster. 1999. Phylogenetically independent comparisons and primate phylogeny. In *Comparative primate socioecology*, edited by P. C. Lee, pp. 44–70. Cambridge: Cambridge University Press.

Purvis, A. A., J. L. Gittleman, G. Cowlishaw, and G. M. Mace. 2000. Predicting extinction risk in declining species. *Proc R Soc Lond B* 267:1947–52.

Purvis, A. A., G. M. Mace, and J. L. Gittleman. 2001. Past and future carnivore extinctions: A phylogenetic perspective. In *Carnivore conservation*, edited by J. L. Gittleman, S. Funk, D. Macdonald, and R. K. Wayne, pp. 11–34. Cambridge: Cambridge University Press.

Purvis, A. A., A. J. Webster, P.-M. Agapow, K. E. Jones, and N. J. B. Isaac. In press. Primate life histories and phylogeny. In *Primate life histories and socioecology*, edited by P. M. Kappeler and M. Pereira. Chicago: University of Chicago Press.

Ragan, M. A. 1992. Phylogenetic inference based on matrix representation of trees. *Mol Phylogenet Evol* 1:53–58.

Ross, C., and K. E. Jones. 1999. Socioecology and the evolution of primate reproductive rates. In *Comparative primate socioecology*, edited by P. C. Lee, pp. 73–110. Cambridge: Cambridge University Press.

Sanderson, M. J., A. A. Purvis, and C. Henze. 1998. Phylogenetic supertrees: Assembling the trees of life. *Trends Ecol Evol* 13:105–9.

Schaller, G. B., J. Hu, W. Pan, and J. Zhu. 1985. *The giant pandas of Wolong*. Chicago: University of Chicago Press.

Sechrest, W., T. M. Brooks, G. A. B. da Fonseca, W. R. Konstant, R. A. Mittermeier, A. A. Purvis, A. B. Rylands, and J. L. Gittleman. 2002. Hotspots and the conservation of evolutionary history. *Proc Nat Acad Sci USA* 99:2067–71.

Simberloff, D. 1998. Small and declining populations. In *Conservation science and action*, edited by W. J. Sutherland, pp. 116–34. Oxford: Blackwell Science.

Springer, M. S., and W. W. de Jong. 2001. Which mammalian supertree to bark up? *Science* 291:1709–11.

Vázquez, D. P., and J. L. Gittleman. 1998. Biodiversity conservation: Does phylogeny matter? *Curr Biol* 8:R379–81.

Webster, A. J., J. L. Gittleman, and A. A. Purvis. 2003. The life history legacy of evolutionary body size change in carnivores. *J Evol Biol* 17:1–12.

Zhu, X., D. G. Lindburg, W. Pan, K. A. Forney, and D. Wang. 2001. The reproductive strategy of giant pandas *(Ailuropoda melanoleuca)*: Infant growth and development and mother-infant relationships. *J Zool (Lond)* 253:141–55.

PANEL REPORT 16.1

Reintroduction of Giant Pandas

AN UPDATE

Sue Mainka, Wenshi Pan, Devra Kleiman, and Zhi Lü

THE IUCN–Species Survival Commission's Reintroduction Specialist Group recognizes two reasons to undertake a reintroduction program for endangered species: (1) to augment the species population in the wild or (2) to increase genetic diversity in the wild. Reintroduction is not an appropriate strategy to deal with reducing a surplus of animals in captivity.

Part of the rationale for maintaining giant pandas *(Ailuropoda melanoleuca)* in captivity is to provide a source of animals for eventual reintroduction to the wild. In 1991, He and Gipps (1991: 19) evaluated the situation and concluded "it would be inappropriate to release captive pandas . . . at this time." A workshop held in China in 1997 reevaluated the situation and again could not recommend giant panda reintroductions. The main reasons were insufficient knowledge for developing an effective giant panda reintroduction program, especially with regard to the status of wild pandas and their habitat and key elements of a successful carnivore reintroduction. In addition, the workshop participants noted a lack of financial resources to implement such a program, insufficient information on suitable habitat, and no captive rearing regimes that would provide viable candidates for release. The participants identified urgent needs to promote habitat conservation and in situ research (Mainka and Lü 1999). Here, we provide a further review of reintroduction as a conservation strategy for giant pandas. Our assessment of pandas as reintroduction candidates is summarized in table PR.16.1.

REVIEW OF REINTRODUCTION AS A CONSERVATION STRATEGY

STATUS OF THE SPECIES

1. Is there a need to increase wild populations of pandas?

An up-to-date survey is a prerequisite for development of further reintroduction planning. A national survey completed in 2002 is acknowledged to be by far the most comprehensive, in both geographical coverage and data collected, of any survey to date. Initial results indicate that the number of pandas in the wild is higher than previously reported, although it is uncertain if this represents a real increase in wild populations or is the result of different census methods and/or increased census effort. It is also unclear

TABLE PR.16.1
Review of the IUCN Reintroduction Specialist Group Release Criteria with Respect to Giant Pandas in 2000

CRITERION	IN 1991	IN 1997	IN 2000
Status of the species			
1. Is there a need to increase wild populations of pandas?	Yes	Yes?	?[1]
2. Are animals available for reintroduction and is there a self-sustaining captive population?	No	No	No[2]
3. Is there a danger to wild populations from a reintroduction program?	?	?	Yes?[3]
Environmental conditions			
4. Have the reasons for the decline in panda numbers been eliminated (e.g., hunting, deforestation, commerce)?	No	No?	Some?
5. Is sufficient habitat protected and secure?	No	?	?[1]
Political conditions			
6. Would a reintroduction program have a negative impact on local people?	No?	No?	?
7. Does community support for reintroductions exist?	2	2	2[4]
8. Is there support from and involvement of government and nongovernment organizations?	Yes?	Yes?	Yes?
9. Would reintroduction activities be in compliance with the laws and regulations of China?	?	Yes	Yes
Biological and other resources			
10. Is sufficient reintroduction technology available at this time?	1	2	3
11. Is critical knowledge of the giant panda's biology available?	2.5	3	3
12. Do sufficient resources exist for establishment and long-term monitoring and evaluation of reintroduction programs?	No	No?	No?
Is reintroduction recommended?	No	No	No[1]

SOURCE: Adapted from Kleiman et al. (1994).

NOTE: Where numbers are reported, they are on a scale from 1 to 5, with 5 being the highest level. A question mark indicates uncertainty with respect to the status of that criterion.

[1] The answer depends on results from the third survey.
[2] Not for a full program, but possibly for an experimental reintroduction.
[3] Concentrate on areas that do not currently contain pandas.
[4] This depends on the area in question.

if the methods used include specific activities that will allow a comparison of the new data with previously collected information. With the release of results from this survey, there will be a better understanding of the current status of wild populations, including whether there is any suitable area in which to release animals, as well as whether we need to augment the wild population.

2. Are animals available for reintroduction and is there a self-sustaining captive population?

Although there has been remarkable progress in the past few years in the captive breeding of giant pandas, this population is not yet self-sustaining. However, there may soon be sufficient animals in captivity to use some individuals for an experimental release program that is designed to provide critical information needed for planning future reintroductions. The results of the national survey completed in 2002 will help to determine whether there are animals in the wild that might be suitable candidates for translocation; as yet, no criteria have been established for identifying an individual as a suitable release candidate.

3. Is there a danger to wild populations from a reintroduction program?

Experience with reintroductions of other carnivores and our increasing knowledge of disease susceptibility in giant pandas suggest that there is a potential danger to both established wild populations and the animals released in the form of disease transmission and difficulties with social interactions. However, until more information becomes available about the current status of wild populations, their demographics, genetic status, health status, and threats, this question cannot be adequately answered.

ENVIRONMENTAL CONDITIONS

4. Have the reasons for the decline in panda numbers been eliminated (e.g., hunting, deforestation, commerce)?

Current information would indicate they have not. However, since the 1997 review, the Chinese government has taken some major steps in this regard, including, most notably, the logging ban that was instituted in 1998. Other forest restoration projects have also been implemented, including the "Farmland to Forest Project." However, it is not clear if the type of habitat that is being restored as a result of these activities will be suitable for pandas. Initial studies indicate that the type of bamboo regrowth that is occurring may not be suitable. In addition, since the logging ban has been instituted, monitoring activities have shown that the incidence of hunting and herbal collection has increased in panda ranges. Both activities have negative effects on the wild population. China's new "Western Development Initiative" could also have profound effects on giant panda habitat, and it is critical that the environmental impact of this Initiative be adequately considered before activities proceed. Continued work in habitat protection, antipoaching, and public education is needed to guarantee the safety of wild pandas.

5. Is sufficient habitat protected and secure?

Implementation of reintroductions depends on the existence of suitable, currently unoccupied or underoccupied habitat. Information concerning such habitats is not currently available.

POLITICAL CONDITIONS

6. Would a reintroduction program have a negative impact on local people?

Giant pandas are not predators that pose a threat to the safety of villagers. However, they may threaten crops and animals in the area if not trained to avoid human settlements. Also, if the release program includes land used by the local people, the potential impact of releases on their livelihood must be taken into consideration and efforts made to mitigate conflict.

7. Does community support for reintroductions exist?

Support will vary, depending on the release site. To gain local government and community support, these entities must participate in the planning process for the release to ensure that the program does not have a negative impact on them. It will also help if they receive some benefit from the release effort. Education programs at the community level will be critical, including instruction on what to do if a panda comes into a village.

8. Is there support from and involvement of government and nongovernment organizations (NGOs)?

The Chinese government is committed to panda conservation and will support release efforts if the efforts contribute to improvement in the status of its giant pandas. NGOs participating in the panel discussion at *Panda 2000* agreed that releases may have a role to play as part of a conservation strategy in metapopulation management for panda conservation. However, both the State Forestry Administration (SFA) of China and the NGOs involved believe that other aspects of panda conservation, such as habitat protection, antipoaching activity, and conservation education and research leading to development of a suitable source population, require more attention before there is public support for release programs.

9. Would reintroduction activities be in compliance with the laws and regulations of China?

Release activities would conform to current Chinese wildlife laws and regulations. As wild pandas are found only in China, no international laws would be a factor, unless captive animals are brought in from a foreign country for this purpose. In such cases, a CITES permit would be required. Also, any release that involves moving a panda from one location to another within China requires a permit from SFA. The actual release would also require a permit from SFA. Currently, none of the reserves includes release projects as part of their management plan, although Wolong Nature Reserve has announced plans for a release in 2005.

BIOLOGICAL AND OTHER RESOURCES

10. Is sufficient reintroduction technology available at this time?

Reintroduction technology is still in the development phase for the majority of species, although some experiences with other carnivore taxa provide important information that could be used in the case of the giant panda.

11. Is critical knowledge of the giant panda's biology available?

Since 1997, when the last review of these criteria was completed, a ban by the Chinese government on the use of radiocollars for studying wild pandas has severely impacted the ability to improve our understanding of the animal's biology and ecology. Knowledge of wild pandas is still minimal, and long-term projects studying social behavior and community structure, diseases and the causes of mortality, population trends, habitat preferences, and many other such topics are essential to planning a successful release.

12. Do sufficient resources exist for establishment and long-term monitoring and evaluation of reintroduction programs?

Resources specifically targeted for a panda release program are not available. It must be emphasized that a successful reintroduction program requires a long-term commitment of people and money—not just to implement the release but also to monitor and evaluate its results. At this time, given the available resources and the current situation for giant pandas, release projects should not be a priority. It is logical to assume, however, that reintroductions will eventually form part of an overall metapopulation management strategy. Therefore, experimental release studies would be an appropriate way to acquire some of the knowledge needed to plan a full-scale program. The 1997 workshop recommended that an international advisory committee be established to begin planning for such a program, but this has not yet been done.

Our conclusion is that a full-scale release program for giant pandas cannot be recommended at this time. However, the latest national survey of the wild population will provide valuable information for future planning of such a program.

REFERENCES

He, G., and J. Gipps. 1991. Working group report on the captive management of giant pandas. In *Giant panda and red panda conservation workshop working group reports,* edited by D. Kleiman and M. Roberts, pp. 13–22. Washington, D.C.: National Zoological Park, Smithsonian Institution.

Kleiman, D. G., M. Stanley Price, and B. B. Beck. 1994. Criteria for reintroductions. In *Creative conservation: Interactive management of wild and captive animals,* edited by P. J. S. Olney, G. M. Mace, and A. T. C. Feistner, pp. 287–303. London: Chapman and Hall.

Mainka, S., and Z. Lü, eds. 1999. *International workshop on the feasibility of giant panda re-introduction.* Beijing: China Forestry Publishing House.

17

Biomedical Survey of Captive Giant Pandas

A CATALYST FOR CONSERVATION PARTNERSHIPS IN CHINA

*Susie Ellis, Anju Zhang, Hemin Zhang, Jinguo Zhang,
Zhihe Zhang, Mabel Lam, Mark Edwards, JoGayle Howard,
Donald Janssen, Eric Miller, and David Wildt*

BECAUSE OF THE precarious status of wild populations, giant pandas *(Ailuropoda melanoleuca)* in zoos and breeding centers play a crucial role in educating the public about the plight of their wild counterparts. Giant pandas existing in captivity also function as a critical "hedge" against extinction and serve as a potential resource for future reintroduction efforts. Additionally, captive giant pandas are important as a research resource and a means for attracting substantial public support for conservation of the species living in the wild (as well as for other endangered species endemic to China). At present, there are approximately 150 giant pandas living in captive facilities in China, some managed under the Ministry of Construction's (MOC) Chinese Association of Zoological Gardens (CAZG) and others under the authority of the State Forestry Administration (SFA) at the China Wildlife Conservation and Research Center for Giant Pandas, located in Wolong Nature Reserve.

There has been a steep and challenging learning curve in developing a consistently successful captive breeding program for giant pandas. The first successful birth by natural mating occurred in 1963 at the Beijing Zoo (Ouyang and Tung 1964). The first successful artificial insemination (AI) with fresh semen occurred in 1978 at the Beijing Zoo; successful AI with frozen semen occurred in 1980 at the Chengdu Zoo (Z. Zhang et al. 2000). Through 1989, giant pandas were successfully bred at zoos in Kunming, Shanghai, Hangzhou, Chengdu, Chongqing, Fuzhou, Xian, and the Wolong Conservation and Research Center, as well as a few zoos outside of China (Zhong and Gipps 1999). Records published in the international studbook reveal that for that first 26-year period, a total of seventy-five litters comprising 119 cubs (with ~50% twin births) were produced (Z. Zhang et al. 2000). Thirty-seven of these infants survived for at least 6 months, for an average survival rate of 31%. This low percentage was a prime reason for the slow growth rate of the captive population. Simply put, husbandry was inadequate to ensure consistent success. Many infants died, and although multiple births were common within a litter, the survival of twins did not occur until 1990.

Inertia has never been associated with the challenge of creating a self-sustaining captive population of giant pandas. Over the years, many research initiatives were carried out, resulting in substantial new information on panda biology and some success in enhanced propagation. For example, in 1990, supplemented with hand rearing, a female giant panda at the Chengdu Zoo reared twins successfully for the first time (Z. Zhang et al. 2000). In 1992, the Beijing Zoo succeeded in hand-rearing a neonate that had never had the opportunity to nurse (Z. Zhang et al. 2000).

CAZG has always realized the importance of the captive giant panda population as a conservation resource. It was also sufficiently visionary to recognize the need to develop an organized plan that would allow a consistent breeding program for giant pandas in China. Therefore, CAZG provided an opportunity for the Conservation Breeding Specialist Group (CBSG) of IUCN's Species Survival Commission to facilitate a Captive Management Planning Workshop in Chengdu in December 1996. Approximately fifty Chinese stakeholders attended this workshop. Working together, participants created a plan for managing the captive population (Zheng et al. 1997). Computer simulation modeling revealed that, if the triggers to reproduction could be identified, then the captive population of giant pandas in China could double within the next 10–14 years. Action-based recommendations were made that would begin addressing the poor reproduction and health problems in all panda age classes. The most significant recommendation was for a biomedical survey of the extant population. The reasoning was simple: developing a self-sustaining population would require maximizing the use of healthy, reproductively fit individuals, which then could be intensively managed to retain existing genetic diversity. This could only be achieved if the health and reproductive status of the existing population was known. The primary focus of this initiative would be to ascertain which animals were healthy and reproductively fit and could participate in a more effective breeding program. Animals with medical problems perhaps could be treated and returned to breeding status. Biological materials (blood and tissue) could be harvested to allow genetic studies (e.g., determining relatedness among individuals or establishing paternity). Finally, a hands-on examination would allow each animal to be identified by tattoo and electronic transponder chip, thereby providing unambiguous identification to help avoid inappropriate matings that could result in inbreeding.

The final negotiations between CAZG and CBSG to conduct the first year of the Giant Panda Biomedical Survey occurred in September 1997. All parties agreed to work together as scientific teams. It was agreed that the survey would begin with selected giant pandas housed at the Chengdu Research Base of Giant Panda Breeding, the Chengdu Zoo, and the Beijing Zoo. It also was agreed that CBSG would assemble a diverse team of U.S.-based scientists and secure funding for their travel and per diem, as well as collect donations of needed equipment items. During the second and third years of the survey, activities were extended to colleagues in the SFA, specifically including evaluation of giant pandas housed at the Wolong Breeding Center.

Upon deciding to act on the Chinese initiative for a survey, CBSG prepared a scientific proposal that was forwarded to various U.S. zoos for funding. In the three years during which the survey was conducted, the CBSG–U.S. team consisted of twenty specialists from seven institutions, who represented the disciplines of veterinary medicine, reproductive physiology, endocrinology, animal behavior, genetics, nutrition, and pathology. This group was complemented by more than fifty Chinese counterpart specialists from MOC and SFA. There was strong political support from the Chinese government, and funding was provided by the U.S. zoo community, complemented with biomedical equipment donations from corporations. All details of the proposed project, including animals to be evaluated, were finalized in Memoranda of

Understanding (MOU) between CBSG and each of the respective breeding facilities or zoos.

The Survey began in March 1998. Data for each individual panda were recorded on forms according to respective disciplines (e.g., anesthesia, medical, reproduction, behavior) in a centralized database. Succinct summaries for each animal also were prepared for the purpose of classifying individuals as "Prime Breeder," "Potential Breeder" (healthy, but prepubertal), "Questionable Breeding Prospect," or "Poor Breeding Prospect." All data were discussed between the CBSG and Chinese teams in several technical meetings, including a two-day meeting held in Beijing in March 1998 that was used to develop the next steps in the Captive Management Plan for giant pandas (Ellis and Wildt 1998).

A second request was issued by CAZG to continue the survey in 1999, including completion of assessments in Chengdu and Beijing and adding a new partner, the Chongqing Zoo. Additionally, CBSG received an invitation from SFA to extend the survey to include some of the giant pandas at the China Wildlife Conservation and Research Center for Giant Pandas in Wolong Nature Reserve. This was greatly facilitated by an existing MOU between the Zoological Society of San Diego and the Wolong Breeding Center. In the 2000 mating season, the CBSG team extended the survey to include Wolong's pandas. By the end of the 2000 season, sixty-one giant pandas had been included in the survey, resulting in massive amounts of new information on species biology.

SURVEY METHODOLOGY

Methods were consistent during all three years of the survey (see A. Zhang et al. [2000] for further details). Anesthesia was induced using primarily ketamine hydrochloride (and additional anesthetic agents in some cases) delivered intramuscularly. Once each animal was tractable, the following procedures were performed:

- An electronic transponder chip (Trovan Electronic Identification Systems, MID FingerPrint, Weymouth, United Kingdom) was inserted subcutaneously in the interscapular area at the dorsal midline at the cranial aspect of the black-and-white hair interface of each animal.

- The animal was tattooed with its respective studbook number in the mucosa of the upper left lip.

- A 0.5×0.5-cm^2 skin biopsy was incised from the inner thigh of a hind limb, and twenty hairs and 10 cc of heparinized blood were collected for preparing samples for future genetic analysis. All samples were stored on site at their respective collection locations.

- Selected healthy males of breeding age were subjected to an approximately 20-min electroejaculation procedure. Ejaculate volume was recorded, and fresh sperm were evaluated for percentage motility, forward progressive status, normal/abnormal morphology and sperm acrosomal integrity. Fresh semen also was diluted in various diluents used historically by the Chinese and CBSG team members. All cryopreserved sperm samples were stored on site at their respective collection locations.

- A general physical examination was performed that included body weight, body measurements, oral/dental examinations, and assessment of limbs and external genitalia. A total of 25–30 cc of blood (including that used for the genetic analysis) was collected from the jugular vein (in Chengdu) or from the cephalic vein (in Beijing). Portions of these samples were used for hematology and serum chemistry analyses. The testes of males were measured and palpated for tone and consistency. A vaginal smear for cytology was prepared from each female and a fecal smear for cytology and Gram stain (from many individuals). An ultrasound examination of the abdominal cavity was made on each animal. During anesthesia, the animal was monitored by noninvasive blood pressure, transcutaneous

TABLE 17.1
Cumulative Results of the Three-Year Biomedical Survey and Assigned Categories by Institution

INSTITUTION	TOTAL NUMBER OF ANIMALS	PRIME BREEDER	POTENTIAL BREEDER	QUESTIONABLE BREEDER	POOR BREEDING PROSPECT	NOT CLASSIFIED
Chengdu Zoo and Research Base	6/14	2/5	3/6	0/1	1/2	0/0
Beijing Zoo	4/5	1/2	0/2	1/0	0/1	2/0
Chongqing Zoo	0/3	0/2	0/0	0/1	0/0	0/0
Wolong Center	14/15	6/5	7/6	1/3	0/1	0/0
Total	24/37	9/14	10/14	2/5	1/4	2/0

NOTES: The entries give numbers of male/female individuals, so that, for example, 2/3 = 2 males and 3 females. The total number of animals at all listed institutions was sixty-one.

pulse oximetry, body temperature, and direct observation. Blood parameters were assessed on-site using a portable analyzer; other blood parameters not assessed by the portable analyzer were assessed by staff from a local human hospital. All remaining blood products were stored on site at their respective collection locations.

- A nutrition survey form was distributed at each location and used to obtain dietary information for evaluation.

- Historical and behavioral data were collected that included origin (wild-born or captive-born), date of birth, and reproductive history, including past opportunities to breed and reproductive success. Information collected also included evaluation by keepers of various behavioral characteristics for each animal, such as calmness, shyness, or aggressiveness.

Following the completion of all evaluations at a given location, all data were discussed between the CBSG team and the various Chinese teams in technical meetings to reach consensus on findings and interpretation. Management recommendations and the data on which they were based were provided to all participants.

SURVEY RESULTS INDICATING FACTORS THAT LIMIT REPRODUCTIVE SUCCESS

Cumulative results from the three years of survey, conducted at the Chengdu Zoo and the Chengdu Research Base of Giant Panda Breeding, the Beijing Zoo, the Chongqing Zoo, and the China Research and Conservation Center for the Giant Panda (Wolong), are shown in table 17.1 (A. Zhang et al. 2000).

Approximately 39% of the animals were prime breeders, 40% potential breeders, 12% questionable breeders, and 8% poor breeding prospects (two were not classified). Although the majority of the animals were categorized as prime or potential breeders, almost 20% of the population experienced one or more problems that prohibited reproductive success. From the survey, six important issues emerged that were influencing the health and reproductive fitness of captive giant pandas in China. Some of these factors were already known to be important, but now explicit data were produced that revealed the incidence of these factors and indentified specific animals afflicted by specific conditions. Furthermore, new factors were identified that required additional attention in terms of research and remediation.

The variables most influencing the health and reproductive status of captive giant pandas in China were the following:

1. Unknown paternities resulting from the common practice of combining natural mating and artificial insemination (in the absence of routine use of molecular identification);
2. Genetic overrepresentation of certain individuals;
3. Behavioral problems, especially aggressiveness, that prevented the safe introduction of males to females;
4. Suboptimal nutrition, especially with regard to fiber intake;
5. "Stunted growth syndrome"; and
6. Males with testicular hypoplasia or atrophy.

We discuss these factors and related findings in detail below.

UNKNOWN PATERNITIES AND GENETIC OVERREPRESENTATION

The ultimate goal of developing a self-sustaining captive population of giant pandas depends on successful reproduction and the ability to manage the pedigree of the population by ensuring specific animal pairings. The original 1996 Masterplanning Workshop in Chengdu revealed that there already was significant genetic diversity in the captive population, with no need to import new founders from the wild (Zheng et al. 1997). However, the survey indicated that there were two significant problems. First, pedigree construction revealed that many of the wild-born founders had not reproduced in captivity. In general, only a fraction of the animals in a given institution were reproducing naturally. For example, the Beijing Zoo was highly successful in producing offspring, but the preponderance of births resulted from the matings of only one male and two females. This observation strongly suggested the need for developing genetic exchange programs (of animals or semen) among institutions to maintain high levels of genetic diversity in the captive population.

Second, it became clear early in the survey that there were questions about the validity of pedigrees, largely due to questions regarding paternity. In fact, the Working Group on Genetics and Reproductive Technologies in the technical workshop that was held following the first year of the survey (Ellis and Wildt 1998) determined that establishing the genetic profile of every giant panda in captivity was of highest priority. This remains a major issue that can only be resolved by molecular genetic technology (i.e., DNA microsatellite analysis). This assessment will ensure that studbook information is correct, so that inadvertent pairing of related individuals is avoided. Paternity determination also will establish the success of past breeding attempts using AI. This is important because it is common practice to combine natural mating with AI, often with sperm from multiple sires. Finally, in cases of twinning and the use of multiple males, paternity analysis will provide new information on the phenomenon of sperm competition.

Studbook records were used to create partial pedigrees of individuals at all four facilities. From the pedigrees and studbooks, it became clear that parentage is further complicated because many key animals (both past potential sires and dams) are dead, and samples are now unavailable. Because resolution of paternity is of highest priority, it would be useful for managers to adopt a policy that involves collecting biomaterials for genetic analyses from every captive giant panda, both wild-born and captive-born. Technical protocols have been generated to accomplish this task by the Working Group on Genetics and Reproductive Technologies in 1998 (Ellis and Wildt 1998). The thoughtful collection of samples when animals are alive will provide the essential biomaterials for conducting many studies that will promote improved management, as well as address scholarly issues.

A plan is in place to analyze samples to sort out paternity questions in the extant giant panda population. A collaboration has been developed with the laboratories of Drs. Stephen O'Brien (National Cancer Institute's Laboratory of Genomic Diversity) and Ya-Ping Zhang (Kunming Institute of Zoology, China) that is aimed at identifying the paternity of all giant pandas.

EVALUATION OF BEHAVIOR

Many giant pandas were discovered to express varying degrees of behavioral aberrancies. For males, aggression often prevents the safe introduction of males to females for mating. To begin to determine if particular behavioral or "personality" traits contribute to an individual's reproductive success, detailed behavioral data from giant panda keepers at all institutions were collected during the survey. This was a subjective assessment tool, using behaviorally based personality characteristics and their definitions. Methods were based on the work of Feaver et al. (1986) with the domestic cat *(Felis catus)*, that of Gold and Maple (1994) with the gorilla *(Gorilla gorilla)*, of Fagen and Fagen (1996) with the brown bear *(Ursus arctos)*, and of Wielebnowski (1999) with the cheetah *(Acinonyx jubatus)*.

Twenty-three behavioral characteristics were chosen as quantitative measures. The keeper survey was developed in English and translated into Chinese. Thirty-eight keepers took the survey; they were asked not to discuss their ratings of individual pandas prior to or during the scoring process. The actual rating method was based on Feaver et al. (1986). A form with calibrated horizontal lines (number of lines corresponding to the number of giant pandas at that facility) was provided for each behavioral characteristic. Each line was 100 cm long on a continuous scale for a particular trait. The minimum score was situated on the left side of the line, the maximum score on the right. Keepers marked each individual's score by placing an x along the line for each individual under each behavioral characteristic. Distance from the left side of the horizontal line was measured in centimeters, with each animal receiving a score of 0–100 for each trait. Data were analyzed using binary logistics regression.

Of the twenty-three variables examined, origin (wild-born) and aggressiveness (sexes combined) contributed most significantly to successful breeding. There were no sex differences, but wild-born animals were more successful breeders (binary "logit" regression, $N = 54$, $p < 0.01$). Origin was a better predictor of successful reproduction than age, although, in general, older adult pandas tended to be more successful at breeding than younger adults. Finally, although aggressiveness toward other pandas did not predict breeding success for either males or females, when sex data were combined, aggressive animals were the more effective breeders (binary "logit" regression, $N = 54$, $p < 0.01$).

SUBOPTIMAL NUTRITION

Nutrition and nutrient status play essential roles in all aspects of an animal's health, including growth, reproduction, and disease resistance. To assess the effect of nutrition on the health status of giant pandas in the survey, detailed information was collected on feeding, dietary husbandry, and dietary nutrient content. At each institution, nutrition and dietary husbandry information included the following:

1. Identifying and quantifying nutrient content of all food items offered;
2. Quantifying currently offered diets (i.e., by mass);
3. Generating data on food intake (for selected animals);
4. Measuring feeding frequency of various diet components;
5. Characterizing distribution of food within enclosures and sites of food presentation;
6. Determining seasonal variations in diets provided; and
7. Describing food storage and preparation facilities.

In addition, characteristics of fecal volume and consistency were described, including defining the typical fecal output within a 24-h period, frequency of mucous stool excretion, and the association of blood with mucous stools. During the medical examination, body weight, anatomical measurements, and body condition were characterized for each giant panda. Plasma and acid-washed serum samples were collected, which can be used in the future to determine the fat-soluble vitamin and trace mineral status of each individual, respectively.

All diets were evaluated using the Zoo Diet Analysis Program software (D. Baer, © 1989). Data on nutrient composition of bamboo represented the largest potential variable in these analyses. Information on bamboo nutrient content reported in the survey was generated from samples of leaf material from a single species, *Phyllostachys aurea,* collected at the Honolulu Zoo in Hawaii. The most valuable information that could be added to these calculated assessments would include the routine nutrient analysis of bamboos actually fed at Chinese zoos and research centers. This is a high-priority project for the near future. Additionally, specific nutrient requirements for various stages of the giant panda's lifecycle have not been scientifically quantified. As a result, nutrient guidelines have been developed for evaluation of diets fed to both the red *(Ailurus fulgens)* and giant panda using nutrient content information on natural foodstuffs, natural feeding ecology, and nutrient requirements of taxonomically related species (e.g., bears, dogs).

The most significant nutritionally related finding from the survey was that dry matter intake was lower than anticipated. We realized that this might be a function of a higher-energy diet fed routinely in Chinese institutions. Additionally, fecal output was lower than anticipated, which may be a result of higher digestibility produced from the less-fibrous diets fed in China. Mucous stool output may be related to a low-fiber/high-starch diet and/or the effect of feeding food items of widely variable nutrient content (e.g., bamboo versus concentrate). Bamboo was the primary variable of all captive giant panda diets. To ensure that the best material is provided for feeding pandas, it was determined that (1) consistent supplies of multiple bamboo species should be identified and established that allow the animals to consume free-choice quantities of material; (2) supplies should be readily available and easily accessible; and (3) routine nutrient analysis and heavy metal screening should be conducted to better understand the seasonal variability and guarantee the quality of this food item. Finally, the survey indicated that the addition of bamboo powder was an appropriate method to increase the fiber content of a concentrate diet. In sum, the survey suggested that suboptimal nutrition may indeed be limiting reproductive success by not assuring optimal health. Compromised nutrition can be corrected through adjustments to nutritional husbandry that are based on sound, high-priority basic research, especially fundamental studies of bamboo nutritional quality.

"STUNTED GROWTH SYNDROME"

Six of the sixty-one giant pandas examined (9.8%) exhibited a cluster of characteristics that the survey team termed "stunted growth syndrome." This condition was characterized by varying degrees of stunted body development in pandas that had been born in captivity. Animals exhibiting this syndrome often had a history of multiple, chronic diseases (especially gastrointestinal distress) with excessive dental staining and tooth wear. Other common characteristics were a coarse hair coat in poor condition, abdominal ascites (excess fluid), and an absence of sexual development or activity. Five of the six affected animals originated at the Chengdu facilities. The etiology of this syndrome has not yet been determined. However, the onset of this condition appeared to be developing in two young individuals included in the survey, studbooks #453 and #454. These dam-reared twins were born in 1997 in Chengdu and remained with their mother for more than 1 year, as part of an early development study conducted by Rebecca Snyder (Zoo Atlanta). Other subjects of the study were studbooks #452 and #461, who also remained with their dams for more than 1 year. At approximately 300 days of

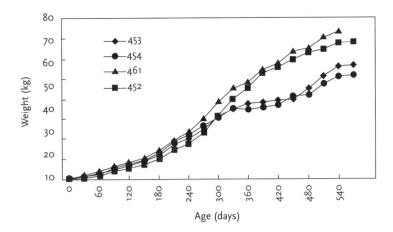

FIGURE 17.1. Weights of age-matched giant panda cubs born in 1997 at the Chengdu facilities, illustrating a distinct divergence in weight gain. Pandas #452 and #453 are age matched, as are #454 and #461.

age, #453 and #454 developed gastrointestinal problems and began to grow more slowly than #452 and #461 (figure 17.1). Examination of the two animals in the 1999 survey revealed that both had excessive dental wear and staining and were stunted in development. Ultrasound examination of #453 also revealed a moderate amount of abdominal fluid. Both weighed less than their age-matched counterparts (#453 weighed 12 kg less than #452 and #454 weighed 21 kg less than #461).

The specific etiology of stunted growth syndrome remains unknown. Perhaps there is a nutritional component. However, the survey team also has speculated about the cause being related to an infectious disease, perhaps a virus. As a result, screening for viral titers has become a high priority, with research being initiated in 2002 and now in progress.

TESTICULAR HYPOPLASIA OR ATROPHY

Abnormally sized testes were discovered in three adult male giant pandas (#323, #181, and #345). In each of these individuals, only one normal-sized testis was located in the scrotal sac. The contralateral testis was hypoplastic, soft in texture, and situated adjacent to the inguinal canal rather than being located in the scrotal sac. There was no obvious cause for this abnormality, although all three individuals were housed (or had been housed) at the Beijing Zoo. For example, #323 was a robust 12.5-year-old male at the Beijing Zoo, the primary breeding sire at this institution. Upon examining the pedigree and photographic records, it was discovered that the male sibling of #323's dam (#181, deceased) also had a unilateral hypoplastic testis. This observation raised speculation that the condition could be heritable, similar to cryptorchidism. This hypothesis was partially rejected upon discovering the same abnormality in #345, a male that appears unrelated to #323 or #181 from current pedigree data. Additionally, there was information from historic curator records indicating that testis size in #345 at one time was normal, with the onset of atrophy at 7.5 years of age.

In summary, testicular hypoplasia in the giant panda is of concern, largely due to its unknown etiology. If heritable, then a significant number of males in the Beijing-produced population may be affected, as #323 was the primary breeding male. The histological evaluation of testicular tissue from affected males should be a high priority. Since the survey was conducted, #323 has died, and detailed histopathology on his affected tissues will yield further useful information. Another priority is to examine all of the male offspring of #323 for the defect (#345 is a genetically unrepresented, nonbreeding male). Until this issue is resolved, using #345 or any descendant of #323 for breeding cannot be recommended.

SEMEN CHARACTERISTICS AND SPERM CRYOPRESERVATION

In addition to determining the variables that were found to be driving reproductive success, it is important to report those factors that were found to be insignificant. Although there were a few males with hypoplastic testes, we identified no physiological causes for male infertility in the population. To allow comparison of seminal traits among males, the same semen collection technique and ejaculator/probe equipment were used on each individual. Following electroejaculation, fresh semen was evaluated as described below. To compare the ability of sperm from each male to survive cryopreservation, a standardized cryomethod was used on each individual. However, throughout the survey, parallel studies were conducted in testing historical sperm cryopreservation methods (especially focusing on diluents and cryoprotectants) developed earlier by the Chinese and CBSG teams. To assess postthaw viability, an aliquot from each sample was thawed and evaluated by both teams together to assess the ability of sperm from each male to survive cryopreservation stress and determine optimal methods of freezing and thawing sperm (which is always specific for each species). All cryopreserved sperm samples were stored on site at their respective collection locations.

Seventeen ejaculates from sixteen males were collected by electroejaculation using an electroejaculator and a 2.6-cm diameter rectal probe. Semen was examined on a heated (37°C) stage using phase-contrast microscopy for a subjective assessment of sperm motility (0–100%) and forward progressive status (on a scale of 0–5, where 0 = no movement and 5 = rapid forward progression). Ejaculate volume and pH were determined, and sperm concentration/ml of ejaculate was calculated using a hemacytometer procedure. Sperm morphology was assessed after fixing a 5-µl aliquot in 100 µl 0.3% glutaraldehyde (prepared in phosphate buffered saline), followed by phase-contrast microscopic examination of 100 sperm/aliquot at 1000×. Sperm were categorized as normal or abnormal due to a coiled flagellum, abnormal acrosome, bent midpiece with cytoplasmic droplet, bent midpiece without cytoplasmic droplet, bent flagellum with cytoplasmic droplet, bent flagellum without cytoplasmic droplet, proximal cytoplasmic droplet, or distal cytoplasmic droplet. Sperm acrosomal integrity was evaluated further by mixing 5-µl diluted semen (diluted in Ham's F10 culture medium) to 45 µl of acrosome stain (rose bengal/fast green) for 90 s. Slide smears were then prepared and acrosomes were divided into four categories: (1) normal apical ridge (uniform staining of the acrosome); (2) damaged apical ridge (nonuniform staining with ruffled or folded acrosome); (3) missing apical ridge (no staining; acrosome not present); and (4) loose acrosomal cap (loose membrane protruding above level of sperm head). A brief summary of the average semen characteristics is presented in table 17.2.

Overall, seminal evaluations in a large cohort of giant pandas indicated that there were good to excellent quality ejaculates collected from all males (with exception of high proportions of abnormal sperm in males <6 years of age). Thus, there was no indication that a limiting factor to reproduction of giant pandas in captivity was the quality of ejaculate. On the contrary, giant pandas maintained in captivity (whether wild- or captive-born) produced prodigious amounts of high-quality spermatozoa. Furthermore, the ability of sperm from all males to survive sperm cryopreservation appeared similar.

Other important observations were made in the male reproductive studies. For example, long-term cold storage at 4°C was effective for maintaining giant panda sperm motility and viability for days. The TEST cryodiluent (Irvine Scientific, Santa Ana, California) and Chinese Sperm Freezing Solution (SFS—made by reproductive labs in China) were comparable and effective for cryopreserving giant panda sperm. Optimum postthaw sperm viability was achieved using rapid freezing rates, and postthaw motil-

TABLE 17.2
Semen Characteristics of Giant Pandas

CHARACTERISTIC	VALUE
Ejaculate volume (ml)	3.9 ± 0.6
Seminal pH	8.1 ± 0.1
Sperm concentration (ml^{-1})	$1182.4 \pm 220.6 \times 10^6$
Sperm motility (%)	68.8 ± 5.2
Sperm progression (0–5, 5 = best)	$2.8 \pm 0.2\%$
Normal sperm structure (%)	66.3 ± 6.4
Intact sperm acrosomes (%)	89.5 ± 5.5

NOTE: Based on seventeen ejaculates from sixteen males.

TABLE 17.3
Summary of Survival Success of Captive-Born Giant Pandas in China

YEAR	NUMBER OF SINGLETON BIRTHS	NUMBER OF TWIN BIRTHS	NUMBER OF TRIPLET BIRTHS	NUMBER OF CUBS	NUMBER OF CUBS SURVIVING TO 6 MONTHS OF AGE	SURVIVAL RATE (%)
1963–1989	32	42	1	119	37	32
1990–2000	40	33	1	109	66	61
1998	4	3	0	10	8	80
1999	4	6	1	19	14	74

ity, status, and intact acrosomes were higher in Hepes-buffered Ham's medium (catalog #12390-035, Life Technologies, Grand Island, New York) compared with TCM medium (made by reproductive labs in China). Washing sperm (to remove seminal fluid) before diluting in cryodiluent did not influence postthaw motility and forward progressive motion. Lastly, variation in postthaw survival was observed using the pelleting method of freezing, perhaps due to inconsistent rates or methods of freezing. A standardized pelleting technique (used later in the survey) during pellet freezing resulted in optimal postthaw sperm quality.

RECENT CAPTIVE BREEDING SUCCESS IN CHINA

The biomedical survey resulted in enormous amounts of new information that has had a direct impact on improving giant panda husbandry. Furthermore, independent scientific studies by Chinese researchers have also exponentially increased knowledge about the biology of this flagship species. The combination of efforts of these scholarly teams has resulted in a remarkably rapid improvement in the success of producing giant panda cubs in the captive program. From 1990 to 2000, seventy-four litters composed of

TABLE 17.4
Number and Distribution of Captive Giant Pandas through May 2000

LOCATION	SOURCE		TOTAL
	WILD-BORN	CAPTIVE-BORN	
In China	44	69	113
Outside China	6	11	17
Total	50	80	130
Percentage	38	62	

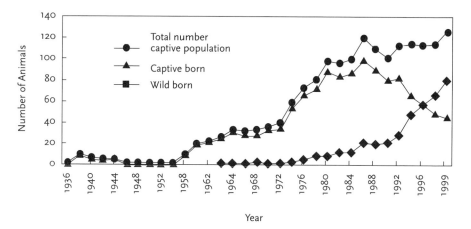

FIGURE 17.2. Changes in the captive population of giant pandas from 1936 to 2000 (only cubs surviving to 6 months are included in the year 2000 figures). At the end of 2000, 130 giant pandas older than 6 months were maintained in Chinese zoos and nature reserve centers, whereas less than twenty individuals were maintained in zoos outside China.

109 cubs were born. Sixty-six cubs survived more than 6 months, and the survival rate reached 61%. From 1998 through 1999, twenty-nine cubs were born in eighteen litters, of which twenty-two survived for at least 6 months, increasing survival success to 76% (table 17.3).

From 1963 through 2000, Chinese facilities exponentially increased their overall ability to successfully breed giant pandas. A total of 228 cubs have been born in 149 litters. One hundred and three cubs have survived more than 6 months. Eighty of these remain alive and account for 62% of the contemporary captive population (table 17.4). Clearly, the captive population has the ability to become self-sustaining with less dependence on animals in nature. For example, figure 17.2 illustrates the decline of giant pandas in the captive population that are of wild-born origin. In contrast, the contribution of captive-born individuals increases, whereas overall captive animal numbers also rise.

CONCLUDING PERSPECTIVE

If challenged with the responsibility of preventing extinction of a given species, what is the most effective way to respond? For most zoo managers and researchers, the first reaction would be "breed it," and by whatever means available. Yet, this apparently most logical of responses is likely to be erroneous, or at the least overly simplistic, because genuine conservation is not linked solely to the propagation of a species (Wildt et al. 2001).

Real conservation can be likened to a thousand-piece jigsaw puzzle that may take decades to solve. Unfortunately, in the case of the giant panda conservation puzzle, time is of the essence. Propagating animals in captivity can and does contribute to wildlife conservation, but only in the context of being addressed simultaneously with other issues (i.e., other puzzle pieces), usually in the context of wildlife and habitat management, ecology, population biology, behavior, nutrition, genetics, veterinary medicine, and social, economic, and ecopolitical factors. In short, captive breeding is only one small component of panda conservation—and we have just begun to piece together the rest of the abundantly complex conservation puzzle involving giant pandas.

Our multidisciplinary approach was key to revealing that no one variable was impeding reproductive fitness in captive giant pandas. Rather, failures appeared to be attributable to multiple, linked factors (e.g., poor nutrition leading to compromised health that, directly or indirectly, decreased reproduction or offspring survivorship). Without interdisciplinary collaboration, some causes and interactions would have gone undiscovered. In some cases, remediation was simple. For example, routine medical evaluations revealed reproductive tract infections that were treated with antibiotics that allowed some previously nonreproductive females to produce offspring. Other cases involved remediation efforts (e.g., modifying diets, sorting out paternities) that were more complex, and detailed systematic studies have been implemented. In all cases, the survey has provided the blueprint for continued management action.

Another benefit of the project was the opportunity to conduct more basic research. For example, a by-product of male fertility evaluations was "surplus" semen for investigating the sensitivity of panda sperm to cooling and cryopreservation. New semen-handling protocols emerged that have been useful for improving artificial insemination. One practical benefit was the production of a surviving cub from a wild-born, underrepresented male with a lethal squamous cell carcinoma. Previously, such an individual would have died without breeding, its genes unrepresented in future generations. Artificial insemination will continue to be important for genetic management, including circumventing sexual incompatibility problems and moving genetic material among breeding centers and from wild to captive populations.

The close partnerships that developed in the intensive working milieu have inspired trust and the development of many friendships between the scientists from divergent disciplines and cultures. Additionally, collaboration between the two Chinese agencies responsible for giant panda conservation (MOC and SFA) has significantly improved. Although the primary purpose of the biomedical survey has been to assess the health and reproductive status of giant pandas in captivity, other valuable results emerged, especially in the areas of knowledge and technology transfer.

High-priority needs have been identified by Chinese panda specialists, including capacity building in animal behavior, enrichment, and veterinary technologies. The first of a series of veterinary training workshops was held in Chengdu in December 1999 and involved training forty-nine veterinarians representing twenty-seven Chinese institutions in various aspects of veterinary medicine, ranging from anesthesia to diagnostic procedures to clinical pathology. Capacity building in behavioral enrichment, veterinary medicine and pathology will continue to improve competence in captive management. Thus, the original project, designed to collect data to solve a management problem, is resulting in range-country capacity building, with the needs for training identified by our Chinese colleagues and implemented by CBSG in collaboration with its partner institutions.

The variety of activities focusing on the captive population has also generated interest among managers of wild giant pandas. For example, CBSG was able to facilitate a workshop at the

China Wildlife Conservation and Research Center at Wolong in October 1999 that focused on the status of giant pandas in nature, research methodologies that could enhance the Chinese Giant Panda National Survey (launched in 1999; see Yu and Liu, panel report 15.1), as well as discussing the potential role of captive-bred individuals for reintroduction. This four-year effort in China with the captive giant panda population has been helpful in facilitating development of a much broader process and dialogue to formulate a better conservation plan for the wild population. Among the outcomes of this undertaking have been identification of high-priority needs, ranging from survey and monitoring of wild populations to infrastructure needs in all reserves and population simulations of the dramatic impacts of habitat fragmentation and the need for improved knowledge of human impacts on panda populations (Yan et al. 2000; see also Dinerstein et al., chapter 15, and Loucks and Wang, panel report 9.1). Most importantly, these high-priority needs have been identified in an inclusive process by Chinese colleagues and not unilaterally by Westerners approaching giant panda conservation in China with a "missionary" (i.e., "we are here to tell you what you need") attitude.

This project, beginning with a simple request from Chinese colleagues for information exchange, has resulted in a remarkable cascade of new biological data, enhanced management practices, and capacity building. The project also illustrates the value of integrative, multidisciplinary research. Whether this is an "ideal" model to be considered for other species may be debatable. The charismatic giant panda is of inordinate interest worldwide—its high profile eased the way for the required approvals and funding. It may be more difficult to stimulate enthusiasm and to secure grants for species with a lower profile, yet are as rare or even more ecologically important than the giant panda. Finally, there was widespread interest in participating by many U.S.-based specialists, thereby allowing the best scientists—as well as those most likely to be team players—to be selected to participate. Not all multidisciplinary projects would have the luxury of unlimited numbers of scientists eager to participate.

Nevertheless, there are other project characteristics that should be considered in formulating similar studies in the future. Clearly, organizing multiple institutions under a neutral entity like CBSG avoided the perception that any one organization was dominating the initiatives. Shunning condescension and focusing on developing collegial and personal relationships and knowledge sharing inspired trust that was essential to working across diverse cultures. The resultant confidence and friendships facilitated later invitations to coordinate the training workshops. Enthusiasm for capacity building was enhanced by having the most skilled Chinese counterparts serving as co-trainers. Finally, it was critical that every priority was put forward by the range-country scientists and managers. The most significant contribution of the Western partners was sharing expertise and transferring tools to the local people who ultimately are responsible for preserving the biodiversity of their country.

The biomedical survey is indeed a useful model for resolving problems in captive breeding programs in a holistic fashion. However, there will never be a simple or quick fix without sustained follow-up that includes partnerships and strong science. Although the survey has assisted in enhancing captive breeding in China, there is much left to be done. Certainly, the highest priority is developing and implementing a long-range genetic management plan that ensures that all wild-born founders and genetically valuable individuals in the contemporary population reproduce. This will require intensive research to resolve problems identified in the survey, as well as continued partnerships. Future success will depend on sharing scientific expertise and animals (or their genetic material), undertaken with an ongoing spirit of true international collaboration.

ACKNOWLEDGMENTS

The giant panda biomedical survey was conducted with a vast array of partners, including

Xie Zhong, He Guangxin, Li Guanghan, Yu Jianqu, Arlene Kumamoto, Lyndsay Phillips, Ray Wack, Meg Sutherland-Smith, Lee Young, Richard Montali, Don Janssen, Bruce Rideout, Barbara Durrant, Rebecca Spindler, and Ulysses Seal. Cooperating organizations include the Beijing Zoo, Chengdu Zoo, Chengdu Research Base for China Panda Breeding, China Research and Conservation Center for the Giant Panda, Chongqing Zoo, National Zoological Park (Washington, D.C.), Zoological Society of San Diego, Zoo Atlanta, St. Louis Zoo, Columbus Zoo, Memphis Zoo, the University of California at Davis, and the American Zoo Association's Giant Panda Conservation Foundation. Corporations providing in-kind support included British Airways, Nellcor Puritan Bennett, Heska, Ohaus, InfoPet Identification Systems, Air-Gas, and Sensory Devices.

REFERENCES

Ellis, S., and D. Wildt. 1998. *Technical meeting for the review of the biomedical survey of giant pandas in captivity in China. Draft workshop report.* Apple Valley, Minn.: Conservation Breeding Specialist Group, IUCN Species Survival Commission.

Fagen, R., and J. M. Fagen. 1996. Individual distinctiveness in brown bears, *Ursus arctos. Ethology* 102: 212–26.

Feaver, J., M. Mendl, and P. Bateson. 1986. A method for rating individual distinctiveness of domestic cats. *Anim Behav* 34:1016–25.

Gold, K. C., and T. L. Maple. 1994. Personality assessment in the gorilla and its utility as a management tool. *Zoo Biol* 13:509–22.

Ouyang, K., and S.-H. Tung. 1964. In the Peking Zoo—The first baby giant panda. *Anim Kingdom* 67:45–46.

Wielebnowski, N. C. 1999. Individual behavioral differences in captive cheetahs as predictors of breeding status. *Zoo Biol* 18:335–49.

Wildt, D. E., S. Ellis, and J. G. Howard. 2001. Linkage of reproductive sciences: From "quick fix" to "integrated" conservation. *J Reprod Fertil* [Supplement] 57:295–307.

Yan, X., X. Deng, H. Zhang, M. Lam, S. Ellis, D. E. Wildt, P. Miller, and U. S. Seal. 2000. *Giant panda conservation assessment and research techniques workshop report.* Apple Valley, Minn.: Conservation Breeding Specialist Group, IUCN Species Survival Commission.

Zhang, A., H. Zhang, J. Zhang, X. Zhou, D. Janssen, D. E. Wildt, and S. Ellis. 2000. *1998–2000 CBSG biomedical survey of giant pandas in captivity in China: Summary.* Apple Valley, Minn.: Conservation Breeding Specialist Group, IUCN Species Survival Commission.

Zhang, Z., A. Zhang, G. Li, L. Fei, Q. Wang, D. E. Wildt, S. Ellis, T. Maple, and R. MacManamon. 2000. *A review of captive breeding success in giant pandas.* Paper presented at *Panda 2000: Conservation priorities for the new millennium,* San Diego, October 15–18.

Zheng, S., Q. Zhao, Z. Xie, D. E. Wildt, and U. S. Seal. 1997. *Report of the giant panda captive management planning workshop.* Apple Valley, Minn.: Conservation Breeding Specialist Group, IUCN Species Survival Commission.

Zhong, X., and J. Gipps. 1999. *The 1999 international studbook for giant panda (Ailuropoda melanoleuca).* Beijing: Chinese Association of Zoological Gardens.

BRIEF REPORT 17.1

Conservation Education Initiatives in China

A COLLABORATIVE PROJECT AMONG ZOO ATLANTA, CHENGDU ZOO,
AND CHENGDU RESEARCH BASE OF GIANT PANDA BREEDING

Sarah M. Bexell, Lan Luo, Yan Hu, Terry L. Maple,
Rita McManamon, Anju Zhang, Zhihe Zhang,
Li Song Fei, and Yuzhong Tian

IN THE FALL OF 1999, as part of its giant panda *(Ailuropoda melanoleuca)* loan agreement with the U.S. Fish and Wildlife Service and China's Association of Zoological Gardens and Ministry of Construction, Zoo Atlanta began to explore the possibility of assisting in the establishment of conservation education programs in China. A formal, long-term plan to establish these programs for the community of Chengdu (Sichuan Province) and its international community was forged in early 2000, with the Chengdu Zoo and Chengdu Research Base of Giant Panda Breeding (hereafter the "Research Base") as partners. Education departments were established at each of these facilities, and initiatives for the Chengdu community began to be developed. A set of visitor-based programs was established at the Chengdu Zoo and Research Base, as well as traveling education programs that could be taken into Chengdu schools. Programming was focused on endangered species of China, with an emphasis on the giant panda. Prekindergarten and kindergarten students, 2½– 6½ years old were a focal point for our initiatives. The most compelling rationale for this focus derives from the higher receptivity to learning at this age. According to Nash (1997) early experiences influence brain development, thus shaping how we learn, think, and behave throughout our lives.

Our goal is to indoctrinate young people with positive attitudes toward wildlife and the environment as their values are in the formative stages. Altering older children's attitudes and beliefs is less easily accomplished (Stapp 1978; Wilson 1996), and attitudes toward animals and wild places differ from one culture to another. Teaching respect for animals and the environment to the very young in China may have far-reaching effects for the future of conservation in their country. The extensive use of kindergarten facilities in China (Tobin et al. 1989; Hindryckx et al. 2001) suggested the means to access a key audience that could be targeted with conservation programming at a critical time in their development. Parents in China rely on kindergarten staff to indoctrinate their children with cultural norms (Tobin et al. 1989), and those surveyed in Chengdu were very supportive of a program of conservation for their children (Bexell et al. 2000).

The kindergarten age group is an underserved group in conservation education in both the United States (Wilson 1996; Ottman 1998) and China (J. Qi, X. Chen, pers. comm.; pers. obs.), underscoring the need for focusing on the early childhood years in instilling respect for nature. Another reason is to be found in the demanding education system in China. As soon as children enter primary school, they begin to compete for advancement through the system, and their success determines their subsequent access to the universities and desirable careers (Pepper 1996; T. Lee, pers. comm.). Teachers and administrators find it difficult to take time from the rigorous study of core subjects that will prepare their students for competitive testing. Therefore, less germane topics are not readily received, and education about conservation falls into this category. However, most Chinese kindergartens are independently governed and therefore not as influenced by government testing requirements as are schools for the older age groups (J. Qi, pers. comm.; Hindryckx et al. 2001).

In 1997, the National Environmental Protection Agency (renamed the State Environmental Protection Administration in March 1998), the Communist Party, and the Department of Education jointly launched an "Action Plan for Environmental Publicity and Education," which emphasizes environmental knowledge in regular curricula (China Consul of International Conservation and Environmental Development 1997). Because of this action plan, the majority of administrators contacted has been receptive to discussions of programming and curricula. Furthermore, teachers have almost no training in conservation and support workshops that would correct this deficit. Other conservation groups from the West (Wildlife Conservation Society, U.S. Peace Corps, Roots and Shoots) have developed environmental education programs in Chengdu. As our programming develops, we will coordinate with these organizations to avoid duplication of effort.

A first step in our program development was to assess the sentiments of leaders at the Research Base and Chengdu Zoo regarding education about giant pandas and their ecosystems. It was important for us to know what had been attempted in the past, and to learn of current needs and desires for conservation education in the surrounding community. We also began to compile information from the Chengdu educational community. Based on what was learned, three schools were visited—one kindergarten, one primary school, and one middle school—to explore the interest level in conservation education, determine where programming would best fit in, and assess the availability of resources.

Learning about the education system in China was an obvious step in our effort. Fashioning effective programs required that we observe how and what children were already being taught. Conservation education methods or content should not be too novel or too difficult to incorporate into existing curricula. It was also necessary for us to learn more about Chinese history and culture in developing material for presentation. A next step was to determine the best methods to use with our targeted audience, and to begin the development of an appropriate curriculum. To the broad task of fostering concern for wildlife and the environment, we sought ways to provide the knowledge, skills, and motivation needed by students that would enable them to address constructively real conservation problems.

Educational programs can have greatly enhanced prospects of being effective if there is family involvement (Furman 1990; Wilson 1993; Bredekamp and Copple 1997). It is well established that the support of significant adults creates a more favorable environment for learning than what schools alone can provide (Carson 1956; Ottman 1998). This is especially true in China, where family values and devotion are particularly strong. Parents and grandparents are therefore invited to participate in workshops that enable them to learn what the children are being taught in our curriculum. They are then positioned to reinforce the value of this new information and model good stewardship behavior outside the schools' sphere of influence.

A report by Vermeer (1998: 229) asserts that "society [in China] at large has little awareness of the threats of environmental pollution to human health and future resource availability, and does little to restrain individual pollutive behavior." Our initial investigations revealed that, indeed, there was almost no instruction in conservation issues in our target audience. Even the life sciences are not extensively taught in Chinese schools. Media coverage of these topics is infrequent, and local cultural and educational opportunities for learning about conservation issues outside of formal schooling are rare. Vermeer (1998: 228) explains "there are clear links [in China] between poverty, lack of awareness or environmental degradation, and threats to human health, limited capacity to sacrifice present consumption on behalf of future generations, and lack of political concern for the environment." An assessment of the effectiveness of our first efforts during the fall of 2000 offered evidence of receptivity to the conservation message and a ready clientele for action-based programs (Bexell et al. 2000).

WHAT HAVE WE ACCOMPLISHED SO FAR?

Training of education managers at the Research Base and Chengdu Zoo was started during the fall of 2000, and will continue indefinitely. Today, the managers are primarily self-sufficient and are building their departments with guidance and new ideas through e-mail correspondence with, and biannual visits to, Zoo Atlanta. Several workshops for teachers, administrators, parents, and grandparents are conducted each year. Long-term education and research relationships with teachers and administrators at the Chengdu City First Governmental Kindergarten and Sichuan Normal University Kindergarten have been established. In addition, several traveling education programs are being used in primary and middle schools and in universities. Lectures are given for staff at the Research Base and Chengdu Zoo in the fall and spring.

During the fall of 2000, we conducted weekly observations in the Chengdu City First Governmental Kindergarten, with the objective of gaining firsthand information about teaching at this age level. Rapid reforms of the education system and cultural changes taking place in China mandate a process of constant updating to maintain the effectiveness of the conservation education program. Provision is made for periodically assessing our progress, effectiveness, and shortcomings. Formal evaluations validate our efforts and provide a basis for assisting other institutions that may want to utilize our strategies.

To fill out the educational program, websites have been developed for use by the Research Base and the Chengdu Zoo as a means of accessing international information on conservation. Each facility has established a volunteer program—the first two such programs in zoological facilities in China. We plan to produce educational videos and other materials and curriculum for circulation among our target audiences. For the benefit of visitors to the Chengdu Zoo and Research Base, graphics have been revised and expanded. Our outreach will eventually extend to communities in and adjacent to five giant panda reserves: Longxi-Hongkou, Wanglang, Anzhihe, Baihe, and Baodinggou. Educational strategies will be modified to accommodate the different needs and resources of each of these communities.

Program success depends on views such as those articulated by Ottman (1998): to have adults who will ultimately make decisions influencing conservation, we must begin by instilling awareness and concern in children. The overheard query of a small child standing with parents in front of a zoo exhibit of the rare giant panda indicates the question that, when it comes to endangered wildlife, kids want to have answered: "How can we save them?"

REFERENCES

Bexell, S. M., L. Luo, and Y. Hu. 2000. Parent and teacher surveys, Chengdu City First Governmental Kindergarten. Unpublished report, Zoo Atlanta.

Bredekamp, S., and C. Copple, eds. 1997. *Developmentally appropriate practice in early childhood programs servicing children from birth through age eight.* Revised edition. Washington, D.C.: National Association for the Education of Young Children.

Carson, R. 1956. *The sense of wonder.* New York: Harper and Row.

China Consul of International Conservation and Environmental Development. 1997. *China Consul Int Cons Env Dev Newslett* 3(1):2.

Furman, E. 1990. Plant a potato—learn about life (and death). *Young Children* 46:15–20.

Hindryckx, G., X. ZongPing, M. Ojala, C. Frangos, S. Opper, D. Harianti, N. Hayes, L. Pusci, O. Onibokun, M. Karwowska-Struczyk, E. Bahovec, M. Montane, and N. Passornsiri. 2001. National profiles of the structural characteristics of pre-primary settings in various nations. In *Early childhood settings in 15 countries,* edited by P. P. Olmsted, J. Montie, J. Claxton, and S. Oden, pp. 55–110. Ypsilanti, Mich.: High Scope Press.

Nash, M. J. 1997. Fertile minds. *Time,* February 3, pp. 49–56.

Ottman, L. 1998. Why teach toddlers and preschoolers? It is not educational and is a waste of time! In *Annual Conference Proceedings,* pp. 170–75. Silver Spring, Md.: Association of Zoos and Aquariums.

Pepper, S. 1996. *Radicalism and education reform in twentieth-century China.* Cambridge: Cambridge University Press.

Stapp, W. 1978. An instructional model for environmental education. *Prospects* 8:495–507.

Tobin, J. J., D. Y. H. Wu, and D. H. Davidson. 1989. *Preschool in three cultures: Japan, China and the United States.* New Haven, Conn.: Yale University Press.

Vermeer, E. B. 1998. Industrial pollution in China and Remedial Policies. In *Managing the Chinese environment,* edited by R. L. Edmonds, pp. 228–61. Oxford: Oxford University Press.

Wilson, R. A. 1993. *Fostering a sense of wonder during the early childhood years.* Columbus, Ohio: Greyden Press.

Wilson, R. A. 1996. Environmental education programs for preschool children. *J Env Ed* 27:28–33.

WORKSHOP REPORT 17.1

International Coordination and Cooperation in the Conservation of Giant Pandas

J. Craig Potter and Kenneth Stansell

COLLABORATIVE RESEARCH and, indeed, the cooperative international exchange of giant pandas for any purpose necessarily rests on the foundation provided by multilateral policy governing the terms and conditions of such exchanges. Because all countries that have been involved in the international exchange of pandas thus far have been signatories to the Convention on International Trade and Endangered Species of Wild Fauna and Flora (CITES), guidance provided by that treaty is fundamental to the development of multilateral policy in this area. The history of CITES involvement with international exchanges and the guidelines that are currently in use must be taken into account when developing loan policies.

Concepts governing policy in the area of international coordination and cooperation must be clarified and discussion stimulated to improve the basis for such coordination and cooperation. A fundamental objective is to encourage exchanges of giant pandas closely aligned with conservation purposes and consistent with the basic purposes of CITES.

Guidelines and concepts in this area require clarification to offset differing perspectives on their applicability and relevance. It is desirable to clarify the concepts that govern policy in this area, so as to foster improved communication among nations of the concepts that provide the basis for the international exchange of giant pandas.

DISCUSSION AND RECOMMENDATIONS

There is general agreement that the governing principles established through CITES for Appendix I animals serve as an essential and necessary framework to guide international coordination. Specifically, Notification to the Parties no. 932 (9/4/96) repealed previous Notification to the Parties no. 477 (5/23/88) and provided updated and clarified recommendations regarding loans of giant pandas. In particular, item 3(d) of the Notification states that every effort must be made to ensure that all the financial benefits from exchanges are devoted to the conservation of the species in China, nearly all such benefits being reserved for in situ conservation. Accepting this framework implies a long-term commitment by all parties concerned, and thus, by its nature requires close coordination and collaboration to be successful.

It is further agreed that improved communication, coordination, and collaboration at all levels and among a wide array of participants both inside and outside China are required to utilize the strengths and specialties of those concerned with giant panda conservation. Although much progress has been made in this direction, the recent expansion of bilateral, international, long-term loan programs for giant pandas involving several countries requires increased willingness to work together in these areas.

From numerous proposed recommendations, we present the following as worthy of further consideration and attention by anyone associated with international exchanges involving pandas:

- Enhanced coordination among the authorities in China and those in importing countries and more thorough discussion, in particular, of in situ projects should be vigorously pursued.
- A more focused process is needed for the exchange of information on research findings, the status of project implementation, and other information related to giant panda conservation.
- There is no substitute for actual on-the-ground experience, and exchange visits offer the potential of gaining a firsthand understanding of the issues associated with panda conservation.
- Although the appropriate central focus is conservation of pandas in the wild, it is necessary to further examine and clarify the relationships of captive programs (i.e., breeding and research) to in situ conservation. This examination will lead to improved understanding and appreciation of the complexities associated with the conservation of giant pandas and their habitats.

Further consideration of these important issues would be an important step in clarifying and focusing the dialogue on international coordination and cooperation.

ACKNOWLEDGMENTS

We thank Gerard Baars, Zhuohua Dai, Meryl Icove, Mabel Lam, Junqing Li, Ling Lin, Timothy Ng, Joel Parrott, Yanling Song, Rebecca Spindler, David Towne, Chanqing Yu, Jinping Yu, Chris West, Don Winstel, Daniel Viederman, Shanning Zhang, Zhihe Zhang, and Alex Zimmerman for their contributions to this report.

Conclusion

CONSENSUS AND CHALLENGE: THE GIANT PANDA'S DAY IS NOW

Donald Lindburg and Karen Baragona

THE WORLD is now dotted with the "flyspecks of wilderness" lamented by Aldo Leopold (1949: 182) as he viewed the U.S. landscape half a century ago. One must search at great length today to find a place that can be called pristine, for even the trails to the top of Everest are littered with human debris. The home of the giant panda is no exception. For starters, unlike national parks or wildlife reserves in the West, many such facilities in China are also the ancestral homes of a significant human population (Liu et al., chapter 14). Whereas in times past, human and animal may have cohabited relatively peacefully, the balance is now tipped significantly against the long-term survival of many wild forms. A familiar refrain, the need for human space, has pushed the giant panda into the remotest regions of the mountains of western China, but even there, it teeters on the brink of a final descent into oblivion.

Recent history speaks of species in a state of such dire threat that only last resort measures stand between extinction and survival. In the western United States, for example, the very last California condor had to be "rescued" from the wild in 1987, as the number of birds had spiraled downward year after year. Fortunately, the gamble paid off: after a brief sojourn in captivity, the numbers are now over two hundred, and this magnificent bird once again soars on the thermals in portions of its Pleistocene range (Snyder 2000). But, has the giant panda had its day, as some have wondered (see, e.g., Entwistle and Dunstone 2000)? We think not. Remarkably, a confluence of streams of concern into a tidal wave of effort often arises at fortuitous moments to foster new optimism on the survival chances of a species, regardless of where a nation stands on the scale of "resource development." This volume captures both the hopes and the concerns of just such a tide of sentiment on behalf of the giant panda. The book's theme—the intertwined strands of biological study on the one hand and conservation action on the other—derives from the latest findings from field and laboratory and shows their complementarity. Readers of this volume are more likely than not to be committed, even passionately so, to making the giant panda's future a secure one. The questions that the committed face are mainly those of process—the "how" rather than the "what" of its conservation. In looking into the decades ahead, we raise six questions that all who join ranks in this effort might ponder.

Can giant pandas and people share the same landscape?

What do we know of the human-animal connection in the protected areas of the panda's range? Can a relationship that will serve the needs of both people and animals be realized? It would appear that relatively few scientists are conducting such investigations in panda ranges today, but Liu et al. (chapter 14) are a notable exception. Among the more salient conclusions of these authors is that reduction of the human presence in panda habitat can be realized rather painlessly through demographic management; that is, by providing incentives for the younger generation to relocate. Discussions of habitat restoration (Zhu and Ouyang, panel report 11.1), reserve management (Yu and Deng, panel report 13.1), China's national planning for panda conservation (Lü and Liu, panel report 14.1), and approaches that take in the broadest possible landscape (Dinerstein et al., chapter 15) call for solutions to the conflicts arising from the competing needs of humans and wildlife. These authors also recognize the need for research in which the human population is the primary focus. Clearly, securing the panda's future will require in-depth studies of human attitudes toward wildlife, human dependence on forest products for income, how this income can be generated by other types of livelihoods that exert a lower impact on the environment, and yet other kinds of analyses by those trained in the social sciences (Western et al. 1994).

How can China's panda conservation community become self-sufficient?

The World Wildlife Fund (WWF) has been pursuing conservation strategies in China since the early 1980s, beginning with cosponsorship of the early research by Schaller et al. (1985) and early surveys of pandas and their habitats. Further evidence of this important collaboration was the publication of China's first National Plan for giant panda conservation (Chinese Ministry of Forestry and World Wildlife Fund 1989). One of the plan's recommendations was to expand the number of protected areas (reserves), and today these number about forty. However, only about 50% of the wild panda population's habitat falls within these reserves, underscoring the need for the broadened landscape approach advocated by Dinerstein et al. (chapter 15), and the assessment techniques described by Loucks and Wang (panel report 9.1).

The creation of new reserves greatly increases the need for infrastructure and reserve staff training. Some progress in meeting these needs is being realized through the United States–China international loan program (see Potter and Stansell, workshop report 17.1), but adequate protection will require the infusion of massive financial support by China itself. WWF has also initiated a broad-scale program for the training of reserve staff and for monitoring the wild population of giant pandas. In 1997, WWF developed an integrated conservation and development project (ICDP) with its Chinese partners in Pingwu County (Minshan Mountains, Sichuan Province), home to the largest concentration of giant pandas in all of China. According to Lü et al. (2000: 333), this project involves broad issues, "including land and resource tenure, decentralization of government power, people participation, public awareness, policy change toward sustainable forest resource use, sustainable land use and alternative livelihood development." In part 4 of this volume, we have given space to additional international collaborations; for example, the development of educational programs for schools (Bexell et al., brief report 17.1) and the first ever in-depth investigation of panda husbandry and health issues (Ellis et al., chapter 17) in an attempt to convey the broad scope of present-day efforts. These and other strategic alliances in the making auger well for future work on giant pandas, particularly at a time when China is increasingly receptive to input from and cooperation with nongovernmental organizations (NGO)—both international ones like WWF, Conservation International, and the Nature Conservancy, and a rapidly expanding legion of homegrown NGOs that now number over 230,000. In the end, it will be those who have direct and long-term involvement with pandas that will determine the fate of the animals, and there are many indications that a vibrant cadre of conservationists is emerging in China today.

How can further field work benefit the conservation effort?

This volume includes reports from investigations in various parts of the giant panda's range, beginning with the work of Hu in the late 1970s (Hu and Wei, chapter 9), and continuing up to the present, particularly in the Qinling Mountains of Shaanxi Province (Yong et al., chapter 10; Liu et al., chapter 11; Wei et al., chapter 13) and the Xiangling Mountains of Sichuan Province (Wei et al., chapter 12). As a result of this new information, where pandas find essential resources and how these are affected by variables of season or human impacts is now becoming evident. From chapter 5 (Pan et al.), chapter 6 (Long et al.), and brief report 4.1 (Wang et al.), we gain insights into a community of giant pandas that was followed with the aid of radiotelemetry for more than a decade. Yet that highly successful effort is now terminated, and today there is no site in China where similar in-depth studies are occurring. Until additional data become available, including samplings of pandas in other parts of their range, we cannot speak with confidence of the demographics, community structure, communication systems, reproduction, dietary requirements, and additional biological information that would ensure more enlightened conservation initiatives. A point-in-time census (Yu and Liu, panel report 15.1) is useful in telling us how many pandas there may be in the wild and where they are. But it does not address the all-important matter of population trends. The prospect of addressing this critical issue awaits information on birth rates and mortality and its causes. What is the disease status of the wild population, for example, especially given a significant presence of domestic animals in many parts of the panda's range? How important might these domestics be as reservoirs of diseases to which the panda is susceptible? Given the rugged terrain in which giant pandas live today, their low density, wariness, and cryptic behavior, we can appreciate the years of effort required to gain an understanding of community structure and mating systems. And, as ranges shrink and become increasingly fragmented, the employment of noninvasive techniques for tracking genetic diversity (Zhang et al., brief report 9.1) or community demographics (Durnin et al., brief report 10.1) appears to lack only the resources needed for their implementation. The connectivity of biological knowledge and effective conservation is clearly evident. It is only as specific answers from long-term and in-depth field investigations are realized that the best strategies for protecting endangered species can be developed.

How does the giant panda's unique constellation of traits affect its survival odds?

The point is often made that extinction is a natural process in the history of life. In that light, it is said—even by some of the most ardent conservationists—that we might consider cutting our losses and allow relicts of a past epoch or other highly disadvantaged species to go the way of the dodo (see, e.g., Snyder 2000). Is the giant panda one such species? Has it had its day? Only recently has an initial summary of the giant panda's life history been published (Zhu et al. 2001), and in-depth analyses of reproductive parameters (Snyder et al., chapter 8; Wang et al., brief report 4.1), how pandas communicate (Swaisgood et al., chapter 7; Hagey and MacDonald, brief report 7.1) or how pandas find and assimilate food (Long et al., chapter 6; Tarou et al., brief report 6.1) have barely begun. As Gittleman and Webster point out (chapter 16), certain of its traits may indeed reduce the odds for the panda's long-term survival. Yet these authors and other contributors to this volume make the equally salient point that the place of a species on the evolutionary tree should offer critical insights into the ways in which conservation strategies need to be structured. Questions of phylogeny continue to occupy those for whom such questions are an academic passion, yet at some point we must ask if a compelling consensus has been attained. To that end, we opted to give space to the latest views on panda phylogeny (Bininda-Emonds, chapter 1; Waits, brief report 1.1) and have added a new perspective from recent work on the evolution of bile salts (Hagey and MacDonald, chapter 2). An up-to-date summary of the all-important fossil record (Hunt, chapter 3) is seen to be consistent with the other lines of evidence in concluding that the giant panda's affinities lie with the ursids. In his broadly drawn report, Garshelis (chapter 4) places the giant panda's life history within the context of its presumptive nearest relatives, and in so doing, provides the raw material from which planning for the panda's future may yet alter the odds of its survival in a more favorable direction.

How might pandas in the wild be more amenable to study?

Central to studies focused on specific populations is the necessity of locating and identifying pandas on a regular basis. At the present, the only known way to do so is by radiotelemetry, and several chapters in this volume demonstrate how critical this approach is to the collection of salient data. Radiotelemetry has a long history of use in wildlife studies around the world (Pride and Swift 1992; Millspaugh and Marzluff 2001). It is a technology that is routinely used in studies of the bear family (e.g., Miller et al. 1996) and the record of usage reveals that bears have been safely collared over long periods of time (Garshelis and McLaughlin 1998), paying huge dividends in knowledge and in enlightened management. Yet telemetry is not widely used in wildlife studies in China today. We do not believe that conservation of the giant panda can be assured, nor can its scientific study go forward, without recourse to radiotelemetry. What role might other technologies have in conserving giant pandas? As yet, the extraction of DNA from feces as a way of identifying community members (Zhang et al., brief report 9.1) or the potential of hair collection for characterizing and monitoring the pandas of a region (Durnin et al., brief report 10.1) are non-invasive techniques whose applications are still to be successfully applied. Technology development using pandas loaned to U.S. zoos and its transfer to appropriate situations in both wild and captive sectors in China has increased, as a result of the research mandates imposed on U.S. recipients by the U.S. Fish and Wildlife Service (1998). Not to be overlooked in increasing amenability is the need for one or two centers where scholars from China and around the world can engage in fieldwork, thereby accumulating long-term scientific information that increases in value with every passing year.

Does the keeping of giant pandas in captivity enhance the well-being of the wild population?

Views on the significance of captive populations to the conservation effort are varied and controversial (DeBlieu 1991; Norton et al. 1995; Lindburg 1999; Balmford 2000). Here, we address this issue solely on the grounds of conservation benefits, and conclude that these populations play at best a supportive role. From a research standpoint, access to captives often generates knowledge that would be unavailable from pursuit of similar questions in the wild. A good example is the work on olfactory communication. Schaller's team (1985) was able to gain a sense of the importance of scent marking during the months of snow cover, when signs of marking were evident, but not at other times of the year. Little is known about the different postures used in marking except from study of captive subjects (Kleiman 1983), nor of the message content of the chemical signals themselves (Swaisgood et al., chapter 7). Hagey and MacDonald (brief report 7.1) used samples from both captive and wild subjects in analyzing the bioactive components of scent marks and urine, but applications of such knowledge as the differential use of these scents by males and females depends on observations of behavior in a captive context. Benign experimentation on captive pandas is further evident in the attempt by Tarou et al. (brief report 6.1) to discover how pandas locate food. A video camera inserted into the den of a wild dam has provided unusual insights into the early relationship of giant panda dams with their infants (Zhu et al. 2001), but routine use of this form of data collection is reserved for captive studies. This brief sampling points to the value of knowledge that may only be uncovered in a captive context, but what is its relevance to the study or management of wild populations? In their chapter on olfactory communication, Swaisgood et al. (chapter 7) suggest that appropriate and strategically placed scents might be used to encourage migrations, such as the passage of individuals through corridors connecting isolated fragments of forest. They further suggest that the introduction of a male into a new population for genetic reasons might benefit from the deposition of his scent in the new community before the animal itself is introduced, as a way for the community to gain familiarity and thus ameliorate aggressive responses. None of these notions have been tested, and we cite them only to indicate how studies of captive pandas might be structured to address in situ situations.

We conclude that the prospects for saving giant pandas are today unequaled. The spirit of *Panda 2000*, conveyed by the keynote addresses and a Memorandum of Consensus signed by representatives of China and the United States (see the appendixes), strongly indicates a flowering of collaborations that are advancing our understanding of the species and promoting strategies for its conservation. Yet these advances can only flourish to the extent that governments lend them support. It is noteworthy that an event beyond human control or predictive powers interceded in 1998 to alter forever the conservation landscape in China. Massive flooding in the Yangtze River basin brought a sobering message to the leaders of a country that had formerly promoted grand-scale logging, thereby denuding watersheds of the protective cover needed to prevent flooding. Much of this watershed falls within the ranges of the giant panda, and China's ban on logging has created new options for these animals.

For decades, the notion of restoring and reconnecting panda habitat was but a cherished pipe dream; today, because of China's forward-looking environmental protection policies, it may well be within reach. For decades, nature conservation was a luxury the Chinese economy could not support; today, with an annual growth rate of nearly 10%, China is not just able but also quite willing to invest billions of dollars to safeguard its natural heritage. These days in China, the only constant is change. What will the next decades bring? We hope the findings, conclusions, and recommendations presented in this volume have helped sketch out a road map along which China—with the support of the Chinese and international research and conservation communities—can plot a course to safety for one of its most prized national treasures, the giant panda.

REFERENCES

Balmford, A. 2000. Priorities for captive breeding—which mammals should board the ark? In *Priorities for the conservation of mammalian diversity*, edited by A. Entwistle and N. Dunstone, pp. 291–307. Cambridge: Cambridge University Press.

Chinese Minisitry of Forestry for China and World Wildlife Fund. 1989. *The national conservation management plan of the giant panda and its habitat.* Beijing: Chinese Ministry of Forestry and World Wildlife Fund.

DeBlieu, J. 1991. *Meant to be wild: The struggle to save endangered species through captive breeding.* Golden, Colo.: Fulcrum.

Entwistle, A., and N. Dunstone, eds. 2000. *Priorities for the conservation of mammalian diversity.* Cambridge: Cambridge University Press.

Garshelis, D. L., and C. R. McLaughlin. 1998. Review and evaluation of breakaway devices for bear radiocollars. *Ursus* 10:459–65.

Kleiman, D. 1983. Ethology and reproduction of captive giant pandas *(Ailuropoda melanoleuca)*. Z Tierpsychol 62:1–46.

Leopold, A. 1949. *A Sand County almanac and sketches here and there.* New York: Oxford University Press.

Lindburg, D. G. 1999. Zoos as arks: Issues in ex situ propagation of endangered wildlife. In *The new physical anthropology: Science, humanism, and critical reflection*, edited by S. C. Strum, D. G. Lindburg, and D. Hamburg, pp. 201–13. New York: Prentice-Hall.

Lü, Z., W. Pan, X. Zhu, D. Wang, and H. Wang. 2000. What has the giant panda taught us? In *Priorities for the conservation of mammalian diversity*, edited by A. Entwistle and N. Dunstone, pp. 325–34. Cambridge: Cambridge University Press.

Miller, S. D., G. C. White, R. A. Sellers, H. V. Reynolds, J. W. Schoen, K. Titus, V. G. Barnes Jr., R. B. Smith, R. R. Nelson, W. B. Ballard, and C. C. Schwartz. 1996. Brown and black bear density estimation in Alaska using radiotelemetry and replication mark-resight techniques. *Wildl Monogr* 133:1–55.

Millspaugh, J. J., and J. M. Marzluff, eds. 2001. *Radio tracking and animal populations*. San Diego: Academic.

Norton, B. G., M. Hutchins, E. F. Stevens, and T. L. Maple, eds. 1995. *Ethics on the ark: Zoos, animal welfare, and wildlife conservation*. Washington, D.C.: Smithsonian Institution Press.

Pride, G., and S. M. Swift. 1992. *Wildlife telemetry: Remote monitoring and tracking animals*. New York: Ellis Horwood.

Schaller, G. B., J. Hu, W. Pan, and J. Zhu. 1985. *The giant pandas of Wolong*. Chicago: University of Chicago Press.

Snyder, N. F. R. 2000. *The California condor: A saga of natural history and conservation*. San Diego: Academic.

U.S. Fish and Wildlife Service. 1998. Policy on giant panda permits. *Fed Reg* 63(166):45839–54.

Western, D., R. M. Wright, and S. C. Strum, eds. 1994. *Natural connections: Perspectives in community-based conservation*. Washington, D.C.: Island.

Zhu, X., D. Lindburg, W. Pan, K. Forney, and D. Wang. 2001. The reproductive strategy of giant pandas (*Ailuropoda melanoleuca*): Infant growth and development and mother-infant relationships. *J Zool Lond* 253:141–55.

Appendixes

APPENDIX A

Keynote Address

Fu Ma
Vice Director, State Forestry Administration, and
Vice President, China Wildlife Conservation Association

Honorable Guests, Distinguished Delegates, Ladies, Gentlemen and Friends, the 2000 International Symposium on the Giant Panda, organized under the joint auspices of the World Wildlife Fund—United States, the San Diego Zoological Society, the China Wildlife Conservation Association, and the Chinese Association of Zoological Gardens, is now in session. It is my pleasure to participate in this conference on the conservation management of giant pandas in this scenic city of San Diego. First of all, on behalf of China's State Forestry Administration, I congratulate the organizers and countries around the world for their concerns, enthusiasm, and support in giant panda conservation management.

The giant panda, a beautiful, affable, and docile creature, is an invaluable heritage, which Nature has left to us human beings. It has survived hardships throughout history, persevered since ancient times, and persisted on this planet to this date. Once upon a time, there was a widespread distribution of the giant panda. It has even withstood the critical test of glaciers during the Fourth Era. In the Modern Era, however, its living environment has been destroyed to a worse extent than ever before. As a result, the giant panda has been forced to retreat to an area high above sea level, in the Chinese provinces of Sichuan, Shaanxi, and Gansu. Today, around one thousand giant pandas live in remote mountain regions of nearly 20,000 km^2, spanning fifty-five counties in fifteen prefectures in three western provinces in China. The shrinking and deterioration of giant panda habitats and the serious threats to panda populations in the wild have become a world concern. Saving and protecting the giant panda has become a common bond for people in China and all over the world. As this conference is being held at the turn of the century and the start of a new millennium, its place in history is given an extra dimension. At this historical moment, we realize that it is precisely due to our concern and protection of giant pandas in the past that this endangered species can cross the threshold of a new millennium together with human beings. If we do not persist in our endeavor, we could, at any moment in time, lose a dear friend on this planet forever.

China is fortunate to be able to protect and breed the endangered giant panda. It is a challenge as well, a glorious and formidable mission, which history has bestowed upon the Chinese.

At the outset, the Chinese government has placed the giant panda in the "priority class of rare and endangered wildlife animal species under protection." Giant pandas were the focus of protection in four of the five earliest nature reserves in China. Since bamboo is the main staple of the giant panda, the bamboo flowering in the early 1980s created a universal wave across China, as well as in international communities, to save the giant panda.

In 1992, the Chinese government officially launched "China's Project for the Protection of Giant Panda and Their Habitats" to ensure protection at a fundamental level. This was the first time in the history of wildlife conservation in China that a species has been designated in a project. The project's objective was to protect existing giant panda habitats, gradually restore habitats that had been destroyed, and continuously expand the wild population. Since the project's implementation, the various levels of government in the three provinces of Sichuan, Shaanxi, and Gansu have suspended production in several forestry industries and relocated residents out of the area of giant panda distribution. A number of nature reserves with the giant panda as focus have been established. So far, thirty-three giant panda nature reserve locations have been established across China. Panda habitats under the umbrella of nature reserves conservation management have a total area of 17,000 km^2. As of the end of 1999, more than 120 cases of rescue of injured and sick giant pandas have occurred across China. Among them, over eighty cases have been successful, with more than seventy having been returned to the wild.

Encouraging results have also been achieved in giant panda research. Extensive research has been conducted in the bioecology, wild habitat, genetics, reproduction, rearing, and the prevention and treatment of diseases of giant pandas by Chinese and international scientists. Remarkable results in artificial reproduction, hand-rearing, and population relocation have been accomplished by the China Research Center for the Protection of the Giant Panda in Wolong, the Beijing Zoo, and the Giant Panda Breeding Center in Chengdu. According to statistics, over 220 panda cubs from more than 160 pregnancies have been produced in China since the 1960s. More than ninety of these cubs have survived to age 6 months. In 1999 and 2000 alone, twenty panda cubs from twelve pregnancies were born in the China Research Center for the Protection of the Giant Panda in Wolong, with nineteen still alive. A miraculous feat has been achieved, in which two out of a set of triplets, the smallest weighing only 53 g, survived.

In order to comprehensively and systematically understand and control the situations regarding giant panda habitats, the level of wildlife populations, and their composition, two giant panda surveys have been conducted—in 1974–1977 and in 1985–1988—with the help of the World Wildlife Fund. We are now in the process of conducting a third national survey on the giant panda. The purpose is to further investigate the number of wild giant pandas and current habitat conditions.

We will review what we have done in the realm of giant panda conservation over the years and identify our strengths as well as problems. The objectives are to accelerate the upgrading and setting up of an ongoing protection and rearing management structure, formulate policies for giant panda conservation management in the future, design plans for conservation and development, and establish scientifically accurate information. With the launching of the "Western Exploration Strategy" currently in China, the implementation of the National Natural Forests Conservation Project offers a unique opportunity to build the momentum and work together to raise giant panda conservation management to a new plateau.

The giant panda is a darling of the world. As an ambassador of the Chinese, it has visited several countries and played an important role in promoting friendships between the people of China and peoples around the world. Since the 1990s, China has engaged in long-term research partnerships with the United States, Germany, and Japan. These activities not only facilitate friendly relations among nations, but also produce fruitful results in research through the joint efforts of Chinese and international scientists.

In 1999, the hard work and seamless teamwork among Chinese and American experts in joint research conducted by China's Association for the Protection of Wildlife Animals and the San Diego Zoo have successfully produced a panda cub, Hua Mei. The name means Sino-American. At present, Hua Mei is enjoying a healthy and active life at the San Diego Zoo. It is a bundle of joy to the people visiting the area. Indeed, this is the primary objective in conducting joint research on giant pandas. I believe that encouraging results will also come out of the joint research between relevant Chinese institutes and, respectively, the Atlanta Zoo; the Washington, D.C., Zoo; and the Prince Zoo in Kobe, Japan.

Over the years, many international organizations have shared their concerns and given their support to the conservation of China's giant panda. The World Wildlife Fund has engaged in long-term research partnership projects with the nature reserves located in regions of giant panda distribution in Sichuan province in China and has played an active role in giant panda conservation. The Global Environment Fund's (GEF) support of the implementation of the Management Program of Nature Reserves in China in regions of giant panda distribution in Shaanxi province was critical for facilitating the relocation of the Shaanxi Evergreen Forestry Bureau and establishing a giant panda reserve. We are in the process of filing an application for a biological diversity conservation program in natural forests with the GEF, which will add mileage to enhancing conservation management of the giant panda reserves in Sichuan and the nature reserves in Baishuijiang in Gansu. All of these will be instrumental in promoting giant panda conservation. I hope that governments and relevant organizations around the world will continue to share their concerns and support for giant panda conservation management, as well as engage in extensive cooperation and exchanges.

The achievements in giant panda conservation, research, and management listed above form the foundation for success and growth of the giant panda conservation effort. They are also the fruits of joint efforts by the Chinese government and other countries and relevant organizations around the world. They are the harvests of your hard labor and aspirations. Once again, I express my gratitude to all of you.

Despite the successes in giant panda conservation, the fact remains that saving the giant panda will always be a long-term and formidable mission. The top priority given by the various levels of government in China, the sustained enthusiasm in every sector of the community, and the concerns and support at the international level will be critical factors in gradually pulling giant pandas out of dire straits and putting them on the road to recovery and expansion. Today, experts and conservation management professionals are gathered here to look back at the strategies, measures, and techniques in giant panda conservation in the past and look ahead to more effective means, tactics, and the application of new technologies to future giant panda conservation. This is both necessary and timely. I invite all of you—delegates, experts, and learned professionals in the field—to feel free to share your wisdom, ideas, and suggestions for the cause of giant panda conservation. As the ministry in charge of conservation of wildlife flora and fauna in China, the Chinese National Forestry Ministry will seriously consider the views of experts from China and around the world in formulating scientifically sound policies for the proper protection and management of the species.

Distinguished delegates, experts, and scholars, thanks to the support and patronage of the Zoological Society of San Diego and the World Wildlife Fund—United States, we are able to gather here in San Diego at the turn of the new century to discuss giant panda conservation and management. I believe that, as long as international communities share their concerns, and as long as our experts continue to work hard, we can expect to do an even better job in protecting the giant panda. Let us join hands, work together, and collaborate in propelling the giant panda conservation effort into the new millennium and achieve even greater success. Finally, best wishes to the conference.

APPENDIX B

Keynote Address

Marshall Jones
Deputy Director, U.S. Fish and Wildlife Service

On behalf of Interior Secretary Bruce Babitt, and U.S. Fish and Wildlife Service Director Jamie Rappaport Clark, it is a tremendous pleasure for me to be here with all of you today. And let me give special thanks to Dr. Kurt Benirschke and Douglas Myers and the entire staff of the Zoological Society of San Diego for the warm welcome and for hosting this meeting here, and to the World Wildlife Fund for supplying so much of the energy and the assistance needed to make this possible. A special thanks is due our partners from China—the China Wildlife Conservation Association and the Chinese Association of Zoological Gardens—for joining with us to help make this conference possible.

Finally and most especially, thanks to Deputy Director of the State Forestry Association Ma Fu, who emphasized the tremendous commitment that has been made by the government of China to this extraordinary program and his hope that we could all join together—nations, private organizations, scientists, and individuals—in this common goal.

This is truly an amazing gathering, seeing so many eminent scientists, conservationists, and leaders from all over the world assembled in one place. Think about it—a meeting like this would not have been possible even five years ago, to have so many leaders, from not only the United States but from China and from other countries together in this way.

The organizers of the conference, I believe, have done an amazing job in developing a broad and deep agenda—a chance to educate one another, a chance to share ideas, and a chance to develop new strategies about how to conserve the giant panda.

Before we break into the papers, panel sessions, and workshops, where all of those things will happen, however, I think it is important that we take a moment and reflect on why we are all here. The giant panda is truly a unique animal. It is so much more than an evolutionarily extraordinary bear that eats bamboo, although that by itself would certainly be enough. It is so much more than a precious symbol of China and its natural heritage, although that too would be more than enough. It is so much more than a logo or a symbol for the largest conservation association, although that too would be enough to make the panda something special. It is so much more than the most valuable animal that any zoological garden can put on display, although that too would be enough. It is so much more than just a symbol of all of our precious and disappearing endangered and threatened species, but that too would be enough. But the panda is all these things and so much more. To me, the giant panda is an incredible force of nature, an animal that has the power to not only bring all of us here together today, but an animal that has the power to mobilize the public, to mobilize everyone to participate in their own conservation interests. There is no other animal, no other natural phenomenon that I can think of that has such power to bring people together or to motivate people across all sectors of society, across all countries, across all kinds of organizations, to be united in one goal. And that goal is to do the things that Deputy Director Ma Fu talked about today: to save the giant panda, in it natural habitat for future generations to experience.

And that extraordinary power means that the panda also has the power to generate money, which can be used for its own salvation. That is why we have pandas here today in the United States at the San Diego Zoo, which has developed such a model program; at Zoo Atlanta, where they are well on the way to developing a similar program; and, we hope in the near future, at the National Zoo in Washington, D.C., following that same model. But it has not always been that way in the United States. Pandas have not always been in this country for the purpose of conservation.

When I first became involved in panda conservation activities in 1987, the giant panda already had a fifty-year history of display in the United States, from the first panda that came to the Brookfield Zoo in 1937, and then a succession of other zoos in the next few years. Those pandas were extraordinary creatures, and those pandas represented the state-of-the-art of what we knew and what we could do with pandas at that time in this country. But in a way, those pandas might also be thought of as a dead end. Pandas that came from China to this country—some of them lived short lives, some of them lived long lives—were not making the contribution to conservation in China that we would hope.

In 1972, after a long hiatus, we had a breakthrough again, and the government of China gave to the people of the United States a pair of pandas that went to the National Zoo, and became, in their own right, beloved symbols. But once again, those pandas were here for limited purposes. They were the best purposes that we all could think of at the time. Valuable research was done, and millions of Americans had the opportunity to experience them, but there still was a missing link. Harnessing those pandas to benefit the pandas in the wild suggests that there was so much more that could be done. The National Zoo started many programs, but there was still much more that remained to be done.

Soon after that gift, the world changed. The Convention on International Trade and Endangered Species (CITES) and then the U.S. Endangered Species Act, an act that was approved just a year after those pandas arrived in Washington D.C., brought to bear a host of new requirements and responsibilities, not just for governments but for all of us. These were to ensure that no endangered species involved in international trade was in that trade solely for commercial purposes, or solely for the benefit of those who were making the transaction and not the animals themselves. These acts were also meant to ensure that no endangered species in trade was taken in a way that was detrimental to the wild population. These were the goals of CITES, with the U.S. Endangered Species Act adding a further layer on top of that, to ensure that not only were transactions not commercial, and that they were not detrimental to the species, but in addition, that they actually would enhance the conservation of the species in the wild, in their native ecosystems.

Research, conservation, display, captive breeding—all of those things were possible, but if and only if they contributed to the conservation of the species in the wild. Unfortunately, we struggled here in the United States on how to implement the requirements of CITES and the Endangered Species Act. When I became involved with pandas in 1987, we had seen a succession of short-term loans to U.S. institutions, each one of them well meaning in intent, but each also representing potentially a lost opportunity in China and a lost opportunity in this country for pandas to really be harnessed to power their own conservation.

And we wound up in a situation where there was a frenzy, as U.S. zoos attempted to negotiate with China for additional loans. It was at that time that the American Zoo and Aquarium Association (AZA) stepped in to develop its own giant panda policy and plan. It was also the time when we in the U.S. government, realizing that we were not living up to our obligations, imposed our first moratorium on panda loans to the United States. We developed our first policy in an attempt to use the permission process as something more than just a piece of paper. In 1991, we developed, as our first policy, the requirement that we use the permission process as an organizing principle for conservation. That same year, I attended a meeting on loan policy that was organized by the National Zoo, counterparts from China, and others, at Front Royal, Virginia. And as many of you that were there told me in no uncertain terms, we had missed the boat again. We still were not there, because the policy that we had developed still failed to really harness the power of the panda for its own conservation.

Ultimately, we heard from the IUCN, from WWF, from AZA, and finally from the Secretary General of CITES, all of whom told us that we needed to look again at policy. We were informed that our standards were not stringent enough and would not do enough to ensure that pandas would survive. We needed a new paradigm, a new model, a new way to approach panda conservation and the contribution that we in the United States could make to it.

And so the U.S. Fish and Wildlife Service imposed another moratorium on panda loans in 1992. At that time, we began to think seriously about a new answer that had come to us from China; namely, long-term panda loans that could contribute, not just a small amount of revenue for a short period of time, but loans that could be put in the context of a comprehensive conservation program. From these deliberations came our new giant panda policy, published in 1998, based on the pilot program, which was developed right here at the San Diego Zoo. This policy was based on more than just having a permit, as you will hear tomorrow, when Ken Stansell and Craig Potter conduct one of the workshops about the regulatory program. The permit is only a vehicle for achieving conservation, and it has to be put in the larger framework of an entire program that will guarantee that giant pandas do not merely come to the United States to go onto public display, although that is a worthy activity. Nor are they to come here to be the subjects of interesting scientific research, or merely to be the mechanism for raising money to go back to China. Rather, loans of pandas to the United States must be all of these, and more, and must be woven together into a transparent program, so that all of us can understand the research goals and how that research contributes to panda conservation, and how the funds that are raised from loans will be used in China to contribute to the goals of China's own conservation plans. Most of those funds must go to in situ conservation, but we also recognize that there is a role for ex situ programs. And so we welcome the opportunities to work with all institutions in the United States and in China.

But we also recognized that the U.S. Government's giant panda policy is only one small piece of the entire puzzle. Deputy Director Ma Fu has detailed the extraordinary breakthroughs that have been made in China. We applaud the efforts of the State Forestry Administration and the government of China in establishing panda reserves, enhancing the breeding of pandas, developing a national plan, and banning logging activities in national reserves.

We in the U.S. Fish and Wildlife Service look back on two decades of close cooperation with the State Forestry Administration and its predecessors, and we look forward to many decades more of cooperation between our two agencies. We also applaud the efforts of the Chinese Association of Zoological Gardens and the Ministry of Construction, new partners for us, but ones with whom we hope to build strong relationships in the future, led by institution like Zoo Atlanta, which, through its relationship with the

Chinese Association of Zoological Gardens, is teaching us more about how we can all work together. We also applaud the efforts of the AZA in developing a giant panda conservation action plan and in establishing the Giant Panda Conservation Foundation, led by David Towne.

The pioneering work that is being carried out here at the San Diego Zoo, and coming soon from Zoo Atlanta (and, we hope, at the National Zoo and perhaps other institutions in the United States), is only a small piece of the entire effort. The work being done in China must always be our goal and focus. There is also valuable work that can be done in other countries—in Mexico, in Europe, and in Asia. We applaud the zoological institutions that are here today and look forward to opportunities for international conservation organizations to become involved in this effort.

So for all of these reasons, I believe we are really poised on the threshold of an extraordinary opportunity, a pivotal point in panda conservation. The question is, will we seize this opportunity? Will we harness the power of the panda to contribute to its own conservation? I know that we can, I am sure that we must, and I am hopeful that we will.

APPENDIX C

Memorandum of Consensus

International cooperation between China and the West is essential to ensuring the survival of the giant panda population and the preservation of its habitat. This conference symbolizes the growing spirit of collaboration between Chinese and Western panda conservation stakeholders, and the vital link between captive-based and field-based initiatives. Both wild and captive giant panda populations are important, but conservation of the wild population and its habitat is the foremost priority.

- Having convened in San Diego at *Panda 2000*, having heard the latest research on what is agreed to be a national treasure, the giant panda;
- Agreeing that conservation of giant panda habitat conservation is the overreaching theme of *Panda 2000* and that efforts to conserve the giant panda in the wild will serve to protect its habitat as well as the species that share it;
- Recognizing the contribution made through ex situ efforts;
- Understanding that the complexities and difficulties involved in giant panda conservation require the joint effort of many organizations and disciplines; and
- Commending the Chinese government for its efforts to conserve wild giant pandas and their habitats and for recent successes in breeding and management of the captive giant panda;

We, the cosponsors of *Panda 2000*, have agreed on the following points:

- We encourage increased international cooperation, the building of mutual trust and the provision, by China and other countries, of policy supports that assist giant panda conservation.
- We recommend that the Chinese government include in the tenth five-year national plan the protection of the environment and biodiversity in the three provinces where giant pandas are found: Sichuan, Gansu and Shaanxi.
- We strongly support the relevant authorities in China as they update China's national plan for conservation of the giant panda and call upon other conservation organizations to offer their knowledge and provide input into the development of this plan, including the prioritizing of critical areas and activities.

- We encourage the Chinese government and all other organizations involved in giant panda conservation to utilize information gained from the national panda survey in making giant panda conservation efforts more strategic.
- We recommend organizations continue and expand efforts to build panda conservation capacity in China, including training in reserve management and other essential skills and the establishment of centers in the reserve system which can undertake long-term study of the giant panda.
- We advise conservation organizations to continue their support and efforts (including scientific research, captive breeding, and population management) toward better understanding of captive and wild giant panda populations.

Li Yuming, Vice Secretary General
China Wildlife Conservation Association

Liu Shanghua, President
Chinese Association of Zoological Gardens

Ginette Hemley, Vice President for Species Conservation
World Wildlife Fund—United States

Kurt Benirschke, President
Zoological Society of San Diego

CONTRIBUTORS

AN LI, research associate at the University of Michigan, has research interests primarily in integrating spatially explicit dynamic modeling with socioeconomics and landscape ecology. He has published in *Ecological Modelling, Ecological Economics,* and *Science.*

KAREN BARAGONA, deputy director of species conservation, World Wildlife Fund—United States, manages the Fund's conservation programs for giant pandas, whales, dolphins, and porpoises. She chairs a steering group for all Fund offices involved in giant panda conservation and oversees its mountain gorilla program.

SCOTT BEARER, Ph.D. candidate in Fisheries and Wildlife at Michigan State University, conducts research on the influence of forestry practices on giant panda habitat use, particularly at Wolong Reserve in China.

SARAH M. BEXELL, conservation programs manager at Zoo Atlanta, manages the zoo's China Conservation Education Program. She is a member of the American Zoo and Aquarium Association's Conservation Education Committee and education advisor to its Giant Panda Species Survival Plan.

OLAF R. P. BININDA-EMONDS is head of the Bioinformatics Research Group at the Technical University of Munich. Using supertrees, he has constructed the first complete species-level phylogeny of the mammalian order Carnivora and is part of a group attempting to build a species-level tree of all extant mammals.

MOLLIE A. BLOOMSMITH, director of research at Zoo Atlanta, received her training in experimental psychology from the Georgia Institute of Technology, Atlanta. Her research interests are in applying knowledge of animals' behavior to improving their welfare and care in captivity.

M. C. BRONSVELD, assistant professor at ITC, Enschede, The Netherlands, has experience in the spatial analysis of environmental factors in relation to their use and conservation.

NANCY M. CZEKALA, senior research endocrinologist at the San Diego Zoo, studies the reproduction of endangered species, focusing on gorillas, giant pandas, and rhinoceroses.

DANG GAODI has been working in Foping Nature Reserve since 1985. His working background is in plant ecology and he is very familiar with the plant species growing in the nature reserves.

DENG XIANGSUI, head of wildlife protection in the Sichuan Forestry Department and board chairman of the Sichuan Wild Animal Protection Council, was trained at Sichuan Forestry College.

ERIC DINERSTEIN, chief scientist, World Wildlife Fund—United States, has undertaken worldwide studies of global biodiversity priorities and also large mammal conservation throughout Asia. He is the author of the recently published *The Return of the Unicorns: The Natural History and Conservation of the Great One-Horned Rhinoceros*.

MATTHEW E. DURNIN, Ph.D. candidate, University of California at Berkeley, has been studying wild giant pandas in Wolong Nature Reserve, Sichuan, China, since 1998. He is a research fellow at the Zoological Society of San Diego and founder of the nonprofit organization Collaborative Research for Endangered Wildlife (CREW).

MARK EDWARDS, nutritionist for the Zoological Society of San Diego, is the nutrition advisor on giant and red pandas for the American Zoo and Aquarium Association. He is also a member of its Nutrition Advisory Group and of the American Association of Zoo Veterinarians.

SUSIE ELLIS, vice president for Indonesia and Philippines programs at Conservation International, formerly directed the IUCN/SSC Conservation Breeding Specialist Group's work with giant pandas in China from 1996 to 2001.

FEI LI SONG, vice president of the Chengdu Zoo, is a member of the China Association of Zoological Gardens and its committee on South China Tiger Conservation and Coordination.

FENG ZUOJIAN, senior mammalogist at the Institute of Zoology, Chinese Academy of Science, is the author of *Mammals in Tibet* and other books. He is secretary general of the Chinese Zoological Society and editor of *Acta Zootaxonomica Sinica*.

DEBRA L. FORTHMAN has held positions in research and conservation at Arizona State University, the Los Angeles Zoo, and Zoo Atlanta and has authored over thirty publications on a range of mammals, including giant pandas.

DAVID L. GARSHELIS, adjunct associate professor at the University of Minnesota, has been a bear researcher with the Minnesota Department of Natural Resources since 1983. He also advises graduate students at the University of Minnesota, who study bears around the world. He has been involved in field studies of five of the eight species of bears.

JOHN L. GITTLEMAN, professor of biology at the University of Virginia, is editor of *Carnivore Behavior, Ecology, and Evolution* (volumes 1 and 2) and co-editor of *Carnivore Conservation*. He is a scientific fellow of the Zoological Society of London.

LEE R. HAGEY, analytical chemist with the Zoological Society of San Diego, is the author of scientific publications in the area of comparative biochemistry.

HE GUANGMING, Ph.D. candidate at Michigan State University, works on the integration of advanced computer technology and system modeling methodology into natural resources management.

JOGAYLE HOWARD, theriogenologist at the National Zoo, Washington, D.C., specializes in the use of reproductive technologies, including artificial insemination and sperm cryopreservation, for conservation and species management.

HU JINCHU, professor of ecology at Sichuan Normal College, has studied giant pandas across their range for over twenty years. He is the author or co-author of many journal articles and several books on giant pandas and other Chinese mammals.

HU YAN, conservation educator at the Chengdu Zoo, has translated many conservation education documents, including *The Asian*

Turtle Crisis and *Awakening the Mass Media to Build a Sustainable Future.*

HUANG JIN YAN, field biologist at the Wolong Nature Reserve, holds a masters degree in forestry from South Central Forestry University, Sichuan, and is a member of the China Wildlife Association. His research is focused on the vegetation of Wolong Nature Reserve.

HUANG XIANGMING, curator of animal management at the Chendgu Research Base of Giant Panda Breeding, works in the areas of giant panda husbandry, conservation, and reproduction. He has published in *Acta Theriologica Sinica* and the *Sichuan Journal of Zoology.*

ROBERT M. HUNT, JR., professor of geosciences at the University of Nebraska, researches the paleobiology and evolution of mammalian carnivores and is a research associate in vertebrate paleontology at the American Museum of Natural History, New York.

DONALD JANSSEN, director of veterinary services for the San Diego Wild Animal Park, is a past president of the American Association of Zoo Veterinarians, the Association of Avian Veterinarians, the American College of Zoological Medicine, and the American Association of Wildlife Veterinarians. He also chaired the Animal Health Committee of the American Zoo and Aquarium Association.

DEVRA KLEIMAN, adjunct professor at the University of Maryland and research associate at the Smithsonian National Zoological Park, has received awards for scientific and conservation achievements from the Society for Conservation Biology and the American Zoo Association. She is primate section chair for the IUCN/SSC Reintroduction Specialist Group.

MABEL LAM, director of M.L. Associates, specialize in wildlife business management and has over fifteen years of experience in dealing with the Chinese government and NGO agencies on conservation programs. She is an advisor to the Giant Panda Conservation Foundation, United States, and to North American zoos on giant panda conservation.

DWIGHT P. LAWSON, adjunct professor of biology at Georgia State University, is general curator at Zoo Atlanta.

LI MING, associate professor of molecular ecology at the Institute of Zoology, Chinese Academy of Science, has studied the conservation genetics of endangered mammals in China for over ten years, and has published numerous papers on the topic.

DONALD LINDBURG is head of the Office of Giant Panda Conservation, Zoological Society of San Diego. He is species coordinator and studbook keeper for giant pandas in North American zoos and chairs the Science and Technology Committee of the Giant Panda Conservation Foundation. He conducts research on giant panda behavior and reproduction.

MARC LINDERMAN, National Science Foundation postdoctoral fellow at Michigan State University, has co-authored several journal articles, including "Ecological degradation in protected areas: The case of Wolong Nature Reserve for giant pandas," published in *Science*.

LIU JIANGUO (JACK), professor of fisheries and wildlife at Michigan State University, conducts research on conservation and landscape ecology, human-environment interactions, systems modeling and simulation, and the ecological impacts of human population and activity.

LIU SHAOYING, head of the Forest Protection Institute, Sichuan Academy of Forestry, and adjunct professor at Sichuan University, performs research on wildlife protection and the systematics of rodents, bats, and insectivores.

LIU XUEHUA, associate professor at Tsinghua University, is the author of *Mapping and Modeling the Habitat of Giant Pandas in Foping Nature Reserve, China,* and has published scientific papers in both English and Chinese.

LIU YONGFAN, deputy director-general and senior engineer for the Department of Wildlife and Plants Conservation, Chinese State Forestry Administration, holds a degree from Northeast Forestry University, Harbin, China, and is engaged in forest and wildlife conservation. The third national survey of giant pandas was launched under his guidance.

LONG YU, lecturer at Peking University, has conducted research on the feeding and nutrition of giant pandas in the Qinling Mountains of China. She is a co-author of *A Chance for Lasting Survival*.

COLBY LOUCKS, senior conservation specialist, World Wildlife Fund—United States, has published articles on giant panda conservation and has helped coordinate a number of priority-setting exercises throughout Asia, including several in the forests that support pandas.

LÜ ZHI, associate professor at Peking University, has studied giant pandas for over fifteen years. As the World Wildlife Fund—China's species and protected areas manager, she developed and implemented all projects related to panda conservation. She is a co-author of *A Chance for Lasting Survival*, a conservation educator at the Chengdu Research Base of Giant Panda Breeding, and a member of the Chengdu Association of Science and Technology. She has published several articles on educational issues in China.

LUO LAN, vice-curator of education at the Chengdu Research Base of Giant Panda Breeding, has assisted in the development of a conservation education program for visitors to the base and for use in local schools. Her programs are also promoted via the Internet.

EDITH A. MACDONALD, research fellow at the San Diego Zoo, performs research on several aspects of giant pandas, including behavior, scent analysis, and endocrinology.

SUE MAINKA, head of the Species Programme for IUCN—World Conservation Union, has worked at Wolong Reserve in China as a veterinarian and conservationist. She has also worked on captive management of wildlife in several countries.

TERRY L. MAPLE, professor of psychology and Elizabeth Smithgall Watts Professor of Conservation and Behavior at the Georgia Institute of Technology, is president and chief executive officer of Zoo Atlanta. He is an elected fellow of the American Psychological Association and a past president of the American Zoo and Aquarium Association.

RITA MCMANAMON, senior veterinarian and director of China conservation programs at Zoo Atlanta, has served on the Ethics Board of the American Zoo Association, and as an associate editor for the *Journal of Zoo and Wildlife Medicine*. Her publications include journal articles and book chapters in the fields of animal health, animal welfare, and behavioral enrichment.

ERIC MILLER, director of animal health and conservation at the Saint Louis Zoo and adjunct assistant professor at the University of Missouri College of Veterinary Medicine, is the author of numerous scientific articles and co-editor of the fourth and fifth editions of *Zoo and Wild Animal Medicine*.

OUYANG ZHIYUN, professor and director of the Key Lab of Systems Ecology, Research Center for Eco-Environmental Sciences, Chinese Academy of Sciences, has written for and edited *Ecosystem Services: Assessment and Evaluation*, *Regional and Ecological Planning*, and *Ecological Integration for Sustainable Development*.

PAN WENSHI, professor of zoology at Peking University and dean of the Giant Panda and Wildlife Conservation and Research Center, is a research associate at the Smithsonian Institution and the Zoological Society of San Diego and a trustee of the China Zoological Society and the China Ecology Society. He is co-author of *The Giant Pandas of Wolong*, *The Giant Panda's Natural Refuge in the Qinling Mountains*, and *A Chance for Lasting Survival*.

J. CRAIG POTTER, a former high-level government official at the U.S. Department of Interior and the U.S. Environmental Protection Agency, practices environmental law in Washington, D.C. He has been a leading proponent of the conservation of endangered species and was significantly involved in the rewriting and reauthorization of the Endangered Species Act of 1982.

HERBERT H. PRINS, professor of tropical nature conservation and vertebrate ecology at Wageningen University, Wageningen, The Netherlands, is a member of the executive board of the Netherlands Nature Conservancy Natuurmonumenten. He was appointed officer in the Order of Oranje Nassau (2002) by royal decree of Her Majesty the Queen Beatrix and as officer in the Order of the Golden Ark (2001), bestowed by His Royal Highness Prince Bernhardhas.

OLIVER A. RYDER, Kleberg Chair in Genetics at the Zoological Society of San Diego, participates in research pertinent to planning and management for selected mammalian, reptilian, and avian species of international conservation concern. His publications on conservation genetics include accounts of collaborations with scientists at the Kunming Institute of Zoology, China.

ANDREW K. SKIDMORE is professor of spatial ecology at the International Institute of Aerospace Survey, The Netherlands, and professor of land cover mapping and monitoring at Wageningen University. Vegetation mapping and monitoring have been the ongoing theme of his research. Recent research has focused on wildlife habitat assessment in east Africa, hyperspectral remote sensing, artificial intelligence techniques for handling geoinformation, and accuracy assessment.

REBECCA J. SNYDER, curator of giant panda research and management at Zoo Atlanta, is also a part-time instructor in the psychology department at Georgia State University.

KENNETH STANSELL, assistant director for international affairs, U.S. Fish and Wildlife Service, oversees the divisions of International Conservation, Scientific Authority, and Management Authority. He works multilaterally with many partners and nations in the implementation of international treaties, conventions, and on-the-ground projects for conservation of foreign species and their habitats. His training is as a research biologist.

STEPHEN STRINGHAM, adjunct professor at the University of Alaska's MatSu campus, is the author of numerous technical papers on the population ecology and behavior of bears, and of the popular book *Beauty within the Beast: Kinship with Bears in the Alaska Wilderness*.

RONALD R. SWAISGOOD, field biologist at the Zoological Society of San Diego, has worked extensively with giant pandas in China and rhinoceroses in Africa. His publications are concerned with the application of behavioral research to conservation efforts in captivity and in the wild.

LORAINE R. TAROU is research assistant for giant panda behavior studies at the Smithsonian National Zoological Park. She is also a graduate student in the School of Psychology at the Georgia Institute of Technology and a member of TECHlab at Zoo Atlanta.

TIAN YUZHONG, president of the Chengdu Zoo, is a member of the Committee of the Breeding Technique of Giant Pandas of China.

LISETTE P. WAITS, assistant professor of wildlife resources at the University of Idaho, has carried out extensive research on the evolution, systematics, and population genetics of bears. She has provided important data for the endangered species recovery programs of the grizzly bear and red wolf.

WANG DAJUN, Ph.D. candidate at Peking University, conducted research on giant pandas in China's Qinling Mountains from 1993 to 1999. He is a co-author of *A Chance for Lasting Survival*.

WANG HAO, lecturer at Peking University, has studied pandas for eight years. He is a co-author of *A Chance for Lasting Survival*.

WANG TIEJUN, M.Sc. degree candidate at the International Institute for Aerospace Survey and Earth Science, Enschede, The Netherlands, has conducted research on giant pandas at Foping Nature Reserve.

WANG ZUWANG, professor of physiological ecology at the Institute of Zoology, Chinese Academy of Sciences, has written several books on small mammals, including *Ecology and Management of Rodent Pests in Agriculture*. He is president of the Mammalogical Society of China and editor of *Acta Zoological Sinica*.

ANDREA J. WEBSTER, research fellow at the University of Reading, conducts research on evolutionary biology and phylogenetics, and has co-authored chapters in the books *Comparative Primate Socioecology and Morphology* and *Shape and Phylogeny*.

WEI FUWEN, professor of ecology at the Institute of Zoology, Chinese Academy of Sciences, has been engaged in giant panda and red panda research for twenty years; his works have been widely published in both English and Chinese. He was the National Science Fund's Distinguished Young Scholar in 2001, and he serves as secretary general of the Mammalogical Society of China.

ANGELA M. WHITE, field consultant for the Zoological Society of San Diego, conducted research on giant panda chemical communication at Wolong Breeding Center, based on which she wrote her masters thesis. Her current research focuses on reproduction of the white rhinoceros at a site in South Africa.

DAVID WILDT, senior scientist and head of the Department of Reproductive Sciences at the Smithsonian National Zoological Park and its Conservation and Research Center, works primarily on the reproductive biology of carnivores. He has recently co-edited a textbook entitled *Reproductive Sciences and Integrated Conservation*.

YANG GUANG, professor of zoology and director of the Institute of Genetic Resources at Nanjing Normal University, is a member of the editorial committee of *Acta Theriologica Sinica* and of the expert committee of the Forestry Ministry of China for the Third Integrated Survey of Giant Pandas. He has published nearly sixty papers on the ecology, ecological genetics, and conservation of endangered animals.

YONG YANGE, director of the Giant Panda Research Center at Foping Nature Reserve, has been involved in panda research for over twenty years and has published several papers on the panda in Chinese journals.

YU CHANGQING, species program director and coordinator of giant panda programs for the World Wildlife Fund's China office, has been a consultant for the Global Environmental Facility Protected Areas Program in China. He has conducted research on endangered species conservation in China for the past fifteen years.

ZHANG ANJU, professor and vice president of the Chinese Association of Zoological Gardens, and vice president and secretary general of the Chengdu Giant Panda Breeding Research Foundation, is a founder of and senior consultant to the Chengdu Research Base of Giant Panda Breeding and a member of the Chinese National People's Congress. He has written and edited numerous books and academic papers on the topics of giant panda reproduction and veterinary medicine.

ZHANG GUIQUAN, assistant director and senior engineer at the China Conservation and Research Center for the Giant Panda at Wolong, holds a bachelors degree in wildlife protection from Northeast Forestry University, Harbin, China.

ZHANG HEMIN, director of the Wolong Nature Reserve and China Conservation and Research

Center for the Giant Panda, holds a masters degree in wildlife management from Idaho State University. He is senior author of a recent book (in Chinese) entitled *Reproductive Studies of the Giant Panda*.

ZHANG JINGUO, deputy director of the Beijing Zoo and senior veterinarian, also holds an adjunct professorship at Capital Normal University and is the associate editor of the recent book *Color Atlas of Hemocytology of Wildlife*.

ZHANG YAPING, vice director of the Kunming Institute of Zoology, conducts research on molecular genetics and is a professor at Yunnan University. He received one of China's Biodiversity Leadership Awards in 2002.

ZHANG YINGYI, program officer for China for Fauna and Flora International, holds a Ph.D. in conservation biology from Peking University. Her research has focused on the indigenous primate fauna of China. She has also conducted research on giant pandas in the Qinling Mountains of Shaanxi Province.

ZHANG YUNWU, laboratory research coordinator in the Genetics Division at the San Diego Zoo, holds a doctorate in genetics.

ZHANG ZHIHE, senior veterinarian and deputy director of the Chengdu Research Base of Giant Panda Breeding, is also deputy secretary general of the Chengdu Giant Panda Breeding and Research Foundation, and has written many articles on the veterinary medicine, reproduction, and genetics of the giant panda.

ZHONG WEI, Ph.D. candidate in bioinformatics at Georgia State University, Atlanta, was coordinator of research at the Chengdu Zoo for eight years.

ZHOU XIAOPING, director of research at the China Research and Conservation Center for the Giant Panda, holds a bachelors degree in zoology from Sichuan University. He conducted fieldwork on the giant panda and several species of pheasant in Wolong Nature Reserve. His current research interests include reproductive behavior and behavioral development in captivity.

ZHU CHUNQUAN, forest program officer for the World Wildlife Fund—China and board member of the Chinese Society of Forestry, conducts research on forest ecology and sustainable forest management. He is one of the key writers and an editor of *Forestry Action Plan for Chinese Agenda 21 and Modern Forestry of China*.

ZHU XIAOJIAN, lecturer at Peking University, has researched and written on the mating system and mother-cub relationships of giant pandas in the Qinling Mountains of China.

INDEX

Page numbers followed by *f* and *t* indicate figures and tables, respectively.

Abies fabri, consumption of, 91
Acinonyx jubatus (cheetah), chemical communication in, 107
"Action Plan for Environmental Publicity and Education" (1997), 265
Age
  chemical communication of, 107, 111–13
  at first birthing. *See* Reproductive age
  and reproductive rates, 63
  in captivity, 255
Age structure, of Qinling Mountain population, 83, 84t
AI. *See* Artificial insemination; Artificial intelligence
Ailurarctos lufengensis
  phylogeny of, 51f
  teeth of, 48–50, 49f, 50f
Ailuridae, 11
Ailurinae, 189
Ailuropoda spp., creation of genus, 45
Ailuropoda microta, fossils of, 48
Ailuropodidae, 11
Ailuropodinae, 189
Ailurus fulgens (red panda)
  behavior of, 12t, 13
  bile of, 39, 44
  diet of, 90, 189–98
  geographic range of, 189
  phylogeny of, 11–29, 29f
    evidence of, 11–14, 12t
    literature on, 14–29, 15t–19t, 20t
    uncertainty in, 25–27
  resource competition with, 7
  sympatry with *Ailuropoda melanoleuca*, 7, 189–98
American black bear. *See Ursus americanus*
American Zoo and Aquarium Association (AZA), 3
Amphicynodontinae, radiations in, 47
Andean bear. *See Tremarctos ornatus* (spectacled bear)
Angelica spp. (wild parsnip), 91
Animal foods, consumption of, 91–93, 92t, 141
Anogenital gland secretions. *See also* Chemical communication
  chemical composition of, 106, 121, 122f
  persistence of, 115
Arctodus spp., 46
Artificial insemination (AI)
  first successful, 250
  paternity determination with, 254
  role of, 261
Artificial intelligence (AI), in habitat assessment, 177
Asia, ursid species in, 54
Asian golden cat. *See Felis temmincki*
Asiatic black bear. *See Ursus thibetanus*
Audio recorder systems, 81
AZA. *See* American Zoo and Aquarium Association

Backbone constraint trees, 21, 21f
Bamboo
  density of, 203t, 204
    equilibrium, 205–7, 206f, 207f
    maximum, 204
    rate of increase in, 203–4, 204t
  die-off of (1970s), 2
  distribution of, mapping of, 150, 222–23
  human harvest of, 201–8
    current rate of, 205
    sustainable rate of, 206–8, 206f, 207f
  modeling use of, 201–8
  nutrients in
    and captive populations, 256
    digestibility of, 98–99, 99t, 141–44
Bamboo consumption, 90–100
  by *Ailurus fulgens*, 90, 189–98

Bamboo consumption *(continued)*
  behavioral adaptations to, 95–98, 96t, 97t
  and bile synthesis, 44
  in captivity, 256
  efficiency of, 97–98
  leaves in, 95, 144, 192, 198
  as major food source, 55, 90
  methodology for studying, 190–91
  nutrients from, 98–99, 99t, 141–44, 256
  predator-prey model of, 202–3
  rarity of, 90
  rates of, 204–5
    sustainable, 206–8, 206f, 207f
  regional differences in, 90–91, 140–44, 142t–143t
  seasonality of, 93–95, 94t, 141, 192
  spatial memory in, 101–5
  species of bamboo in, 55, 140, 142t–143t
  stems in, 95, 144, 192, 194t, 195t, 198
  success of, 99–100
Bamboo rat. *See Rhizomys sinense*
Banqiao Village, 83
Barbed-wire hair snares, 170, 171–75, 173t, 174t
Bartlett-Box test, 191, 194–95
Bashania spp., consumption of, 91
  leaves versus stems in, 95
  nutrients from, digestibility of, 98–99, 99t
  regional differences in, 141–44
  seasonality of, 93–95, 94t
Bashania faberi, nutrients in, 141, 144
Bashania fargesii, nutrients in, 144
Bashania spanostachya
  abundance of, 102
  consumption of, 141, 144, 192, 193t, 194t, 195t
Bear. *See* Ursidae
Behavior
  adaptations to bamboo consumption in, 95–98, 96t, 97t
  in captivity, and reproductive fitness, 255
  during estrus, 126
  human, in new conservation paradigm, 218, 219
  maternal, 60–61, 85

noninvasive monitoring of, 170–75
in phylogenetics, 12t, 13
literature on, 21–22
in Qinling Mountain population, 85
reproductive, 125–31
  correlation with hormones, 128, 129–31
  female, 127–29
  male, 129–30
  methodology for studying, 126–27
Beijing Zoo
  biomedical survey at, 252, 253t
  breeding at
    first successful, 2, 250
    genetic overrepresentation in, 254
    hand-rearing in, 251
    testicular hypoplasia or atrophy at, 257
"Bet-hedging," 63
Bile salts, 38–44
  bacterial degradation of, in intestines, 41
  collection of samples, 38–39
  definition of, 38
  EMI-MS profile of, 39, 40–41, 42f
  HPLC profile of, 41, 43f
  phylogenetics and, 38, 41–44, 43f
  treatment of samples, 38–39
  of ursids versus other carnivores, 39–41, 40t, 41t
Biomedical survey, of captive populations, 250–62
  benefits of results of, 259–60, 261–62
  methodology of, 252–53
  need for, 250–51
  planning for, 251–52
  and reproductive success
    limitations on, 253–59
    recent changes in, 259–60
Birthing
  age of first. *See* Reproductive age
  in dens, 60–61, 145
  seasonality of, 58–60, 59f, 60f, 64
  in Qinling Mountains, 75, 76, 76t
"Bite size technique," 83–84, 152–53
Black bear. *See Ursus americanus* (American black bear); *Ursus thibetanus* (Asiatic black bear)

Bleating. *See* Vocalizations
Blood samples, 81, 252–53
Body plan, and phylogenetics, 13–14
Body size
  of cubs, 60, 85
  and migration, 56
  as refuge, 57
  variation among ursids in, 53
Bones, consumption of, 91–93, 92t
Breeding. *See* Captive breeding
Bremer decay index, 20
Bronx Zoo, 2
Brown bear. *See Ursus arctos*

California condor, 271
Camera systems, 81, 170–71, 172t, 173–75, 173t
Canis familiaris (dog), phylogeny of, 36, 37f
Capacity building, 261
Captive breeding
  artificial insemination in, 250, 254, 261
  biomedical survey and, 253–60
  birthing in, seasonality of, 59–60, 60f
  chemical communication in, 107, 110–11, 117
  as conservation strategy, 260–61
  cub survival rates in, 250–51, 259–60, 259t, 260t
  difficulty of, 2, 250
  first successful, 2, 250
  genetic overrepresentation in, 254
  litter size in, 61, 65
  mating in, seasonality of, 59, 75–76, 76f
  paternity in, establishment of, 88, 254–55
  sex ratios in, 63
Captive Management Plan, 251, 252
Captive Management Planning Workshop (1996), 251
Captivity, *Ailuropoda melanoleuca* in
  biomedical survey of, 250–62
  chemical communication in, 116–17, 274
  conservation benefits of, 260–61, 274–75
  diet of, 91, 255–56
  first in West, 2
  genetic diversity of, 88

chemical communication and, 117
genetic overrepresentation and, 254
international exchange of, 3, 268–69
number of, 250
wild-born versus captive-born, 260, 260f
reintroduction of, 246–49
as self-sustaining population, 247–48, 251, 260
semen of, 258–59, 258t
stunted growth syndrome in, 256–57, 257f
testicular hypoplasia or atrophy in, 257
Carnassial teeth, 46
Carnivora
versus carnivores, 46
supertree of, 237, 238f
teeth of, 46
Carnivores
bile salts of, 38–44, 40t
evolution of herbivory in, 42–43, 90
evolutionary rates in, 242, 243t
extinction risk in, 239–42, 239t, 240f, 241t
phylogeny of, 14, 237, 238f
Carrion, consumption of, 91, 92t, 141
Cattle, spatial memory in, 105
Caves, dens in, 60, 145
CAZG. *See* Chinese Association of Zoological Gardens
CBSG. *See* Conservation Breeding Specialist Group
Cellulose, in bamboo, utilization of, 98–99, 99t
Changqing Nature Reserve (CNR), migration in, 159
Character weighting, in phylogenetics, 14
Cheek teeth, 46, 48, 49, 50f
Cheetah. *See* Acinonyx jubatus
Chemical communication, 106–18
in captivity, 116–17, 274
deposition of, 106
age assessment of, 107, 115–16
chemical composition of, 106, 121–24, 122f
height of, 112–13
persistence of, 115
postures for, 112–13

distribution of scent on pelage, 123–24, 123f
functions of, 107
information conveyed in, 107, 109–15
age and competitive ability, 111–13
individual identity, 114–15
mother-infant identity, 113–14
sex and reproductive condition, 109–11
in mating, 107, 109–11, 114–15, 129–30
methodology for studying, 108, 274–75
noninvasive techniques in, 170–71, 172t, 173t
prevalence among ursids, 58
responses to, 108–10, 112, 113, 114
at scent stations, 106–7, 171
urine in, 106, 110, 113, 115, 124
Chen, X., 265
Chen, Y., 210
Chengdu Research Base of Giant Panda Breeding, conservation education at, 264–66
Chengdu Zoo
biomedical survey at, 252, 253t
breeding at, 251
conservation education at, 264–66
stunted growth syndrome at, 256–57, 257f
Children, conservation education for, 264–66
Chimonobambus spp., consumption of, 141
China
conservation education initiatives in, 264–66
first captive pandas offered by, 2
government conservation programs in, 4, 133–34, 275
national conservation plan of, 226–27, 228, 234, 272
national surveys by, 134, 150–54, 234–35
and reintroduction efforts, 248–49
ursid species in, 54
U.S. collaboration with, xiii–xiv
China Wildlife Conservation and Research Center for Giant Pandas, 250
biomedical survey at, 253t

Chinese Association of Zoological Gardens (CAZG), 250, 251, 252
Chongqing Zoo, biomedical survey at, 252, 253t
CITES. *See* Convention on International Trade and Endangered Species of Wild Fauna and Flora
Clouded leopard. *See* Neofelis nebulosa
CNR. *See* Changqing Nature Reserve
Coloration, and extinction risk, 241
Communication, chemical. *See* Chemical communication
Community support. *See also* Humans
for conservation, 212
for education initiatives, 264–66
for reintroduction, 248
Competition, interspecific, resource partitioning and, 197–98
Competitive ability, chemical communication of, 111–13, 114
Condor, California, 271
Conservation Breeding Specialist Group (CBSG), xiii, 251, 252, 261–62
Conservation strategies, 213–69
biological studies in, 217, 229
captive populations in, 260–61, 274–75
chemical communication in, 117–18
community-based, 212
education initiatives in, 264–66
evaluation of, biological framework for, 228–32
genetic diversity in, 88
habitat in, 146, 217–24
international coordination and cooperation in, 268–69
life histories in, variation of, 53, 65–66
new paradigm for, 217–24
conceptual framework of, 218–19, 219f
methodology of, 219–21
phylogenetics in, 7, 27–29
for Qinling Mountain populations, 82
reintroduction as, 246–49
scales of, multiple, 229–32
timeline for, 229

INDEX 299

Convention on International Trade and Endangered Species of Wild Fauna and Flora (CITES), 249, 268
Convergence, and phylogenetics, 13–14, 27
Corona photographs, 220
Cubs. *See also* Litters
　behavioral development of, 85
　chemical communication in composition of, 121
　　mother-infant recognition through, 113–14
　maternal transport of, 61
　in maternity dens, 60–61, 85
　number per litter, 61–63, 62t
　in Qinling Mountain population, 85–86
　sex ratio of, 63, 75, 75t, 85
　size of, 60, 85
　survival rates for, 86, 201
　　in captivity, 250–51, 259–60, 259t, 260t
　　variation among ursids in, 62t, 63, 64, 65
Cuon alpinus (dhole), predation by, 145

Darwin, C., xii
David, A., 7, 14, 45
Decision support system (DSS), 223
Deer, spatial memory in, 105
Delayed implantation, 58–59
Demographics, human, in new conservation paradigm, 218, 219, 220
Deng, X., 211
Dens
　maternity, 60–61
　　in Qinling Mountains, 85
　　regional differences in, 145
　sociality and, 58
　switching of, 60–61
　types of, variation in, 60–61
DFA. *See* Discriminant function analysis
Dhole. *See Cuon alpinus*
Diet, 90–100. *See also* Bamboo; Food; Nutrition
　animal foods in, 91–93, 92t, 141
　availability of food and, 55, 56
　of captive populations, 91, 255–56
　evolution of, 90
　　bile salts and, 42–43
　fasting in, 56, 60–61

flexibility in, 55, 56
methodology for studying, 190–91
nonbamboo plants in, 140–41
regional differences in, 90–91, 140–44
soil in, 91, 93
variation among ursids in, 55
Diploid number, in phylogenetics, 13
Discriminant function analysis (DFA), 192, 195, 197t
Disease, reintroduction and, 248
Distribution. *See* Geographic distribution
Diversity. *See* Genetic diversity
DNA. *See also* Genetic diversity
　from fecal matter, extraction of, 155–57, 157f
　from hair, for population monitoring, 171–73, 175
　phylogenetics based on, 36, 37f
Dog. *See Canis familiaris*
Dominance
　chemical communication of, 111–13
　in mating, 85

Ears, scent distribution on, 124
Eco-hydropower station program, 224
Ecoregion scale
　conservation at, 229–30, 231t
　definition of, 229
Ecotourism, 212
Education initiatives, in China, 264–66
Electricity, 224
Electrospray mass spectrometry (EMI-MS), of bile, 39, 40–41, 42f
Elevational range
　migration within, 144, 159, 163, 163f, 165f
　variation in, 54, 55f
EMI-MS. *See* Electrospray mass spectrometry
Enamel patterns, on teeth, 49, 50
Endangered status, 155, 159
　range reduction and, 228
Enhydra lutris (sea otter), extinction risk for, 240
Environmental education, 264–66
Environmental events, and life history strategies, 64
ESC. *See* Expert system classifier
ESNNC, 177–85, 179f, 183t

Estrous cycle
　chemical communication of, 107, 109–11
　female hormones during, 126, 127–28
Evolution. *See also* Phylogenetics
　"bear school" theory of, 13, 27
　of herbivory in carnivores, 13, 42–43, 90
　"raccoon school" theory of, 13
　rate of
　　for biological traits, 242–43, 243t
　　extinction risk and, 242–43
Expert system classifier (ESC), 177, 183, 183t
Extinction risk, 236–43, 273–74
　biological characteristics correlated with, 236, 237–43, 239t, 241t
　evolutionary rates and, 242–43
　modern versus historical, 236
　supertree analysis and, 236–39
Eye patches, scent distribution on, 124

Fargesia spp., consumption of, 91
　leaves versus stems in, 95
　nutrients from, digestibility of, 98–99, 99t
　seasonality of, 93–95, 94t
Fargesia qinlingensis, consumption of, 141
Fargesia robusta, consumption of, 141
Fargesia scabrida, consumption of, 141
Fargesia spathecea, consumption of, regional differences in, 90
Farmlands, in Qinling Mountains, 82–83
Fasting, 56
　maternal, 60–61
Fecal matter
　biochemical composition of, 81
　deposited during feeding, 191
　diet data from, 190
　DNA extracted from, 155–57, 157f
Feeding behavior. *See also* Foraging
　methodology for studying, 190–91
　regional differences in, 141–44
Feet
　scraping of, in response to chemical communication, 110
　of *Ursus*, similarities among, 45

Felis temmincki (Asian golden cat), predation by, 145
Field studies
  early, 2
  further need for, 273
Fish, in ursid diet, 55
Fish and Wildlife Service, U.S., international loan programs under, 3
Flehmen, 108–9, 110
FNR. *See* Foping Nature Reserve
Food. *See also* Diet; Foraging
  availability of
    and diet composition, 55, 56
    and hibernation, 56
    and home ranges, 56
    mobility and, 56
    and sociality, 58
    variation in, 55, 56
  selection of, seasonality of, 93–95, 94t, 192
  spatial distribution of, 101
Foot. *See* Feet
Foping Nature Reserve (FNR), 159–68
  habitat of, 160–61
    mapping suitability of, 176–85
    utilization of, 166–68, 167f
  literature on, review of, 160
  migration in, 159–68, 160f, 161t, 162f
Foraging
  behavioral adaptations to, 95–98, 96t, 97t
  efficiency of, 101–5, 103f
  regional differences in, 141–44
  selectivity in, 93–95
  spatial memory in, 101–5
Fossils
  geographic range of, 48, 48f
  phylogenetic evidence in, 12, 12t, 45–51
Fruit
  availability of, 56
    crop failures and, 64
  consumption of, 55, 56, 141, 192, 198
Fuelwood consumption, 222

Genetic diversity, 88–89
  applications for studies of, 88
  assessment of, 86–87
  of captive populations, 88
    chemical communication and, 117
  genetic overrepresentation and, 254
  in conservation, role of, 88
  methodologies for studying, 81, 86–87
    with DNA from fecal matter, 155–57, 157f
    with DNA from hair, 171–73, 175
  of Qinling Mountain population, 86–87, 86t, 145
  recommendations on, 89
  regional differences in, 145–46
  reintroduction and, 246
Geographic distribution, 137, 138f
  in Qinling Mountains, 82, 83f
  third national survey of, 134, 150–54
Geographic information system (GIS)
  in habitat assessment, 161–62, 176–77, 220
  in migration studies, 159, 160
  national survey results in, 150
  recommendations for use of, 153–54
Geographic ranges
  of *Ailurus fulgens*, 189
  current, 137, 138f, 146, 149, 160f, 228
  genetic diversity studies and, 88
  historical, 88, 137, 138f, 149, 160f, 228
  of Qinling Mountain population, 83
  variation among ursids in, 53, 54, 55f, 64
Gestation, in Qinling Mountains, 74, 75t, 76, 85
The Giant Panda (Morris and Morris), vi
GIS. *See* Geographic information system
Global Environmental Facility, 211
Global positioning system (GPS)
  habitat mapping with, 150
  in new conservation paradigm, 220
  recommendations for use of, 153–54
Golden monkey. *See Rhinopithecus roxellana*
GPS. *See* Global positioning system
"Grain-to-Green" program, 133, 188, 223, 230
Grasses, consumption of, 91
Grazing animals, spatial memory in, 105
Grizzly bear. *See Ursus arctos* (brown bear)
Gulo gulo (wolverine), extinction risk for, 241

Habitat(s), 133–212. *See also specific locations*
  assessment of
    GIS in, 161–62, 176–77, 220
    through mapping, 149–50, 176–85, 178f, 179f
    methodology for, 149–54, 191–92
    recommendations on, 153–54, 178, 179f
    remote sensing in, 81, 149–50, 176–77, 178, 179f, 220
  in conservation efforts, 146
    new paradigm for, 217–24
  current status of, 133, 149
  government programs protecting, 4, 133–34
  human impacts on, 217–24, 272
  land-use planning in, 227
  loss and fragmentation of, rate of, 133, 187
  minimum requirements for, 133
  regional differences in, 137–40, 139t, 145–46
  and reintroduction, 248
  restoration of, 187–88, 188f
  selection of, chemical communication in, 117
  suitable
    availability of, 179, 183t
    classification of, 178, 179, 180t, 181t
    definition of, 178, 179, 180t
    mapping of, 176–85, 183t
    spatial distribution of, 179, 181f, 182f
    transition zones between, 179, 184
  survey of, third national, 134, 150–54
  tropical versus temperate, species in, 54
  utilization of
    noninvasive monitoring of, 170–75
    by red versus giant pandas, 194–97, 196t, 197t, 198
    seasonality of, 166–68, 167f
    variation among ursids in, 54–55

INDEX   301

Hair samples, 81
  from barbed-wire snares, 170, 171–75, 173t, 174t
Harkness, R., 2
Health, in captive populations, assessment of, 250–62
Helarctos malayanus. See *Ursus malayanus* (sun bear)
Hemicellulose, in bamboo, utilization of, 98, 99, 99t
Hemicyoninae
  gait of, 47
  radiations in, 47
  teeth of, 47
Herbivory, evolution in carnivores of, 90
  in "bear school" theory, 13, 27
  bile salts and, 42–43
Hibernation
  and cub size, 60
  den switching during, 61
  in maternity dens, 60–61
  among ursids, 56
High-performance liquid chromatograph (HPLC), of bile, 41, 43f
Home ranges
  assessment of, methodology for, 153
  chemical communication in, 117–18
  core areas of, 58, 125, 163
  female versus male, 125–26, 164, 166t
  food abundance and distribution in, 56, 125
  overlaps in, 58, 125, 163
  seasonality of, 162f, 163–66, 163f, 165f
  size of, 56, 57t, 164–66, 166t
  territoriality in, 58, 117–18, 126
  variation among ursids in, 56, 57t, 64
Hormones, reproductive, 125–31
  correlation with behavior, 128, 129–31
  female, 127–29
  male, 129–30
  methodology for studying, 127
HPLC. See High-performance liquid chromatograph
Hu, J., xi, xiii
Humans
  bamboo harvest by, 201–8
  current rate of, 205
  sustainable rate of, 206–8, 206f, 207f
  ecotourism for, 212
  and habitats, 217–24, 272
    loss of, 217–18
    restoration of, 187–88
  hunting by, 2, 248
  mortality caused by, 66
  in new conservation paradigm, 217–24
  population size of, 221, 222f
  public opinion among, 1, 2
  in Qinling Mountains, 82–83, 184, 185
  and reintroduction efforts, 248–49
  relocation of, 217–18, 222, 272
  in Wolong Nature Reserve, 217–24, 222f
Hunting
  by early Westerners, 2
  and reintroduction, 248
Hyaenarctos spp., 12

Implantation, delayed, 58–59
India, ursid species in, 54
Indocalamus longiauritus
  animal consumption of, 202, 203t, 204–5
  human harvest of, 205, 206f, 207, 207f, 208
  vital rates for, 203–4, 203t
Insects
  bamboo consumption by, 202, 203t, 206
  in ursid diet, 55
International coordination and cooperation, xiii–xiv, 268–69
International loan programs, 3, 268–69
International Union for the Conservation of Nature (IUCN)
  Red List of, 239, 240, 240f
  Species Survival Commission of
    Conservation Breeding Specialist Group of, xiii, 251, 252, 261–62
    Reintroduction Specialist Group of, 246, 247t
IUCN. See International Union for the Conservation of Nature

Jones, M., 282–85

Karyotype, in phylogenetics, 12t, 13
Keynote addresses, at *Panda 2000* conference, 279–85
Kleiman, D., 3

Lairs, cubs in, 85
Land-use planning, 227
LANDSAT MSS imagery, 149, 220
LANDSAT TM imagery, 150, 220
Landscape scale
  conservation at, 229–32, 231t
  definition of, 230
Latitudinal range, variation in, 54, 55f
Laws, related to reintroduction, 249
Lee, T., 265
Leopard. See *Panthera pardus*
Lesser panda. See *Ailurus fulgens*
Liang Mountains, bamboo species consumed in, 90
Liangshan Mountains, 138f. See also Mabian-Dafengding Natural Reserve
  bamboo consumption in, 140, 142t–143t
  comparative ecology of, 137–46
  counties of, 140, 140t
  current population status in, 146
  diet in, 140–44
  feeding behavior in, 141–44
  genetic diversity in, 146
  habitat of, 139t, 140
  mating in, 145
  migration in, 144
  predation in, 145
Licking, in response to chemical communication, 110
Life histories. See also *specific strategies*
  conservation implications of, 65–66
  definition of, 53–54
  and extinction risk, 241–42
  in Qinling Mountain population, 74–76
  variation among ursids in, 53–66
Lifespans, 63
Literature, review of, 3
  on Foping Nature Reserve, 160
  in phylogenetics, 14–29, 15t–19t, 20t, 36, 37t
Litters. See also Cubs
  interval between, 62t, 63
  in Qinling Mountains, 74–75

sex ratio of, 63
  in Qinling Mountains, 75, 75t, 85
  size of, variation in, 61–63, 62t, 64, 145
Liu, J., 235
Liu, S., 235
Liu, X., 235
Loan programs, international, 3, 268–69
Logging
  ban on, 4, 133, 230, 248, 275
  in Qinling Mountains, 82
London, zoo in, 2
Lufeng fossil locality, 48–51, 49f
Lycaon pictus (wild dog), extinction risk for, 241

Ma, F., 279–81
Mabian-Dafengding Natural Reserve
  bamboo shoots in, human versus panda utilization of, 201–8
  habitat of, 201–2, 202t
  observation station in, 137
Malayan sun bear. *See Ursus malayanus* (sun bear)
Mammae, number of, and litter size, 61–63, 62t
Mapping
  of bamboo distribution, 150, 222–23
  habitat assessment through, 149–50, 176–85, 178f, 179f
Martes flavigula (yellow-throated marten)
  at hair snares, 171, 173t
  predation by, 145
Mating. *See also* Captive breeding
  chemical communication in, 107, 109–11, 114–15, 129–30
  delayed implantation after, 58–59
  dominance in, 85
  promiscuity in, 58
  and litter sex ratios, 63
  seasonality of, 58–59, 59f, 74, 75f, 126
  in Qinling Mountains, 74–76, 75f, 76f, 85
  regional differences in, 145
Matrix representation with parsimony analysis (MRP), 14–20
Melursus ursinus. *See Ursus ursinus* (sloth bear)

Memorandum of consensus, at *Panda 2000* conference, 286–87
Memorandum of understanding (MOU)
  between CBSG and zoos (1997), 252
  between MOF and WWF (1997), 234
  between SFA and WWF (2000), 234
Memory, spatial, 101–5
Mice, chemical communication in, 109, 112
Migration, 159–68
  food availability and, 56
  habitat utilization and, 166–68, 167f
  methodologies for studying, 153, 159–60, 161–62
  in Qinling Mountains, 82, 144
  regional differences in, 144
  seasonality of, 162f, 163–66, 163f
  time spent in, 164, 165f
Milne-Edwards, A., 7
Ministry of Construction (MOC) (China), 229, 250
Ministry of Forestry (MOF) (China), 234
Minshan Mountains, 138f
  bamboo consumption in, 140, 142t–143t
  comparative ecology of, 137–46
  counties of, 139, 140t
  current population status in, 145
  diet in, 90–91, 140–44
  distribution in, 139
  feeding behavior in, 141–44
  genetic diversity in, 86t, 87, 145
  habitat of, 138–39, 139t
  integrated conservation and development project in, 272
  mating in, 145
  migration in, 144
  predation in, 145
Mitochondrial DNA, phylogenetics based on, 36, 37f
MOC. *See* Ministry of Construction
MOF. *See* Ministry of Forestry
Molars, 46, 49
Molecular phylogenetics, 12t, 14, 36
  conflicting results in, 7, 36
  literature on, 21, 36, 37t

Morphology, in phylogenetics, 12–14, 12t
  literature on, 21
Morris, D., vi, 115
Morris, R., vi, 115
Mortality. *See also* Survivorship
  human-caused, 66
  rates of, and extinction risk, 242
Moschus chrysogaster (musk deer), consumption of, 91
Moscow, zoo in, 2
Mother-infant recognition, chemical communication in, 113–14
MOU. *See* Memorandum of understanding
Mountain islands, in Qinling Mountains, 82, 83f
MRP. *See* Matrix representation with parsimony analysis
Musk deer. *See Moschus chrysogaster*
Mustela sibirica (Siberian weasel), at hair snares, 171, 173t

National Zoo (Washington DC), 2
Neofelis nebulosa (clouded leopard), predation by, 145
Neural network classifier (NNC), 177, 183, 183t
NGOs. *See* Nongovernmental organizations
Niches, breadth and overlap of, 191, 194, 195t
Nixon, Richard, 2
NNC. *See* Neural network classifier
Nongovernmental organizations (NGOs), reintroduction efforts supported by, 248–49
North America, ursid species in, 54
Nutrients, in bamboo, utilization of, 98–99, 99t, 141–44, 256
Nutrition
  in captivity, and reproductive success, 255–56
  and cub survivorship, 63
  and litter intervals, 63
  in Qinling Mountain population, 90–100
  and reproductive age, 63
Nycticebus pygmaeus (pygmy loris), chemical communication in, 117

O'Brien, S., 255
Olfaction. *See* Chemical communication
Optimal foraging theory, 101
The Origin of Species (Darwin), xii

Paleontology, and phylogenetics, 12, 12t, 45–51
Pan, W., xi, xiii
Panda 2000: Conservation Priorities for the New Millennium (conference), xiv, 1–2
    keynote addresses at, 279–85
    memorandum of consensus at, 286–87
Panthera pardus (leopard), predation by, 145
Paternity, in captive breeding, establishment of, 88, 254–55
Pelage
    and extinction risk, 241
    scent distribution on, 123–24, 123f
Pen swapping, chemical communication and, 108, 111
Phoca hispida, 55
Phoca vitulina, phylogeny of, 36, 37f
Phyllostachys nigra, consumption of, 91
Phylogenetics, 11–44
    applications for, 236–37
    "bear school" in, 13, 27
    bile salts and, 38, 41–44, 43f
    controversy over, 7, 36, 38, 237
    convergence and, 13–14, 27
    early classifications in, 7, 45
    evidence of
        behavioral, 12t, 13, 21–22
        in literature, 21–22
        molecular, 7, 12t, 14, 21, 36, 37f, 37t
        morphological, 12–14, 12t, 21
        paleontological, 12, 12t, 45–51
    and extinction risk, 236–43, 273–74
    history of trends in, 7, 11
    literature on, 14–29, 15t–19t, 20t
        cumulative values of, 22–24, 24f
        sliding window analysis of, 20–21, 22, 23f, 24, 25f, 26f
        supertree analysis of, 14–20, 21, 21f, 24–25, 25f, 27f, 28f
    "raccoon school" in, 13, 27

versus taxonomy, 11, 27–28
    teeth in, 47–51
Pinus armandii, 91
Polar bear. *See Ursus maritimus*
Polyporus frondosus, 91
Population(s)
    current status of
        regional differences in, 145–46
        and reintroduction, 246–48
    genetic diversity in, 86–87, 86t
    noninvasive monitoring of, 170–75, 274
Population density
    equilibrium, 203, 205, 207–8
    habitat restoration and, 187
    in Qinling Mountains, 83, 138
    rate of increase in, 205
Population growth
    in Qinling Mountains, 84
    rate of, 205
Population size
    noninvasive monitoring of, 171–73
    of Qinling Mountain population, 83–84
Population viability analyses
    of Qinling Mountain population, 84
    of regional differences, 145–46
    for *Ursus arctos*, 65
Predation
    by *Ailuropoda melanoleuca*, 90, 91
    on *Ailuropoda melanoleuca*, 145
    predator-prey models of, 202–3
        equilibrium in, 205–7
    on ursids
        by other ursids, 55, 58
        refuge from, 57
        and sociality, 58
    by *Ursus maritimus*, 55
"Preference test," 108
Pregnancy, diet during, 91–93
Primates, extinction risk in, 239–41, 239f, 240f, 241t
Procyon lotor (raccoon), phylogeny of, 7, 11
    molecular, 36, 37f
Procyonidae, phylogeny of, 11–29
    evidence of, 11–14, 12t
    literature on, 14–29, 15t–19t, 20t
Pseudothumb, and foraging efficiency, 101
Public opinion, 1, 2
Publications. *See* Literature

Pygmy loris. *See Nycticebus pygmaeus*

Qi, J., 265
Qinling Mountains, 81–87, 138f. *See also* Foping Nature Reserve
    age structure in, 83, 84t
    bamboo selection in, 140, 142t–143t
        seasonality of, 93, 94t
    birthing in, 85
        seasonality of, 75, 76, 76t, 85
    comparative ecology of, 137–46
    conservation strategies for, 82
    counties of, 138, 140t
    current population status in, 145
    diet in, 140–44
    distribution in, 139
    feeding behavior in, 141–44
    genetic diversity in, 86–87, 86t, 145
    habitat of, 82–83, 83f, 138, 139t, 150
    litter intervals in, 74–75
    mating in, 85
        seasonality of, 74–76, 75f, 76f, 85
    metapopulation in, 82
    methodology for studying, 81–82, 83–84
    migration in, 82, 144
    nutritional strategy in, 90–100
    population density in, 83, 84t, 138
    population growth in, 84
    population size in, 83–84
    population viability analysis for, 84
    predation in, 145
    range in, 83
    reproduction in, 74–76
        age at, 84t, 85, 145
        rate of, 84, 84t
    sex ratios in, 75, 75t, 85
Qionglai Mountains, 138f
    bamboo consumption in, 140, 142t–143t
    comparative ecology of, 137–46
    counties of, 139, 140t
    current population status in, 145
    diet in, 140–44
    distribution in, 139
    feeding behavior in, 141–44
    genetic diversity in, 86t, 87, 145
    habitat of, 139, 139t
    migration in, 144
    predation in, 145

Qiongzhuea spp., consumption of, 141
Qiongzhuea macrophylla
   animal consumption of, 202, 203t, 204–5, 206
   equilibrium density of, 206, 206f, 207f
   human harvest of, 205, 206f, 207, 207f, 208
   nutrients in, 141, 144
   vital rates for, 203–4, 203t

Raccoon. *See Procyon lotor*
Radiotracking
   Chinese ban on, xii, 249
   data obtained from, 81, 153, 159
   equipment for, 160f, 161
   in habitat assessment, 177
   in migration studies, 153, 160, 160f, 161
   need for, 154, 274
Ranges. *See* Geographic ranges; Home ranges
Red List, IUCN, 239, 240, 240f
Red panda. *See Ailurus fulgens*
Reforestation, 4
Refuge strategies, 57
Reintroduction efforts, 246–49
   chemical communication in, 117–18
   criteria for, 246–49, 247t
Remote sensing (RS)
   with camera systems, 170–71, 172t, 173–75, 173t
   in habitat assessment, 81, 149–50, 176–77, 178, 179f, 220
Reproduction, 125–31. *See also* Birthing; Captive breeding; Mating
   behavior of, 127–30
   categories of strategies, 54
   hormones in, 127–30
   methodology for studying, 126–27
   in Qinling Mountains, 74–76, 75f, 75t, 76f, 76t
   regional differences in, 145
   variation among ursids in, 58–64
Reproductive age, 62t, 63
   in Qinling Mountains, 84t, 85, 145
   regional differences in, 145
Reproductive rates
   age and, 63, 255
   and conservation, 65–66
   and extinction risk, 241
   in Qinling Mountains, 84, 84t
   variation in, 61–63, 62t, 64–65
Research. *See* Field studies
Reserve Management Project, 211
Reserves. *See also specific reserves*
   design of, 88, 210–12
   ecosystem health in, 210
   expansion of system, 4, 133, 146, 210, 272
   genetic diversity studies and, 88
   limitations of, 149
   management of, 210–12
   monitoring of, 211, 212
   patrolling of, 211, 212
   recommendations on, 211–12
   threats to, 210, 211
Resources. *See also* Food; Habitat(s)
   overlap of, 191, 194, 195t
   partitioning of, 197–98
Restoration, habitat, 187–88, 188f
Rhinopithecus roxellana (golden monkey), 91
Rhizomys sinense (bamboo rat)
   bamboo consumption by, 90, 202, 203t, 206
   panda consumption of, 91
Roads, in Qinling Mountains, 82–83, 83f
Rock caves, dens in, 60
Roosevelt, K., 2
Roosevelt, T., Jr., 2
RS. *See* Remote sensing
Rut, spring, 129

Satellite imagery, habitat assessment with, 81, 149–50, 176
Scent marking. *See* Chemical communication
Scent stations, 106–7, 171
Schaller, G. B., xi–xii, xiii
Schools, in China, conservation education in, 264–66
Sea otter. *See Enhydra lutris*
Selenarctos thibetanus. *See Ursus thibetanus* (Asiatic black bear)
Semen, in captive populations, 258–59, 258t
Sex, chemical communication of, 107, 109–11
Sex allocation theory, 63
Sex ratios, 63
   in Qinling Mountains, 75, 75t, 85

SFA. *See* State Forestry Administration
Shannon-Wiener index of niche breadth, 191
Shao, K., 235
Sheep, spatial memory in, 105
Siberian weasel. *See Mustela sibirica*
Sinarundinaria spp., consumption of, 93
Sinarundinaria fangiana, consumption of, regional differences in, 90
Sinarundinaria nitida, consumption of, regional differences in, 90
Site scale
   conservation at, 229–30, 231t, 232
   definition of, 230
Sliding window analysis, of literature on phylogenetic affinity, 22, 23f, 24, 26f
   procedure for, 20–21
   on supertrees, 24, 25f
Slopes, preferred characteristics of, 138, 139f, 167f, 168, 198
Sloth bear. *See Ursus ursinus*
Snyder, R., 256
Social relations
   chemical communication in, 112
   organization of, 125–26
   and reproduction, 125, 131
   resource availability and, 58, 125
   solitariness in, 58, 106
Socioeconomics, in new conservation paradigm, 218, 219, 220
Soil, consumption of, 91, 93
Solitariness, 58, 106
Song, Y., 210–11
Spatial memory, 101–5
Spectacled bear. *See Tremarctos ornatus*
Sperm cryopreservation, 258–59
SPOT satellite imagery, 150
Spring rut, 129
State Forestry Administration (SFA) (China), 229, 234, 249, 250
Studbook records, paternity determination from, 254
Stunted growth syndrome, 256–57, 257f
Sun bear. *See Ursus malayanus*
Supertree analysis
   backbone constraint trees in, 21, 21f
   for carnivores, 237, 238f

INDEX   305

Supertree analysis *(continued)*
    definition of, 236, 237
    and extinction risk, 236–39
    of family-level relationships, 25, 28f
    procedure of, 14–20, 21f, 237
    results of, 24–25, 27f
    sliding window analysis of, 24, 25f
Surveys
    biomedical, of captive populations, 250–62
    first national (1974–1976), 234
    methodology for, 150–54
        "bite size technique" in, 83–84, 152–53
        in Qinling Mountains, 81–82, 83–84
    second national (1985–1988), 234
    third national (2002), 134, 150–54, 234–35
        application of results of, 134, 150, 153, 235
        form used in, 151f–152f
        review of methods for, 235
Survival strategies
    definition of, 54
    variation among ursids in, 54–58
Survivorship, 55
    for adults, 66
    for cubs, 86, 201
        in captivity, 250–51, 259–60, 259t, 260t
        variation among ursids in, 62t, 63, 64, 65
        in Qinling Mountains, 84t, 86
Sympatry, 7, 189–98

Tangjiahe Nature Reserve, 137
Taurochenodeoxycholic acid
    in nonursid carnivores, 39, 40t
    structure of, 39f
    in tauroursodeoxycholic acid synthesis, 44
    in ursids, 40, 41, 41t, 43, 44
Taurocholic acid
    in nonursid carnivores, 39–40, 40t
    structure of, 39f
    in ursids, 40, 41, 41t
Taurodeoxycholic acid
    in nonursid carnivores, 39, 40t
    in ursids, 41, 41t
Tauroursodeoxycholic acid
    function of, 42–43
    medical use of, 43
    in nonursid carnivores, absence of, 40, 40t
    structure of, 39f
    synthesis of, 44
    in ursids, 38, 40, 41–44, 41t, 43f
Taxonomy
    versus phylogenetics, 11, 27–28
    subjectivity of, 11, 27–28
Teeth, 46–51, 49f, 50f
Temperate habitats, ursid species in, 54
Territoriality
    chemical communication of, 111–13, 117–18
    in reproduction, 126
    variation among ursids in, 58
Testicular hypoplasia or atrophy, in captivity, 257
Thalarctos maritimus. *See Ursus maritimus* (polar bear)
Third National Survey of the Giant Panda and its Habitat (2002), 134, 150–54, 151f–152f, 234–35
Tibetan black bear. *See Ursus thibetanus* (Asiatic black bear)
Tooth. *See* Teeth
Tourism, 212
Trees
    climbing in, 56–57
    dens in, 60, 145
    marking of, 58
        in Qinling Mountains, 85
    nests in, 56–57
Tremarctinae, 46
Tremarctini, 46
Tremarctos ornatus (spectacled bear)
    bile salts of, 41t, 43f
    birthing in, 59–60, 60f
    geographic range of, 43, 54
    home range of, 57t
    karyotype of, 13
    life history of, versus other ursids, 64–65
    litters of
        sex ratio of, 63
        size of, 61–63, 62t
    mating in, 59
    phylogeny of, 13, 43, 43f, 45–46
    tree climbing by, 56
Tropical habitats, ursid species in, 54
Tsuga chinensis, 91
Twins, 61, 65, 251

Ungulates, in ursid diet, 55
United Nations Development Program, 211
United Nations Environment Program, 211
United States, Chinese collaboration with, xiii–xiv
Urine, in chemical communication, 106, 110, 113, 115, 124
Ursavus spp., 46
    fossils of, 46, 48
    geographical distribution of, 46
    phylogeny of, 48, 50, 51f
    size of, 46
    teeth of, 49, 50
Ursidae
    bile salts of, 38–44, 41t, 43f
    classification of, 47t
    evolution to herbivory, bile salts and, 42–43
    fossils of, 45–46
    life histories of, variation in, 53–66
    living species of
        number of, 53
        oldest, 44
    phylogeny of, 11–29
        bile salts and, 38, 41–44, 43f
        evidence of, 11–14, 12t
        literature on, 14–29, 15t–19t, 20t
    predation on, 55, 57, 58
    teeth of, 49
Ursinae
    classification of, 46, 47t
    radiations in, 47
    teeth of, 49–50
Ursus spp.
    bears not in genus, 45
    classification of, 45, 46
    living species of, 45
    radiations of, 46–47
    skeletal similarities among, 45, 46
Ursus americanus (American black bear)
    bile salts of, 41t, 43f
    birthing in, 58–59, 62t, 63
    dens of, 60
    diet of, 55
    geographic range of, 54
    hibernation by, 56
    home range of, 56, 57t
    life history of, versus other ursids, 64–65

306 INDEX

litters of
    interval between, 62t, 63
    size of, 61, 62t
  mating in, 58–59
  reproductive age in, 62t, 63
  tree climbing by, 56–57
Ursus arctos (brown bear)
  bile salts of, 41t, 43f
  birthing in, 58–59, 62t, 63
  diet of, 55
  hibernation by, 56
  home range of, 56, 57t
  life history of, versus other ursids, 64–65
  litter size of, 61, 62t
  mating in, 58–59
  population viability analyses for, 65
  reproductive age in, 62t, 63
  reproductive output of, 65
  size of, 53
  tree climbing by, 56
Ursus malayanus (sun bear)
  bile salts of, 41t, 43f
  birthing in, 59, 59f
  diet of, 55
  genus of, 45
  geographic range of, 54
  home range of, 57t
  life history of, versus other ursids, 64–65
  litters of
    sex ratio of, 63
    size of, 61, 62t
    transport of, 61
  mammae of, 62t, 63
  mating in, 59
  size of, 53
  tree climbing by, 56
Ursus maritimus (polar bear)
  bile salts of, 41t, 43f
  birthing in, 59, 60, 62t, 63
  dens of, 58
  diet of, 55
  fasting by, 56, 60
  genus of, 45
  geographic range of, 54
  habitat of, 55
  home range of, 56, 57t
  life history of, versus other ursids, 64–65
  litter size of, 61, 62t, 63
  mammae of, 62t, 63
  mating in, 58, 59
  reproductive age in, 62t, 63

size of, 53
sociality of, 58
Ursus thibetanus (Asiatic black bear)
  bile salts of, 38, 41t, 43f
  birthing in, 58–59
  dens of, 60
  diet of, 55
  genus of, 45
  geographic range of, 54
  hibernation by, 56
  home range of, 56, 57t
  life history of, versus other ursids, 64–65
  litter size of, 61, 62t
  mating in, 58–59
  resource competition with, 7
  tree climbing by, 56–57
Ursus ursinus (sloth bear)
  bile salts of, 38, 41t, 43f
  birthing in, 59, 60
  dens of, 58, 60
  diet of, 55
  fasting by, 60
  genus of, 45
  geographic range of, 54
  home range of, 57t
  life history of, versus other ursids, 64–65
  litters of
    sex ratio of, 63
    size of, 61–63, 62t
  mating in, 59
  sociality of, 58
  tree climbing by, 57

Vanderploeg and Scavia selectivity index, 190–91
Vegetation coverage
  loss of, rate of, 187
  in Qinling Mountains, 82
Verhulst-Pearl logistic, 203
Vision, in foraging, 105
VNO. *See* Vomeronasal organ
Vocalizations
  in reproductive behavior, 127, 129
  in response to chemical communication, 109–10
Vomeronasal organ (VNO), 108–9, 110, 111

Wang, D., 150
Western China Development Program, 4, 134, 230, 248
Wild dog. *See* Lycaon pictus
Wild parsnip. *See* Angelica spp.

WNR. *See* Wolong Nature Reserve
Wolong Giant Panda Breeding and Research Center, mating at
  chemical communication in, 111
  temporal distribution of, 75–76, 76f
Wolong Nature Reserve (WNR)
  China Wildlife Conservation and Research Center for Giant Pandas in, 250
    biomedical survey at, 253t
  diet in, 90, 91, 93
  establishment of, 221
  habitat of
    assessment of, 150, 221, 221f
    human impact on, 217–24, 222f
  mating in, seasonality of, 145
  migration in, 159
  observation station in, xiii, 137
Wolverine. *See* Gulo gulo
Wood, for fuel, consumption of, 222
Working Group on Genetics and Reproductive Technologies, 254
World Bank, 211
World War II, 2
World Wildlife Fund (WWF), xi, xiii
  conservation programs of, 2–3, 211, 272
  habitat restoration study by, 187
  memoranda of understanding with, 234
  in national surveys, 150, 234
WWF. *See* World Wildlife Fund

Xiangling Mountains, 138f
  bamboo consumption in, 140, 142t–143t
  comparative ecology of, 137–46
  counties of, 139, 140t
  current population status in, 145–46
  diet in, 140–44
  distribution in, 139
  feeding behavior in, 141–44
  genetic diversity in, 146
  Greater, 139
  habitat of, 139–40, 139t
  Lesser, 139
  migration in, 144
  predation in, 145
Xinglongling (Qinling Mountains), 82, 83

Yangtze River basin, flooding in, 275
Yele Natural Reserve
   habitat of, 189–90
   observation station in, 137–38
   sympatry in, 189–98
Yellow-throated marten. *See Martes flavigula*
Yong, Y., 210

Young. *See* Cubs
Yu, C., 211, 235
Yushannia spp., consumption of, 141
Yushannia chungii, consumption of, 141

Zhang, Y.-P., 255
Zhu Jing, xiii

Zoo Atlanta, in conservation education initiative, 264–66
Zoo Diet Analysis Program software, 256
Zoos. *See* Captive breeding; Captivity; *specific zoos*

Project management, copyediting, indexing, and composition:	Princeton Editorial Associates, Inc.
Text:	Scala and Scala Caps
Display:	Scala Sans and Scala Sans Caps
Printer and binder:	Thomson-Shore